Specialist Surfactants

Specialist Surfactants

Edited by

I.D. ROBB
North East Wales Institute
Plas Coch
Mold Road
Wrexham
Wales
U.K.

BLACKIE ACADEMIC & PROFESSIONAL
An Imprint of Chapman & Hall
London · Weinheim · New York · Tokyo · Melbourne · Madras

Published by Blackie Academic and Professional, an imprint of Chapman & Hall, 2–6 Boundary Row, London SE1 8HN, UK

Chapman & Hall, 2–6 Boundary Row, London SE1 8HN, UK

Chapman & Hall GmbH, Pappelallee 3, 69469 Weinheim, Germany

Chapman & Hall USA, 115 Fifth Avenue, New York, NY 10003, USA

Chapman & Hall Japan, ITP-Japan, Kyowa Building, 3F, 2-2-1 Hirakawacho, Chiyoda-ku, Tokyo 102, Japan

DA Book (Aust.) Pty Ltd, 648 Whitehorse Road, Mitcham 3132, Victoria, Australia

Chapman & Hall India, R. Seshadri, 32 Second Main Road, CIT East, Madras 600 035, India

First edition 1997

Typeset in 10/12 Times by Doyle Graphics, Tullamore, Co. Offaly, Republic of Ireland

Printed in Great Britain by The University Press, Cambridge

ISBN 0 7514 0340 7

A catalogue record for this book is available from the British Library

Library of Congress Catalog Card Number: 96–83830

∞ Printed on acid-free text paper, manufactured in accordance with ANSI/NISO Z39.48-1992 (Permanence of Paper)

Contents

Contributors

D. Balzer Talstr.21, D 45721 Haltern, Germany.

P. Fletcher School of Chemistry, University of Hull, Hull HU6 7RX, UK.

R.M. Hill Central Research and Development, Dow Corning Corporation, Midland, MI 48686-0994, USA

A. Khan Department of Physical Chemistry 1, Chemical Centre, Box 124, Lund University, S-221 00 Lund, Sweden.

E. Marques Department of Physical Chemistry 1, Chemical Centre, Box 124, Lund University, S-221 00 Lund, Sweden. Also Department of Chemistry, University of Coimbra, Portugal.

H.N. Patrick School of Chemistry, University of Sydney, NSW 2006, Australia.

I.D. Robb North East Wales Institute of Higher Education, Plas Coch, Mold Road, Wrexham, Clywd, Wales LL11 2AW, UK.

P. Terech Département de Recherche Fondamentale sur la Matière Condensée, SESAM-PCM, C.E.A.-Grenoble, 17, rue des Martyrs, 38054 Grenoble Cédex 09, France.

G.G. Warr School of Chemistry, University of Sydney, NSW 2006, Australia.

R. Zana Institut C. Sadron (CRM-CNRS), 6, rue Boussingault, 67000 Strasbourg, France.

Preface

Throughout the natural biological world sustenance of life depends (among other things) on the fact that colloidal systems have surprisingly large surface areas, allowing interfacial reactions to take place on a large scale. Reactions at cell interfaces are an example. Similarly, many industrial processes utilize this high surface area of colloids or, increasingly, exploit the compartmentalization that colloidal dispersions offer. Reverse phase emulsions are examples of the latter.

Most of these colloidal systems use surfactants to maintain their stability and integrity, often as mixtures or in conjunction with polymers. The natural world has evolved systems with macromolecules that are far more complex and subtle than man has yet been able to achieve, largely due to higher orders of structure and interactions that proteins and polysaccharides (for example) can produce. In contrast, man's achievements with surfactants have been more on a par with those found in nature. The first soaps derived from vegetable oils have been overtaken in complexity by surfactants with a variety of charge – positive, neutral or zwitterionic – and a multitude of types and arrangements of the hydrophobic chain(s) such as unsaturated, non-hydrocarbon or multichained, of even or uneven lengths. Conventional application of surfactants has been in aqueous systems, though it is apparent that surface activity and emulsification can be demonstrated in mixtures of nonaqueous solvents such as hydrocarbons and fluorocarbons.

The variety of surfactants is increasing almost daily, many of the new materials having novel properties that allow their industrial survival until the full benefit of these properties become widely appreciated. One of the obstacles to full utilization of new surfactants is a comprehensive awareness on the part of experimenters of the potential of these new materials. Often one becomes aware of a few remarkable applications without knowing the full picture of the preparation and properties of the materials.

The main purpose of this book was thus to bring together the important information on some of these more novel surfactants. It was thought important to display both their scientific and industrial aspects to aid their further understanding and utilization. The book provides basic information on the preparation, properties and application of some quite interesting surfactants and should be of interest to both academic and industrial researchers. It has been assumed that the fundamental behaviour of surfactants is understood or at least can be obtained from the numerous publications on the subject.

There are many 'conventional' surfactants currently available commercially, indeed the word 'conventional' now covers a wide range of anionic, cationic and neutral surfactants, mainly with single chains. Truly novel surfactants are being produced on a small scale in great numbers. Thus, selecting those surfactants to be included in this book may appear initially as somewhat arbitrary. However, each has some unique aspects that have found application in a limited area and demonstrated properties that may provide the solution to other problems, if only a full picture of their properties were available.

In preparing this book, I would particularly like to thank Dr J. Eastoe of Bristol University, who was most helpful with the formative stages of the book and the authors of the chapters for their helpful cooperation.

I. Robb

1 Introduction to specialist surfactants

I.D. ROBB

1.1. Introduction

Surfactants have long been used to improve the properties of dispersions and films in many industial products and processes, with the simpler (and hence cheaper!) ionic and nonionic surfactants mainly being used. In recent times there have emerged more unusual surfactants that initially are finding use in niche areas, that are expanding rapidly. In order to understand the behaviour of these specialist surfactants it is worth considering some of the principles that govern this behaviour.

The useful and fascinating properties of surfactants are essentially a result of combining into one molecule certain groups that as separate molecules would be incompatible. Thus, surfactants usually have polar and nonpolar components and the resolution of this 'incompatibility' by aggregation or adsorption at interfaces is what makes surfactants the important materials they are. The basic details of surfactant aggregation have been outlined in a number of books including the recent contribution by Clint [1].

Surfactant aggregation is complex but reasonably well understood. The type of aggregate formed is largely determined by the relative cross section areas and shapes of the polar and nonpolar groups. The relations between the geometry of a surfactant and type of mesophase preferred have been derived [2], and are shown in Table 1.1. The geometry of the surfactant molecule is described by taking a as the head group area, v the volume of

Table 1.1 The relation between surfactant geometry and preferred mesophase

Packing factor v/al_c	$<1/3$	$1/3-1/2$	$1/2-1$	~ 1	>1
Effective shape of surfactant molecule	cone	truncated cone	truncated cone	cyclinder	inverted truncated cone
Preferred structure	spherical micelles, isotropic phase	cylindrical micelles, hexagonal phase	flexible bilayers, vesicles	planar, bilayers, lamellar phase	inverted micelles, isotropic phase

the hydrophobic tail and l_c the critical chain length, this being slightly less than the length of a fully extended chain.

These cross section areas are the effective areas and include not only the actual geometrical areas but also any electrical repulsion between head groups. Thus ionic surfactants, in the absence of added electrolyte, have relatively large effective head group areas. At concentrations just above the critical micelle concentration, and in the absence of electrolytes to reduce the electrical repulsion, ionic surfactants have an effective shape of an ice cream cone, the resulting micellar aggregate being spherical. In the presence of high electrolyte, ionic surfactants may form hexagonal or lamellar phases at high surfactant concentrations. These packing arguments apply to the specialist surfactants described in this book as much as to the more widely studied conventional materials.

The 'incompatibility' of the groups in a surfactant manifests itself not only by aggregation but also by adsorption at interfaces. At the air/water (A/W) interface, there is usually a positive adsorption of surfactant molecules, giving a higher concentration of surfactant in the surface layer than in the bulk. This positive adsorption lowers the free energy of the system, as observed by a decrease in the surface tension γ. The relation between this free energy change, the adsorption and the surfactant activity (a_2) is described by the Gibbs equation:

$$\Gamma = (1/RT)\,\partial\gamma/\partial \ln a_2$$

where Γ is the surface excess of the surfactant. Whilst surfactants show positive adsorption, the rarely studied increase in surface tension of water in the presence of some acids is a result of the negative adsorption of ions from the interface. These principles of positive and negative adsorption altering the free energy of the system may be applied to dissolved molecules as well as at macroscopic (A/W) interfaces as discussed below.

1.2 Head group interactions

Since the properties of surfactants are partially controlled by those of the head group it is worth considering how different groups respond, particularly to variations in electrolyte and temperature. Surfactants are often used at variable temperature or electrolyte concentrations and it is important to understand which head groups are most susceptibe to changes in these conditions. Addition of electrolyte often changes the quality of solvent for various polar groups, though these changes depend on the nature of both the polar group and the ion, showing that the effect of the ions on the structure of water is not the primary cause of the effect on solvent quality. Determining the effects of temperature and added electrolyte on the solvent quality is probably more easily done by examining the properties of polymers consisting of these groups since changes in radius of gyration or second virial coefficient can be measured without phase separation needing to take place.

The effects of electrolytes are best considered by analogy with the adsorption of solutes at interfaces, as mentioned above. Addition of solutes to a solution of dissolved polar materials (which may be in the form of a polymer) will result in an uneven distribution of the ions around the polymer, changing the free energy of the system. Hall [3,4] has derived, from the Kirkwood Buff theory, an expression to describe this effect of the uneven distribution of components of a mixed solvent on the free energy of the polymer. He showed that the chemical potential μ^{θ} of a non-electrolyte in a solution of different solvent components i can be represented by the relation:

$$d\mu^{\theta} = -\sum_{i=1}^{c} N_{si}^{*}\, d\mu_i$$

where N_{si}^{*} is the difference in concentration of the ith component in volume elements near to and far from the non-electrolyte. For $d\mu^{\theta}$ to decrease (salting-in), N_{si}^{*} must be positive, indicating adsorption of electrolyte immediately adjacent to the non-electrolyte, directly analogous to lowering of surface tension by positive adsorption of surfactants.

The mechanisms by which this uneven distribution of ions occurs makes no difference to the result that positive adsorption of the ions leads to a lowering of the free energy of the polymer (hence better solvent conditions). The reasons for the different distribution of ions are not always clear. In the case of large singly charged ions such as iodide, it is likely that hydrophobic interactions lead to their positive adsorption on the polymer. Conversely strongly hydrated ions (small or high charged ions, e.g. OH^- or SO_4^{2-}) do not show hydrophobic interactions and are partially excluded from adjacent to some polymers. This exclusion does not happen with all polymers. The data [5–7] show that, generally, polymers containing carbonyl or ether groups in their polar regions such as poly(vinyl pyrrolidone) or poly(ethylene oxide) are salted out by SO_4^{2-} whereas polymers containing –NH or –OH groups such as poly(acrylamide) or unmodified polysaccharides are not so influenced. The reasons for this difference are not obvious though they may be related to the different way the polymeric backbones are hydrated, i.e. the orientation of the water molecules attached to the chain may determine the distance of closest approach of hydrated ions to the polymer chain. The consequences of this are that those nonionic surfactants having head groups containing ether or carbonyl groups (such as the well known ethoxylated types) are strongly influenced by electrolytes, whereas those having saccharide head groups are largely independent of electrolyte.

Polymers such as poly(ethylene oxide) or poly(vinyl pyrrolidone) that are salted out by electrolytes usually show an inverse temperature solubility, i.e.

Table 1.2 Temperature dependence of solution properties of pullulan

Temp. (°C)	Fraction 1		Fraction 2	
	$\langle S_z^2 \rangle$ (nm)	$A_2 \times 10^4$ (ml.mol g^{-2})	$\langle S_z^2 \rangle$ (nm)	$A_2 \times 10^4$ (ml.mol g^{-2})
15	62	1.88	38.0	2.23
25	63.4	1.94	41.0	2.30
35	63.2	1.96	44.2	2.56
45	70.8	1.83	37.6	2.09

$\langle S_z^2 \rangle$ is the mean square radius of gyration (from light scattering) and A_2 is the second osmotic virial coefficient.

they become more insoluble with increasing temperature and exhibit a lower critical solution temperature. Those polymers not so influenced by electrolytes, such as polysaccharides or polyacrylamide [5] are also less affected by temperature as shown by the data [8] in Table 1.2 for two fractions of pullulan, a linear polysaccharide.

Thus surfactants having saccharide (or amide) head groups are more independent of temperature than the ethoxylated equivalent. The fundamental reasons for the difference in response to solution conditions of various polar groups is not well understood and requires more study.

For simplicity chemists usually represent liquid interfaces with straight lines knowing that thermal and electrical interactions will promote undulations, as shown in Figure 1.1. Undulations in surfactant mesophases have been studied for some time, though they have only recently been measured in single layers at interfaces. Lu *et al.* [9] and Lyttle *et al.* [10] using neutron reflection, measured the extent of the undulations to be about 13 Å for a cationic surfactant compared to about zero for a non-ionic surfactant. One may speculate that the lateral repulsions between the cationic head groups

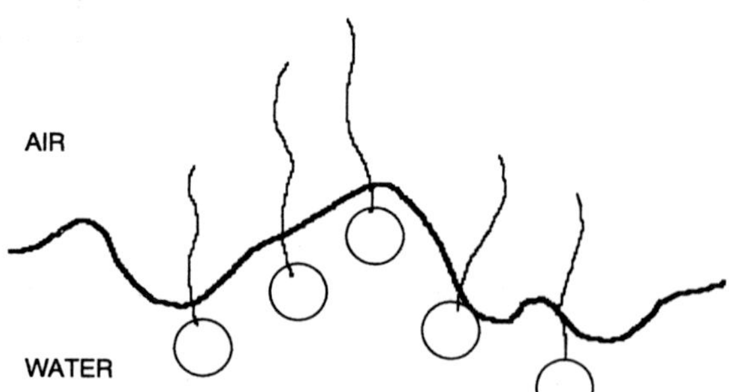

Figure 1.1 Undulations at the air/water interface containing adsorbed surfactants.

contributed significantly to this large undulation and that surfactants having head groups that showed attractive rather than repulsive interactions would have much smaller undulations. The latter might include amide, zwitterionic, mixtures of cationic and anionic (catanionic) or to a lesser extent, bolaform surfactants. The influence of this reduced undulation on interfacial properties such as transport across the surface, surface viscosity or elasticity is unknown but may be significant.

A further speculation on the importance of head group interactions comes from a consideration of the relative stability of soap and detergent foams. It is well known that the foams of most single chain synthetic detergents are broken by the presence of calcium soap precipitates (a simple bathtime experiment) whereas that formed from soaps is not so easily broken. The existence of acid soaps, i.e. protonated carboxylic acid groups that can hydrogen bond within the head group structure, may contribute to this difference, though this is not completely understood.

1.3 Chain interactions

1.3.1 Aqueous systems

The differences in chain types of most interest include chain length, number and unsaturation. The effects of most of these differences can be understood in terms of their ability to pack together. The longer the chains of single chain surfactants the greater the tendency towards flatter, lamellar structures. Doubling the number of chains, if both are the same length, also encourages lamellar structures. However, if the chains are of unequal length, with one being much shorter than the other, the packing is disrupted and the more curved surfaces of discs or rods are preferred. Unsaturation usually introduces more *cis* bonds to chains, breaking the regular array of an all *trans* configuration, thus again disrupting packing and usually lowering the melting point of the lipid array. Nature has exploited these differences to great effect in forming various membranes and cellular structures and whilst commercial surfactants also utilize these differences, there are further benefits yet to be obtained.

1.3.2 Non-aqueous systems

Incompatibility between water and hydrophobic molecules is relatively easy to identify as the result of the strong interactions between the water molecules. However, incompatibility between different hydrocarbon chains is also well known and is manifested especially in polymeric systems where the entropy of mixing is small; mixtures of different polymers (having no mutual attraction) usually phase separate at high enough concentrations.

The fundamental origin of this incompatibility is less easily identified but may be associated with reduced contact between segments of different polymer chains. With small molecules, mixtures of different hydrocarbons are often miscible due to the high entropy of mixing. An extreme case of this difference is shown by hydrocarbon and fluorocarbon chains. Fluorocarbon chains are not only more bulky but also more rigid and hence there is a tendency for these to separate. This opens the way for fluorocarbons and hydrocarbons to show some fascinating aggregation behaviour, as discussed in chapter 5.

1.4 Concluding remarks

The aggregation and adsorption of surfactants are a result of a balance, usually between chains tending to associate and head groups opposing this association. Changes in the geometry, charge or solvation of the chains or head group can have a significant effect on this balance and hence on the properties of the surfactant. Once we are no longer bound by the view of traditional surfactants, novel ones become more feasible. One example is that based on particulates. It is well known that colloidal particles such as clays, having hydrophilic/hydrophobic patches, can stabilize emulsions. These (Pickering) emulsions are very stable yet little work on the synthesis of macroscopic particles as surfactants has been reported. There is still great scope for novel surfactants to be made and exploited.

References

1. Clint, J. *Surfactant aggregation.* Blackie, Glasgow, 1992.
2. Israelachvili, J.N. *Intermolecular and surface forces.* Academic Press, London, 1985.
3. Hall, D.G., *J.C.S. Faraday Trans. II*, 1972, **68**, 2169.
4. Hall, D.G. *J.C.S. Faraday Trans. II*, 1974, **70**, 1526.
5. Garvey, M.J. and Robb, I.D., *J.C.S. Faraday Trans. I*, 1979, **75**, 993.
6. Bailey, F.E. and Callard, R.W., *J. Appl. Polymer Sci.*, 1959, **1**, 56.
7. Clifferi, A. and Orofino, T.A., *J. Phys. Chem.*, 1966, **70**, 32778.
8. Buliga, G.S. and Brant, D.A., *Int. J. Biol. Macromol.*, 1987, **9**, 71.
9. Lu, J.R., Le, Z.X. *et al. J. Phys. Chem.*, 1995, **99**, 8233.
10. Lyttle, D.J., Lu, J.R. *et al. Langmuir*, 1995, **11**, 1001.

2 Self-assembly patterns in double- and triple-chained ionic surfactants

H.N. PATRICK and G.G. WARR

2.1 Introduction

The vast bulk of the scientific literature on amphiphilic compounds deals with single-chained compounds. The adsorption, surface tension-lowering and micellization behaviour of thousands of such surfactants are now documented under a huge variety of conditions of temperature, pressure, and concentration of salt and other additives. The formation of lyotropic liquid crystals, as well as of microemulsions and sponge phases, often with an added oil, co-surfactant (a moderate chain-length alcohol), or both, is also widely reported in the literature.

It is particularly with reference to self-assembly phases containing a co-surfactant that this chapter is written. We consider the use of synthetic multiple chained surfactants instead of surfactant or surfactant/co-surfactant mixtures. These often have many advantages from both a fundamental and applied point of view: reducing the number of components often makes fundamental features and relationships more apparent, simplifies thermodynamic analysis, and at a more quotidian level, reduces the possibilities for formulation errors.

Our scope is deliberately limited to ionic amphiphiles, excluding both zwitterionic and non-ionic compounds. We also exclude lipids, as these are so numerous that they are properly the subject of a separate review.

Below we outline some of the most common kinds of double- and triple-chained ionic surfactants. We then discuss the patterns which emerge from studies of their self-assembly in water and in the presence of an oil. Aqueous phases are taken first, beginning with micelles and moving on to liquid crystalline phases, before adding oil and considering inverse micelles, microemulsions and swollen liquid crystals.

2.2 Cationic surfactants

2.2.1 *Quaternary ammonium surfactants: an archetype for the preparation of multi-chain surfactants*

The most common cationic surfactants are quaternary ammonium salts [1]. The structures of some common single-, double-, and triple-chained quaternary ammonium surfactants are shown in Figure 2.1.

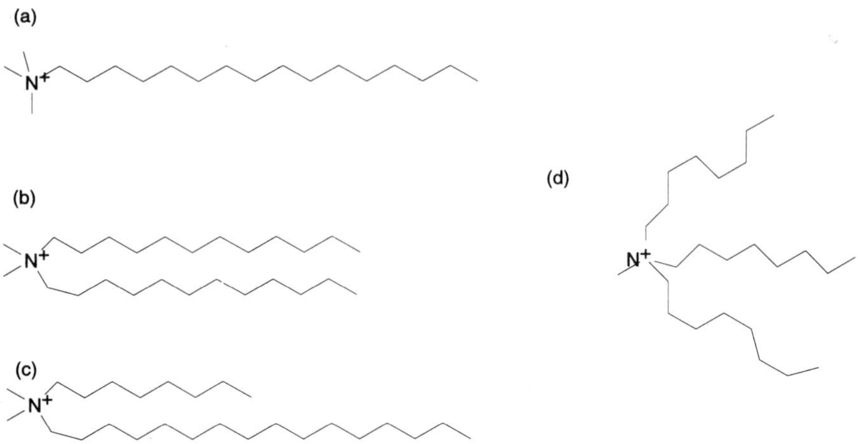

Figure 2.1 Typical structures of cationic quaternary ammonium ionic surfactants. (a) Single-chained hexadecyltrimethylammonium, (b) symmetrical double-chained didodecyldimethylammonium, (c) asymmetric double-chained hexadecyloctyldimethylammonium, (d) triple-chained methyltrioctylammonium.

In single-chained surfactant systems, the conjugate acids of primary, secondary and tertiary amines are also commonly studied. This is not the case for double-chained surfactants; however, protonated trialkylamines have been examined for their flotation response [2].

This class of surfactants has been studied exceedingly widely, largely owing to their ease of synthesis. This means that a large variety of structures are known encompassing single-chained, symmetric and asymmetric double-chained, and triple-chained ions. Two common synthetic routes involve reaction of an amine with a bromoalkane. The simpler of the two is reaction of a bromoalkane R_1Br with a tertiary amine $R_2R_3R_4N$ as shown in Scheme 2.1.

In the other route, common for synthesizing symmetric double-chained surfactants, two moles of bromoalkane are reacted with a dialkylamine (usually dimethylamine, as shown). This is done in basic solution (KOH or K_2CO_3) to ensure that the protonated tertiary amine is not stabilized at the

Scheme 2.1 The synthesis of a quaternary ammonium surfactant by the reaction between a bromoalkane and a tertiary amine.

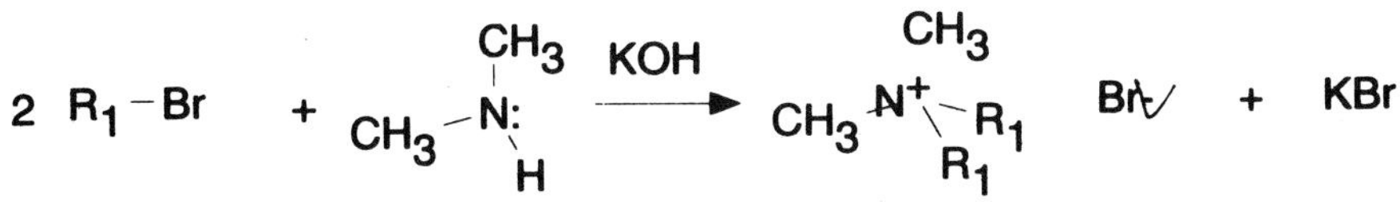

Scheme 2.2 The synthesis of a quaternary ammonium surfactant by the reaction between a bromoalkane and a dialkylamine.

expense of the quaternary ammonium ion. This reaction yields equimolar KBr and quaternary ammonium product, as indicated in Scheme 2.2.

These reactions may produce low yields of the quaternary ammonium salt due to steric hindrance around the nitrogen, particularly if something other than dimethylamine is used.

In both of these syntheses the bromoalkane is favoured as a starting material because bromine is an excellent leaving group; however, chloroalkanes can also be employed. Once the quaternary ammonium salt is prepared any counterion may be substituted for bromide by simple ion exchange. A particularly convenient procedure is to exchange bromide for hydroxide, usually in 90 : 10 methanol : water, and then to titrate the product with the acid of choice. The pH changes in this are useful indicators that exchange has occurred, but direct substitution also proceeds without problems. Among the systems prepared and studied in this way are didodecyldimethylammonium and dihexadecyldimethylammonium fluoride, chloride, iodide, hydroxide, acetate, propionate, sulfate and nitrate [3–6].

Numerous quaternary ammonium surfactants have been produced and studied, including the symmetric double-chained species mentioned above as well as numerous asymmetric double-chained compounds, discussed below. Triple-chained surfactants are typically produced from the symmetric trialkylamines reacted with bromomethane [7–9].

2.2.2 *Other cationic systems*

In addition to these simplest of surfactants, Kunitake [9–11] has produced many double- and triple-chained ammonium surfactants, which also include a connector and a spacer before the cationic head group, by much more elaborate synthetic pathways. The tails are hydrocarbon or fluorocarbon chains. The connector is often an amide or ester linkage, separated from the head group by methylene spacers. Single-chain molecules have also been synthesized and include a hydrocarbon or fluorocarbon tail, a rigid segment, a spacer of methylene units, and a hydrophilic head group. These molecules were all designed to form bilayers on self-assembly. The rigid segment was incorporated to induce the ordering of the molecules into bilayers. It is made of aromatic groups or multiple hydrogen bonding units. Several examples

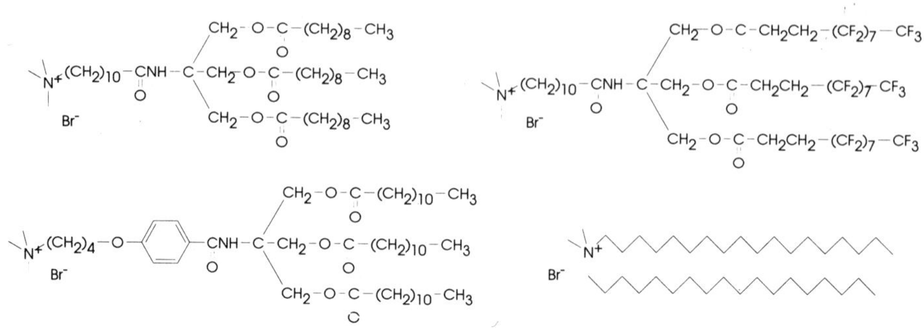

Figure 2.2 Typical examples of multiple-chained surfactants prepared by Kunitake *et al.* [9,11].

are shown in Figure 2.2. The self-assembly phases formed by these surfactants have been identified by electron microscopy and their phase boundaries measured with differential scanning calorimetry [9,12,13].

2.3 Anionic surfactants

2.3.1 AOT: a different archetype

The most common double-chained anionic surfactants provides a different structural model. This is commercially known as Aerosol OT, or just AOT; sodium bis(2-ethylhexyl)sulfosuccinate, shown in Figure 2.3(a). The sodium salt is most commonly used, and is commercially available from numerous sources in varying degrees of purity. Purification of AOT typically involves drying under vacuum and washing to extract organic impurities [14]. Some attention is currently being paid to the role of counterions in microemulsions of AOT, particularly multivalent metal ions. These are easily prepared by ion exchange from the commercial product [15,16].

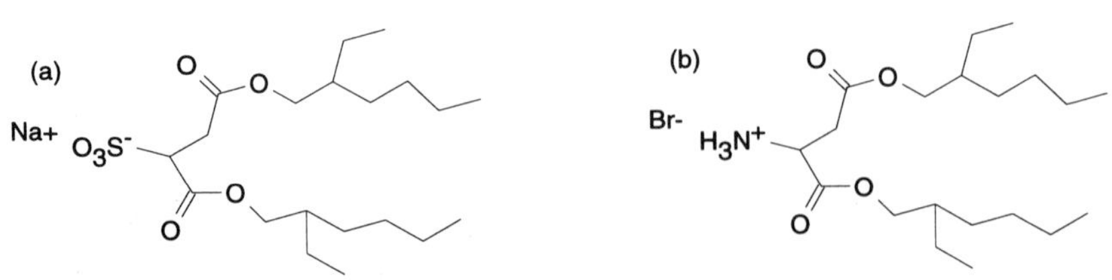

Figure 2.3 Structures of (a) sodium bis-(2-ethylhexyl)sulfosuccinate (AOT), and (b) its cationic analogue, BC8ASP [77].

Unlike quaternary ammonium surfactants, AOT has a branched hydrophobic 'tail' structure which renders it very flexible at an interface (see below). AOT received a great deal of attention in the 1970s in non-aqueous solvents, where it forms reverse micelles, although much of this work was affected by trace amounts of water in the solvent, which leads to the formation of a water-in-oil microemulsion (see below).

Other anionic surfactants based around the AOT structure have been examined, including Aerosol MA, sodium bis-(1,4-dimethylbutyl)sulfosuccinate and numerous *n*-alkylsulfosuccinate anions [17,18].

2.3.2 Sodium alkylbenzene sulfonates (NaABS)

Another common anionic surfactant structure is alkylbenzene sulfonate. This class of surfactants is very widely used industrially, constituting the largest production tonnage of anionic surfactants after soaps. One of the most common is dodecylbenzene sulfonate, which is usually used as the sodium salt, NaDBS. However, as the acid is quite stable and readily available, other counterions may also be studied without difficulty [19]. It is not obvious at first glance that this is a double-chained surfactant at all, until the tail structure is elaborated a little further. Isomerically pure NaDBS is available, but more commonly one obtains a mixture of isomers as shown in Figure 2.4. This shows the branching isomers of dodecyl-*p*-benzenesulfonate, but it should be noted that commercial preparations may also contain *o*-substituted isomers [20]. In one study [21,22], the *para* isomers were prepared and purified individually, and the effect on adsorption and self-assembly behaviour examined. Formation of vesicles and lyotropic phases by isomerically pure and mixed NaDBS in the presence of different added salts has also been examined. We will return to this briefly later.

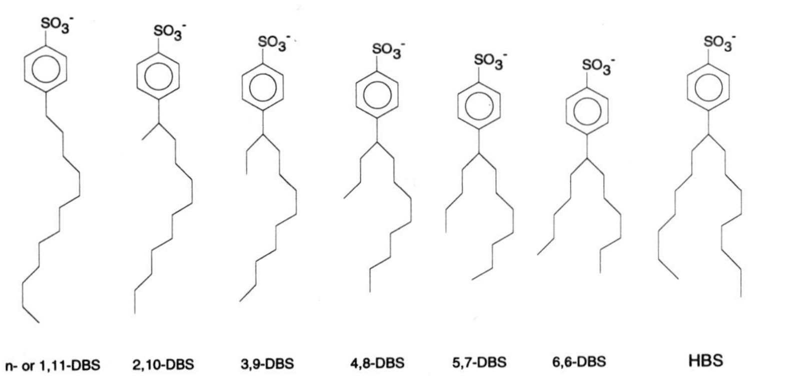

Figure 2.4 Dodecylbenzene sulfonate (DBS) structure, showing straight-chain and branched- or double-chained isomers using the nomenclature of Dick [21,22]. Also shown in the 6,6 isomer of hexadecylbenzenesulfonate [23].

Also widely used is a hexadecyl analog of SDBS, sodium 8-phenyl-*n*-hexadecyl-*p*-sulfonate (SHBS), also shown in Figure 2.4. A preparation for this and other isomerically pure alkylbenzene sulfonates has been given by Gray [23].

2.3.3 Other classes

Among other classes of surfactants, glycerol-based surfactants with two or three long hydrocarbon tails and two hydrophilic ionic head groups are readily prepared. The two head groups are sulfonate, sulfate or carboxylate [24,25]. These have not been studied extensively, but are reported to have much better surface tension lowering ability and lower critical micelle concentrations than their single-chain analogues [24,25].

2.4 Self-assembly behaviour: surfactant geometry

The molecular architectures realized by amphiphile self-assembly are highly varied and complex, but can be unravelled using quaternary ammonium surfactants as a model. The gradual variations in structure which can be achieved using quaternary ammonium surfactants make them useful model systems for understanding the principles which determine structures formed in surfactant self-assembly.

In 1976 Ninham and co-workers [26] detailed the geometric considerations implicit in the self-assembly of surfactants into micelles and vesicles in aqueous solution. Their recognition of a surfactant packing geometry was a watershed in understanding micellization, and subsequently higher order self-assembly processes. Although the original work focussed on single-chained surfactants, and particularly on spherical and rod-like micelles, demolishing the case for micellar disks along the way, its extension to multiple chains and other self-assembly structures was soon to follow.

The essential idea was that the geometry of a surfactant molecule in an aggregate could be encapsulated into a surfactant packing parameter, v/a_0l_C, as illustrated in Figure 2.5.

Spherical micelles are the simplest case, and illustrate the geometrical arguments behind all allowed interfacial shapes. In a micelle of aggregation number N, the volume of the hydrophobic core must be $V = Nv$.

In any self-assembly structure the alkyl chains pack at approximately liquid hydrocarbon density. Gruen [27] has fitted density data to produce the following formula for the volume of a liquid-like hydrocarbon chain as a function of the number of methyl and methylene groups

$$v = 54.30n_{CH_3} + 27.05n_{CH_2}$$

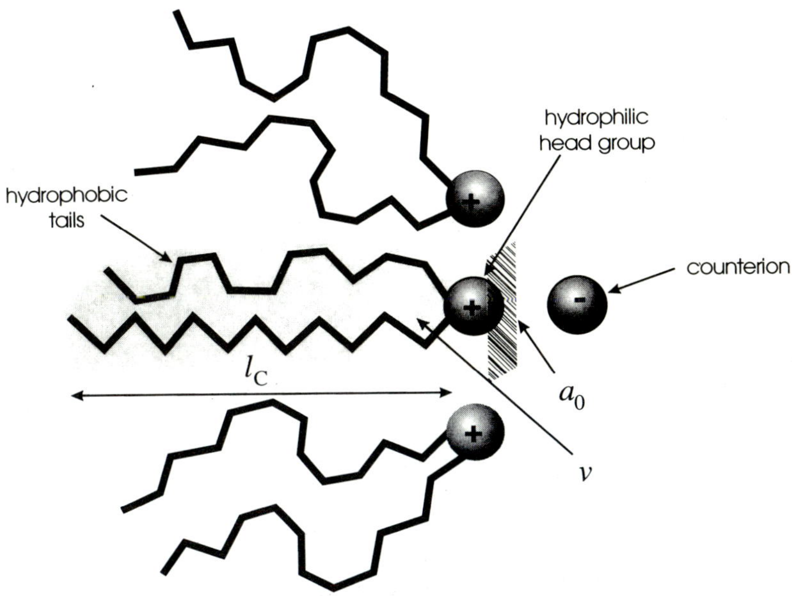

Figure 2.5 Schematic diagram of an amphiphile oriented at an interface, illustrating the concept of the surfactant packing parameter, v/a_0l_C, where v is the molecular chain volume, a_0 the area per surfactant head group, and l_C the alkyl chain length.

(where v is in Å^3) which is almost identical to Tanford's older result for single alkyl chains containing n_C carbon atoms [28]

$$v = 27.5 + 26.9n_C$$

If each surfactant molecule takes up an area a_0 at the micelle/solution interface, then the total surface area of each micelle must similarly be $A = Na_0$. (This is, of course, true for any micellar geometry.)

Now, for spheres the volume and surface area are also given by $V = 4/3\pi r^3$ and $A = 4\pi r^2$ so that the volume to surface ratio is $V/A = v/a_0 = r/3$. As the radius of the micelle must be $\leqslant l_C$, the (longest) chain length of the surfactant, the condition for spherical micelles simply becomes $v/a_0l_C \leqslant 1/3$.

l_C may be calculated for an all-*trans* chain from bond lengths, bond angles, and van der Waals radii, and gives [28]

$$l_C = 1.50 + 1.265n_C$$

(where l_C is in Å).

For single-chained surfactants with, for example, a dodecyl chain, $l_C = 16.7$ Å and $v = 352$ Å^3. In the absence of added salt, the area per molecule at air/solution interfaces just below the critical micelle concentration (CMC) is around 65 Å^2, giving $v/a_0l_C = 0.324$ and, therefore, spherical micelles.

There is an important connection to thermodynamics which must not be overlooked here. v and l_C are molecular properties which cannot easily be changed: liquid hydrocarbons are not easy to compress. a_0, however, arises from a balance of interactions at the interface, and must be regarded as an equilibrium property of the system. One can therefore say that v/a_0l_C describes the *spontaneous curvature* of a surfactant interface.

The curvature, H, of any element of an interface is defined by the two principal radii of curvature as

$$H = 1/r_1 + 1/r_2$$

where r_1 and r_2 are defined in Figure 2.6. By convention, curvature around the oil phase is positive and around the aqueous phase is negative.

Similar arguments for rods can be used to produce analogous geometric conditions for packing. These arise in micellar solutions and also in nematic and hexagonal phases. So it is for planar bilayers, which occur in vesicles, liposomes and lamellar phases.

In the years since Israelachvili, Mitchell and Ninham [26] first established the notion of molecular or packing geometry, several new phases have been postulated or discovered. The geometrical basis for surfactant packing has consequently been expanded to encompass these new conditions, including concentrated liquid crystals and the inverted structures necessary for microemulsions [29]. The general geometrical conditions now widely recognized are summarized as follows:

spheres	$v/a_0l_C \leqslant 1/3$	$H = 2/l_C$
rods	$1/3 < v/a_0l_C \leqslant 1/2$	$H = 1/l_C$
sheets	$1/2 < v/a_0l_C \leqslant 1$	$H = 0$
inverted structures	$v/a_0l_C > 1$	$H < 0$

One of the ways which a_0 of ionic surfactants can be affected is through the counterion. Changing the counterion from chloride to bromide, iodide, or acetate changes the extent of nonelectrostatic ('Stern layer') binding and can alter the repulsions between neighbouring head groups within a micelle. Adding salt can also increase the amount of counterion binding through a simple adsorption equilibrium and have a significant effect on the resulting self-assembly patterns. Addition of strongly binding counterions or high electrolyte concentrations to single-chained surfactant solutions is a common strategy for creating cylindrical or worm-like micelles by adjusting v/a_0l_C [30,31].

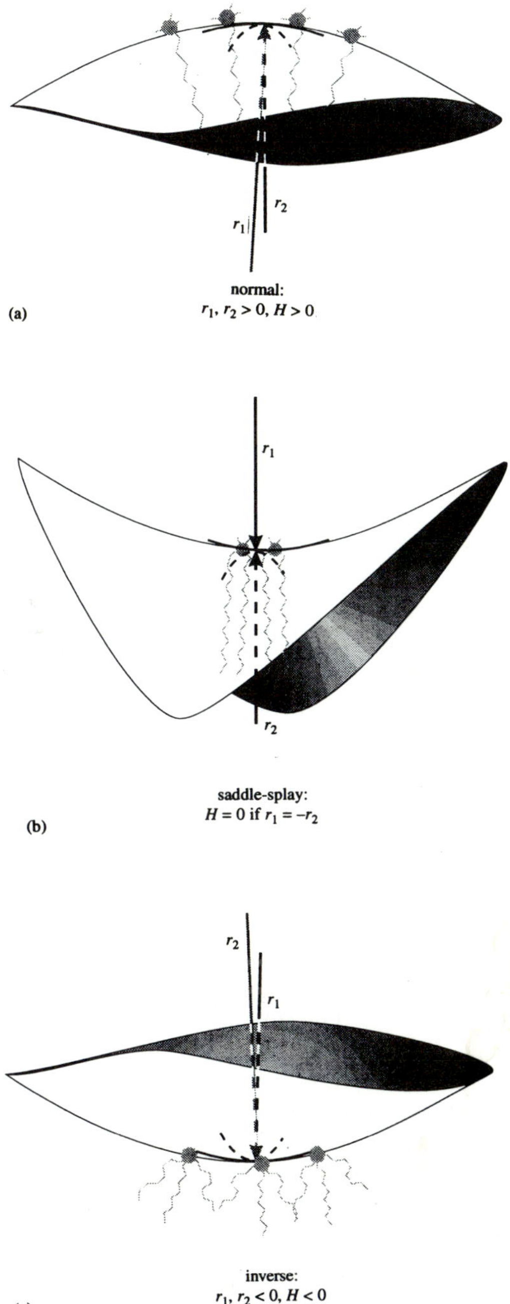

Figure 2.6 Curvature of an element of surfactant-coated interface defined in terms of principal radii of curvature showing (a) normal, or positive, (b) zero, and (c) inverse, or negative curvature.

It is also common in ionic surfactants to see a_0 decrease as surfactant concentration increases. Increasing surfactant concentration generally increases the ionic strength in the system, which better screens electrostatic repulsion between ionic head groups. This shifts the balance of interactions towards lower area per molecule, and hence larger $v/a_0 l_C$. The usual progression of geometrical elements is the same whether salt is added or surfactant concentration is increased [29,32]:

$$\text{spheres} \rightarrow \text{rods} \rightarrow \text{bilayers} \rightarrow \text{inverted rods} \rightarrow \text{inverted spheres}$$

For liquid crystal phases, this commonly leads to the following order with increasing surfactant concentration:

$$\text{micelles } (L_1) \rightarrow \text{hexagonal } (H_1) \rightarrow \text{lamellar } (L_\alpha) \rightarrow \text{inverse hexagonal } (H_2)$$

Inverted structures have proven to be a particularly rich field, with extensive investigations of water-in-oil (w/o) and bicontinuous microemulsions. More recently, self-assembly structures with near-zero net curvature have emerged, including cubic phases, sponges, and postulated branched micelles. The recent work of Hyde [33] relating the surfactant packing parameter to the mean and Gaussian curvatures of an interface has been important in understanding such structures.

Two types of cubic phase are of relevance here. In single-chained surfactants, structures based around an ordered array of spherical micelles have been reported. Such V_1 cubic phases usually appear between a spherical micellar L_1 phase and a hexagonal phase [34]. These will rarely arise in double or triple chained surfactants as these do not normally form spherical micelles. More significant are bilayer-based cubic phases, I_2. These structures are surfactant bilayers which, rather than being planar, lie on a geometric surface of zero mean curvature. Every point on such a surface is saddle point, curving equally in towards either side. An idea of the local curvature in such a system is illustrated in Figure 2.6(b).

Minimal surfaces divide space symmetrically into two interpenetrating labyrinths. In the case of I_2 phases, 'space' is just the majority solvent, usually water. I_2 phases exist in several different symmetries [35–39] including tetrahedral (diamond, or face-centred cubic) and octahedral (body-centred cubic). They may be most easily visualized by imagining swelling the surface of a ball-and-stick model of a unit cell, and then coating that surface with a bilayer of surfactant. The resulting structure is regular and periodic, and has two sets of interpenetrating channels. As the midplane of the bilayer where the tails meet has zero mean curvature, the curvature in the plane of the head groups is slightly negative (unlike Figure 2.6(b)). Such phases are generally observed between the lamellar and inverse hexagonal phases.

2.5 Double-chained surfactants

2.5.1 Aqueous solutions of double-chained surfactants

In aqueous solution, the surfactant packing parameter gives us certain expectations regarding self-assembly patterns. Comparing packing parameters for single- and double-chained surfactants, we do not expect to see micelles in double chained surfactant systems. If the packing parameter for a single-chained surfactant is consistent with spherical micelles as above, say $v/a_0l_C = 0.3$, then two tails effectively doubles the molecular volume without changing the chain length or a_0, giving $v/a_0l_C = 0.6$. This is too high even for rods to form. We therefore anticipate self-assembly structures with near-zero interfacial curvature: vesicles, lamellae and perhaps sponge phases.

One set of systems which exemplifies the transition from micelles to bilayer-based structure is the asymmetric quaternary ammonium bromides, which have been investigated systematically [40]. Various isomers of alkylbenzenesulfonates have also been studied, and these are discussed below.

(a) Quaternary ammonium surfactants. Zana has systematically examined the properties of aqueous solutions of amphiphiles of the general structure $(C_{12}H_{25})(C_nH_{2n+1})N(CH_3)_2Br$, with $n = 1–12$ [40,41]. In this series, increasing n first causes little change in the critical micelle concentration, but it then decreases abruptly above $n = 3$, as shown in Figure 2.7. Once $n > 3$ the alkyl moiety apparently incorporates itself into the hydrophobic part of the aggregate. This has two effects, firstly it increases v without much affecting a_0 or l_C, and is accompanied by a change from spherical micelles through rods ($n = 6–8$) into vesicles ($n = 10–12$). Secondly, the hydrophobic contribution to self-assembly increases as more alkyl groups are incor-

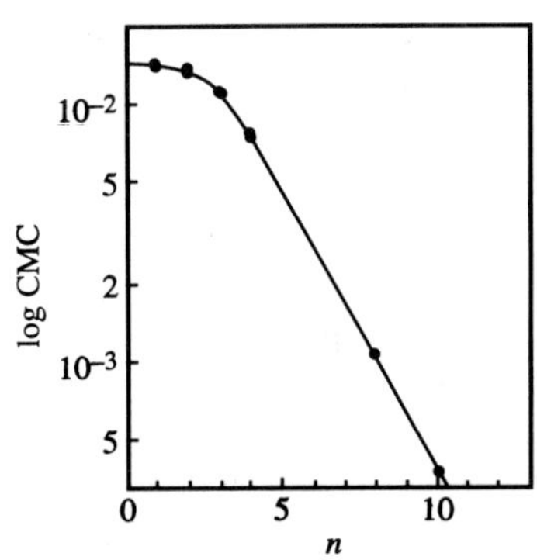

Figure 2.7 Critical micelle concentrations (mol l^{-1}) of $(C_{12}H_{25})(C_nH_{2n+1})N(CH_3)_2Br$ versus length of second alkyl chain, n. Re-drawn from Reference 40.

porated into the micelle. It is this which causes the CMC to decrease precipitously, following a log–linear relationship with chain length as is observed in single-chained surfactants (see Figure 2.7).

Similar and consistent behaviour is seen in aqueous solutions of hexadecyloctyldimethylammonium bromide (C_8C_{16}DAB) [42]. At the CMC cylindrical micelles form immediately which grow and become entangled with increasing concentration. This, too, is consistent with the v/a_0l_C picture. According to Gruen's formula, $v = 54.3 \times 2 + 27.05 \times 22 = 704$ Å^3, the chain length is 21.7 Å, and the area per head group remains approximately 68 Å^2, yielding $v/a_0l_C = 0.48$: rods.

C_8C_{16}DAB is also interesting in that its first liquid crystal phase is lamellar, as shown in Figure 2.8 [43,44]. This can be rationalized considering that its asymmetric tails might easily intercalate in a bilayer structure, particularly at high concentrations where a_0 might be reduced, favouring bilayers over rods. C_8C_{16}DAB continues the progression in packing parameter, forming an inverse hexagonal phase at higher concentrations. The structure of the intervening L_1 phase has not been investigated.

The progression in packing parameter can also be realized by mixing single- and double-chained surfactants, forming prolate ellipsoidal mixed micelles of dodecyltrimethylammonium bromide and didodecyldimethylammonium bromide (DDAB) [45].

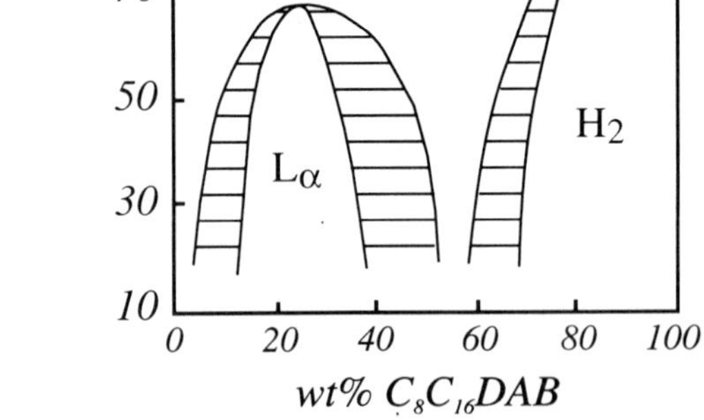

Figure 2.8 Binary phase diagram of hexadecyloctyldimethylammonium bromide (C_8C_{16}DAB). (Re-drawn from Reference 44.) L_1 denotes a liquid isotropic phase, L_α a lamellar phase, and H_2 an inverse hexagonal phase. Two-phase coexistence regions are shown by horizontal tie lines.

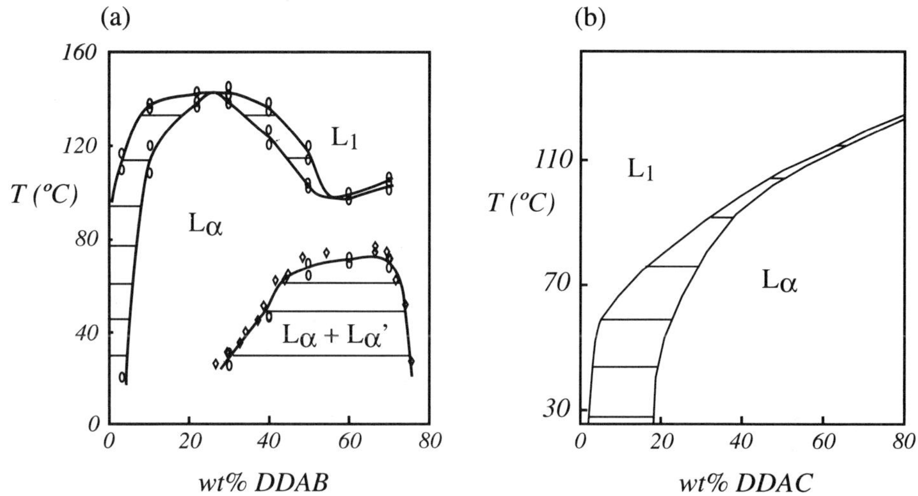

Figure 2.9 Binary phase diagrams of didodecyldimethylammonium bromide (DDAB) and chloride (DEDAC) in water. (a) DDAB. Data points locate the phase boundaries by optical observation (○), Reference 44, and neutron scattering (◇), Reference 47. (b) DDAC. (Redrawn from Reference 6.) L_1 denotes a liquid isotropic phase and L_α a lamellar phase. $L_\alpha + L'_\alpha$ indicates a two phase demixing between lamellar phases.

DDAB is the final compound in the above series. Its self-assembly behaviour has been studied extensively, and in broad view it is easy to understand. Only lamellar phases form, as would be expected for a system with $v/a_0 l_C = 704/(68 \times 16.7) = 0.62$ [6,44,46]. (This is also the case with most other counterions, even though differences between counterion binding can affect a_0 somewhat [6].) The detailed phase behaviour of the DDAB system actually shows a demixing between concentrated and dilute lamellar phases with an upper consolute point at 62.2 wt% and 75°C [47]. Binary phase diagrams of DDAB and DDAC constructed from several studies are shown in Figure 2.9.

(b) Alkylbenzene sulfonates. Sodium hexadecyl-8-*p*-benzenesulfonate (SHBS, Figure 2.4) forms a dispersion of vesicles in aqueous solution above approximately 10^{-3} M [48,49]. Its $v/a_0 l_C$ can be estimated, if we consider the phenyl ring as part of the alkyl tail. The chain length may be estimated using space-filling molecular models, which agree well with Tanford's formula, and the volume estimated similarly. (This will necessarily be approximate, as the packing of rigid phenyl groups will not be at liquid hydrocarbon density, but changing the way the phenyl ring is factored in has little effect on $v/a_0 l_C$.) As the area per molecule is determined mainly

by electrostatic interactions, it should be the same for single- and double-chained surfactants, approximately 65 Å^2. This yields for the chain $v = 54.3 \times 2 + 27.05 \times 14 + 72 = 559$ Å^3, the molecular length is 16.1 Å, and hence $v/a_0 l_C = 559/(65 \times 16.1) = 0.53$. A lamellar phase is therefore expected as the first self-assembly phase, as is widely reported.

All of the isomers of dodecylbenzenesulfonate are reported to form micelles in aqueous solution [21], as are commercially available mixtures of isomers [19,20,50]. For the 6,6 isomer (Figure 2.4) $v/a_0 l_C = 505/(65 \times 13.6) = 0.57$ (c.f. 0.56 as in Reference 51) and we would again expect bilayers, this time contrary to observations.

At higher concentrations, the 6,6 isomer of SDBS does form a lamellar phase as its first liquid crystal. A partial phase diagram is shown in Figure 2.10 [52]. It is interesting to note that, like DDAB in water, 6,6-DBS also exhibits a coexistence region between two lamellar phases with an upper critical temperature. The coexistence of two lamellar phases has yet to be fully explained, but work on the DDAB system has identified a critical role for counterion binding in determining the interactions between adjacent bilayers [47].

The formation of micelles by sodium 6,6-DBS remains unexplained in terms of the surfactant packing parameter. Small angle neutron scattering (SANS) studies confirmed the existence of small, spherical micelles in dilute

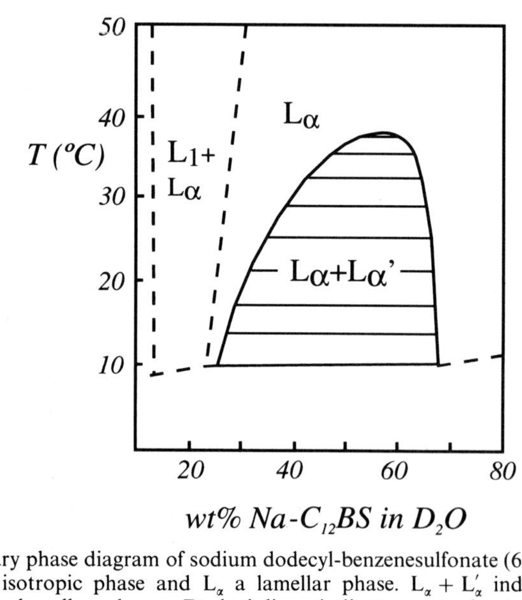

Figure 2.10 Binary phase diagram of sodium dodecyl-benzenesulfonate (6,6-DBS) in D_2O. L_1 denotes a liquid isotropic phase and L_α a lamellar phase. $L_\alpha + L'_\alpha$ indicates a two phase demixing between lamellar phases. Dashed lines indicate approximate phase boundaries. (Re-drawn from Reference 52.)

solution (at 45°C) [53]. Further, the existence of spherical micelles of SHBS has also been demonstrated in a small concentraion range below the concentration where vesicle dispersions have been observed [50].

Comparisons between branched and straight chain isomers of DBS have helped elucidate the micellar structure in these systems. For these, v and a_0 are approximately constant, and only l_C changes between isomers. For the straight chain isomer, $v/a_0l_C = 505/(65 \times 21.1) = 0.37$, and spherical or rod-like micelles are expected.

Two studies comparing single- and double-chained isomers of DBS and decylbenzenesulfonate have produced the same conclusions [53,54]. Namely that straight- or single-chained surfactants form spherical micelles which grow only slowly with concentration. The difference with double-chained isomers, such as 6,6-DBS or HBS, is that they form smaller micelles which grow more rapidly with increasing concentration or addition of salt. SANS studies further suggest a greater degree of hydrocarbon/water contact for branched isomers, providing a driving force for growth and eventual transformation into a bilayer-based structure. The phenyl group acts as a spacer between the alkyl chain and ionic group, and forms part of the polar shell around the hydrophobic micelle core. This is supported by a recent NMR study of chemical shifts of the phenyl protons on micellization [51].

Self-assembly of alkylbenzenesulfonates is driven by the expulsion of hydrocarbon tails from water, but their complex head-group structure leads to a thicker polar layer around the hydrophobic core. This complicates the application of the v/a_0l_C concept, as such aggregates have a less well defined optimal curvature than the more ideal quaternary ammonium species. As a result, double-chained alkylbenzenesulfonates form spherical micelles which deform and grow into aspherical aggregates with increasing concentration or added salt. Only at higher concentrations do they agree with geometrical expectations, forming a planar bilayer with a better-defined interface.

(c) Dialkylsulfosuccinate salts (Aerosol OT). AOT has a low solubility in water, and forms an L_1 isotropic phase up to approximately 1.35 wt% at room temperature [55]. Within this phase AOT forms small, spherical micelles with an aggregation number near the CMC of approximately 15 which grows to around 30 with increasing concentration [56]. In this respect it is like the branched-chain alkylbenzenesulfonates.

Also like the alkylbenzenesulfonates, AOT has a v/a_0l_C which predicts bilayers or inverted structures. Literature data gives a molecular length of 12.6 Å and volume of 611 Å^3 for the hydrophobic region [56]. Using Tanford's formula for the chain length of one hexyl chain gives 9.1 Å, whereas CPK molecular models give 11 Å up to the C–S bond. Using a molecular area of 65 Å^2, this yields $0.75 < v/a_0l_C < 1.0$. Although not as rigid or bulky as a benzenesulfonate, the polar head group of AOT is large enough to cause a similar aggregation pattern.

It has been asserted that AOT forms a 'mesomorphous phase consisting of parallel rod-like aggregates... in hexagonal array' [56] between 1.35 and 29 wt% in water, but we have been unable to substantiate this in any published phase diagrams [18,55,57].

At higher concentrations the aggregation pattern falls into line with geometrical expectations. Between 19 and 72 wt%, AOT forms a lamellar phase, then a cubic phase between approximately 75 and 78 wt%, and finally an inverse hexagonal phase above approximately 80 wt% [55]. The cubic phase is believed to be a bilayer-based structure [39,55]. It should therefore have slightly negative interfacial curvature and lie between the lamellar and inverse hexagonal phases in the expected phase progression.

The aqueous phase behaviour of AOT is sensitive to electrolyte concentration, and addition of NaCl eventually leads to the formation of an L_3 or sponge phase in the same concentration range as the lamellar phase. The L_3 phase structure is also based around a bilayer, but exhibits none of the long-range order seen in liquid crystals [39].

2.5.2 Ternary phase behaviour: the effect of added oil

The presence of an immiscible second solvent leads to a new range of self-assembly behaviour. Isotropic liquids called microemulsions may form, liquid crystals may swell, and new liquid crystal phases may be formed. Recently, isotropic sponge phases have also been reported. In considering ternary mixtures, we look first at the formation of liquid isotropic phases, and then consider liquid crystals.

The earliest studies of ternary, surfactant-containing systems were investigations of the phenomenon of *solubilization*. This is the encapsulation of poorly soluble, i.e. hydrophobic, solutes within swollen micelles, giving these solutes a much greater solubility in surfactant solutions than in water. Numerous studies of solubilization by micelles formed of single-chained surfactants have been reported [58]. The *solubilization capacity* of a solution, the maximum amount of hydrophobic solute which can be added to it, may be interpreted using $v/a_0 l_C$.

Within the $v/a_0 l_C$ approach, addition of oil modifies the interfacial curvature of the surfactant film by increasing the effective volume of the alkyl tail(s). This increases the tendency to form inverted structures, and has been incorporated in an *ad hoc* manner as $v_{eff}/a_0 l_C$, where $v_{eff} = v_{surfactant} + v_{pen}$, where v_{pen} is the volume of oil which penetrates into the film per surfactant molecule. v_{pen} is not something which is always known, so quantitative interpretation of surfactant packing in ternary systems has only been possible in a limited number of cases.

In systems containing added oil, $v_{eff}/a_0 l_C$ describes the curvature of the surfactant film at an oil/water interface, and is equivalent to describing the interfacial curvature using a parallel surface construction with the two surfaces being l_C apart [59].

Recall that v/a_0l_C is a measure of the spontaneous curvature of the interface. If we neglect for the moment any changes in curvature due to oil penetration, it is easy to demonstrate that rod-like micelles will have a substantially greater solubilization capacity than spheres. Consider the spherical micelle discussed earlier, with $v/a_0l_C = 0.30$. The spontaneous curvature of the interface is $2/l_C$, and any addition of oil to the micelle will necessarily increase its radius beyond l_C, shifting the interface away from its preferred curvature at some energy cost. On the other hand, a rod-like system with, say $v/a_0l_C = 0.45$ has spontaneous curvature $1/l_C$. Adding oil to the core of this micelle releases it from the constraint that the radius must be $r \leqslant l_C$. Maintaining the same spontaneous curvature, the system can therefore swell to form spherical droplets with radius satisfying the curvature condition $H = 2/r$ (spheres) $= 1/l_C$(rods). Therefore the droplet radius can increase to $r = 2l_C$ without any change in interfacial curvature. The volume enclosed by a sphere is much greater than by a cylinder with equal surface area, so the solubilization capacity should be quite substantial, and can be determined from straightforward geometry.

The same general arguments apply to inverted structures, including water-in-oil (w/o) and bicontinuous microemulsions with negative interfacial curvature. We begin with the same postulate, that the interface adopts a particular interfacial curvature, or v_{eff}/a_0l_C value, and assume that this remains constant throughout the microemulsion phase.

(a) Quaternary ammonium surfactants. The curvature argument has been applied with great success to DDAB microemulsions assuming that the phase consists of a continuous oil medium containing surfactant-coated water droplets and/or locally cylindrical tubes, tubules, or conduits of water, also coated with surfactant [60,61]. This description, known as the disordered open connected, or DOC model is quite successful at describing the structural changes observed experimentally in inverted microemulsions with a given oil. It predicts the curious observation that DDAB/oil/water microemulsions start out as conducting and bicontinuous, but with the addition of water undergo reverse percolation to a droplet state [46,62–64].

Following the argument above for solubilization of oils into normally curved systems, the same thing follows for reverse systems: as water is added a microemulsion will form when there is enough water to create (approximately) a single, long, conducting cylinder with a particular v_{eff}/a_0l_C characteristic of the system. This also defines a cylinder radius, r_C. (The DOC model actually pictures this as a network of cylinders meeting at branch-points, which are approximated by spheres.) As water is added, the density of spherical branch-points increases and the number of cylindrical conduits decreases, conserving the total interfacial area. Eventually, all of the surfactant is occupied surrounding spheres of radius $2r_C$ and the system has percolated to a discrete structure for purely geometrical reasons.

Specific oil penetration into surfactant films was clearly demonstrated by Gruen and Haydon [65–67] and its effect on self-assembly structure noted by Ninham and Evans. They found that the properties of microemulsions of DDAB + alkane + water depended strongly on the oil used [62–64], and that the behaviour could be rationalized if it was recognized that smaller molecules penetrated further into the tails of DDAB, causing the interface to curve more tightly around water. This is readily seen in Figure 2.11, which shows partial ternary phase diagrams of DDAB + alkane + water systems, including the microemulsion (L_2) regions. Cyclohexane is the

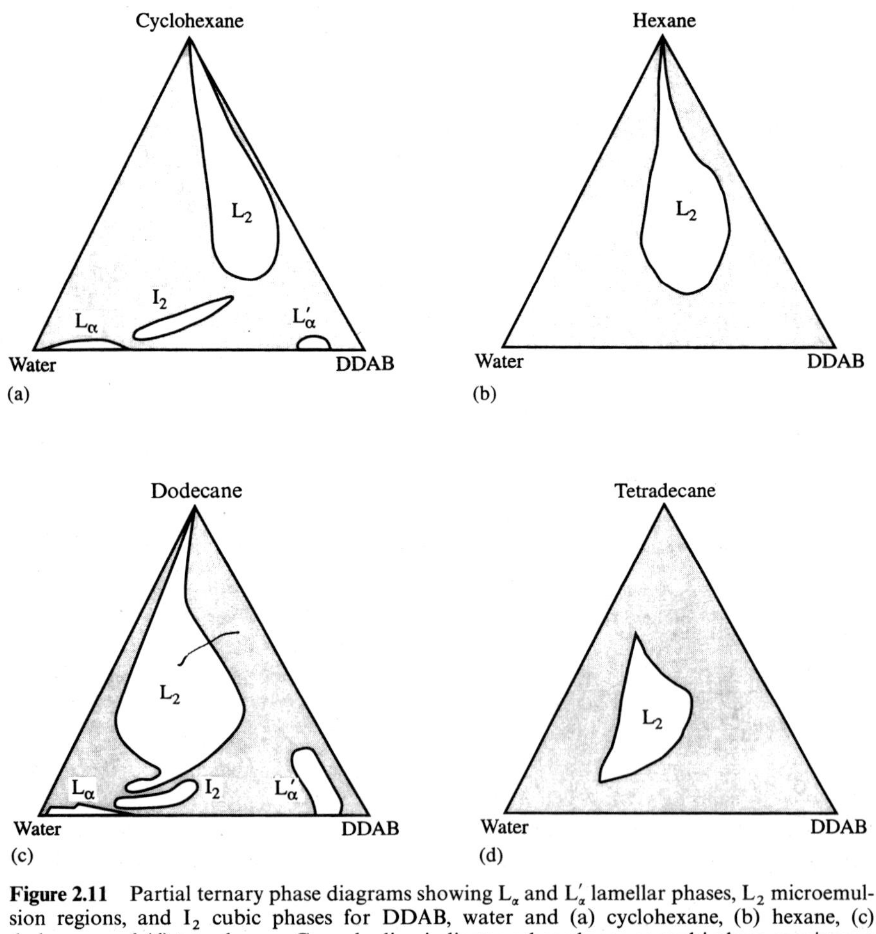

Figure 2.11 Partial ternary phase diagrams showing L_α and L'_α lamellar phases, L_2 microemulsion regions, and I_2 cubic phases for DDAB, water and (a) cyclohexane, (b) hexane, (c) dodecane, and (d) tetradecane. Grey shading indicates other phases or multi-phase coexistence regions. (Re-drawn from References 46, 63 and 64.)

smallest molecule, and penetrates the dodecyl chains of the surfactant to such an extent that only a few weight percent of water is needed to form tightly (negatively) curved water conduits. As we progress towards less penetrating oils — hexane, dodecane and tetradecane — the minimum water needed to form a microemulsion increases and the spontaneous curvature of the interface approaches zero. So it is at the high water side of the phase diagram, where all the water is contained in droplets. The maximum radius which can be achieved by a water droplet is limited by interfacial curvature, which is determined by v_{pen}. Highly penetrating oils like cyclohexane therefore have much lower water capacity than the less highly penetrating oils. Cyclohexane can stabilize microemulsions only up to 21 wt% water, compared with dodecane microemulsions, which can contain over 55% water.

Tetradecane is an unusual case, and is viewed as a totally non-penetrating oil by Ninham and Evans [63,64]. Note that DDAB microemulsions formed with tetradecane cannot form w/o droplets, or any discontinuous structure, as it cannot be diluted infinitely with oil. Small angle scattering supports the view that these microemulsions with tetradecane are unusual, and may best be described by a water-swollen lamellar structure akin to a sponge phase [68].

Much of this oil specificity is also very clearly reflected in the swelling of lamellar phases of DDAB. Figure 2.11 [46] shows both lamellar (L_α) and microemulsion (L_2) regions of ternary mixtures of DDAB, water, and several alkanes. The lamellar phase is the favoured state for DDAB in water, so the maximum swelling of the lamellae by an oil is a useful measure of how much an oil alters the interfacial curvature: more highly penetrating oils increase v_{eff} and hence destabilize the lamellae at lower concentrations. Figure 2.11 clearly shows that dodecdane can swell the lamellar phases to a much greater extent than smaller oils like cyclohexane.

Counterion effects can be important in self-assembly systems, particularly using cationic surfactants. The difference between DDAB and DDAC in water is clearly visible in Figure 2.9. Studies of microemulsions with different counterions are rare, but the phase diagrams of DDAC and DDAI have been studied [69]. Of these ions, chloride binds much less strongly to the DDA ion and iodide much more strongly. Within the $v_{\text{eff}}/a_0 l_C$ interpretation we would expect a greater repulsion between surfactant head groups for DDAC and therefore a larger a_0, hence a smaller $v_{\text{eff}}/a_0 l_C$. The minimum water required for formation of a stable DDAC microemulsion should therefore be greater than for DDAB under the same conditions.

Asymmetric chained surfactants also obey these general principles both in aqueous solution and as microemulsions. This is illustrated by systematically varying tail structure and particularly by comparing DDAB with C_8C_{16}DAB, systems which have the same total alkyl chain volume. Figures 2.12 and 2.13 show the L_2 microemulsion regions of $C_{10}C_{12}$DAB and

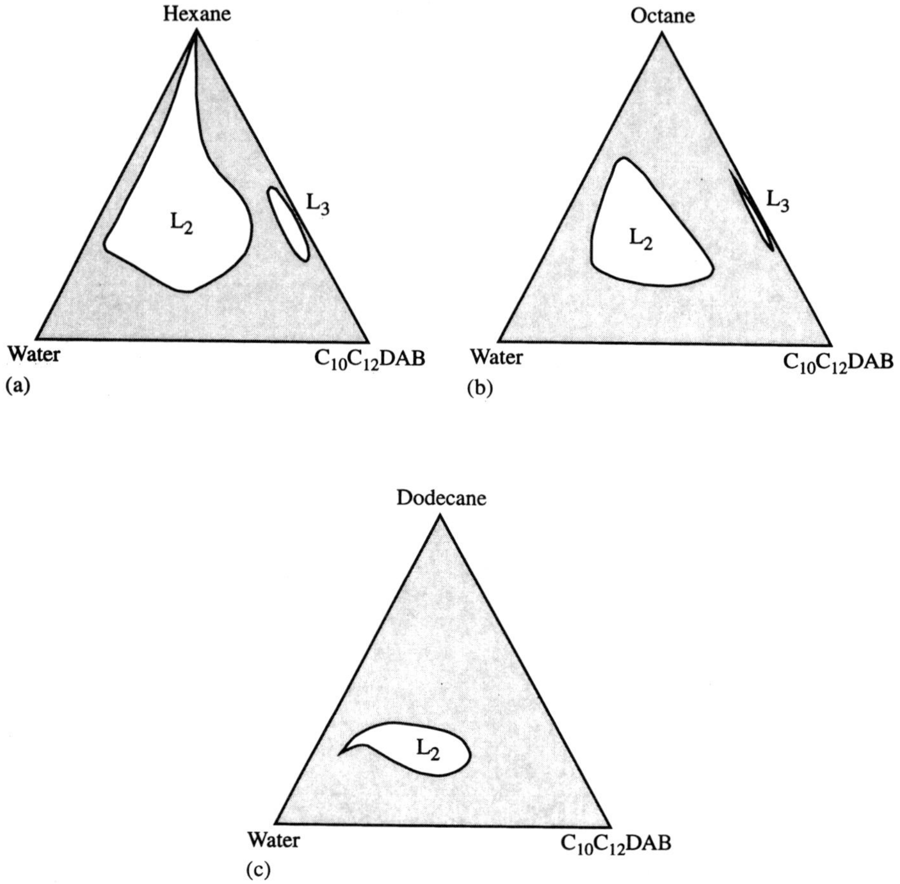

Figure 2.12 Partial ternary phase diagrams showing L_2 microemulsion regions for $C_{10}C_{12}$DAB, water and (a) hexane, (b) octane, and (c) dodecane. (Re-drawn from Reference 44.)

C_8C_{16}DAB in ternary phase maps with several oils, which should be compared with the phase diagrams of DDAB shown in Figure 2.11.

$C_{10}C_{12}$DAB behaves in a manner perfectly analogous to DDAB. The patterns observed for minimum water content, maximum water content and percolation are as expected. Hexane, the smallest and most penetrating of the oils shown, forms microemulsions as the lowest water content, corresponding to the largest (inverse) interfacial curvature. Once the size of the oil molecule is increased even a small amount to octane, the L_2 phase detaches itself from the oil corner and behaves as does tetradecane with DDAB. Octane penetrates the surfactant tails enough to balance the slight tendency towards normal curvature. Further increasing the size of the oil has little effect, moving the microemulsion region towards the water corner

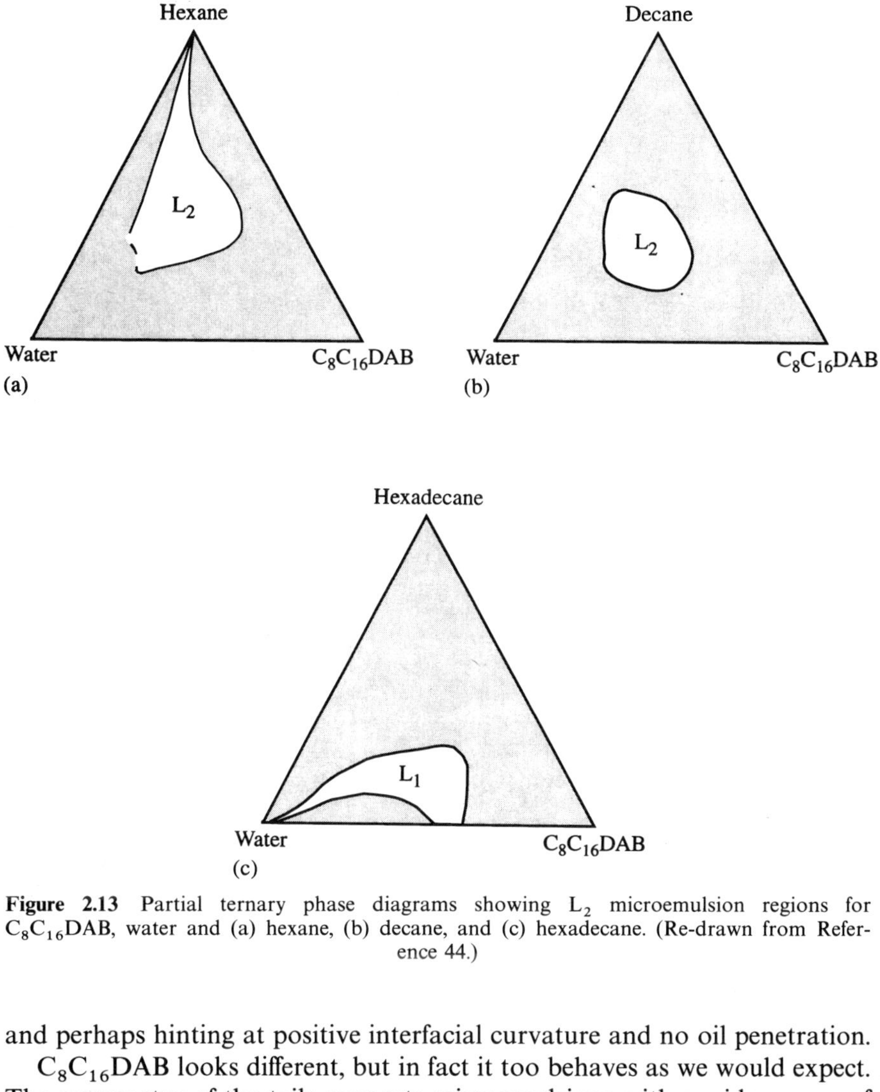

Figure 2.13 Partial ternary phase diagrams showing L_2 microemulsion regions for $C_8C_{16}DAB$, water and (a) hexane, (b) decane, and (c) hexadecane. (Re-drawn from Reference 44.)

and perhaps hinting at positive interfacial curvature and no oil penetration.

$C_8C_{16}DAB$ looks different, but in fact it too behaves as we would expect. The asymmetry of the tails supports microemulsions with a wider range of alkanes than does DDAB. With hexane, a penetrating oil, the observed phase behaviour is the same as for DDAB. Conductivity measurements have shown that the L_2 phase is bicontinuous at low added water, and percolates to w/o droplets as water is added [44]. With decane we see behaviour analogous to that of the DDAB + tetradecane + water system, which is entirely reasonable for an oil too bulky to penetrate between the shorter alkyl tails of $C_8C_{16}DAB$, but can penetrate between the longer chains and flatten the interface towards net zero interfacial curvature. The unusual-looking case is hexadecane, but here we simply see non-penetration of the oil taken to its logical conclusion. Remember that C_8C_{16}DAB has a positive spontaneous curvature: $v/a_0l_C = 0.456$. The rod-like micelles formed in this system are swollen by the non-penetrating hexadecane, as we foreshadowed above. The structure of the isotropic liquid formed along the water/C_8C_{16}DAB axis has not been studied. However, as it lies between lamellar and inverse hexagonal phases (see Figures 2.8 and 2.13) it must have slightly inverse interfacial curvature, and is also easily swollen by hexadecane, forming the curious connected phase which 'wraps around' the swollen lamellar phase of C_8C_{16}DAB/water.

Self-assembly phases of cationic surfactants of the DDAB type are robust against temperature variations [44]. Binary and ternary phase equilibria change only very little between room temperature and nearly 100°C [6,44]. In fact, direct determination of v_{eff}/a_0l_C in $2C_{10}$DAB/hexane/water microemulsions by small-angle neutron scattering at temperatures from 20 to 60°C showed no experimentally detectable change [70]. This is consistent with a large bending energy ($\gg kT$) for the surfactant film, i.e. a highly favoured spontaneous interfacial curvature or v/a_0l_C parameter. Such systems give easily predictable behaviour which can be tuned by changing intermolecular potentials, mainly through electrostatic (changing counterion) or steric means.

Although not present in the didodecyldimethylammonium bromide + water binary phase diagram, I_2 cubic phases are formed by the addition of small amounts of oil to the DDAB lamellar phase. The position of these phases is indicated for cyclohexane and dodecane on Figure 2.11. Each of the two regions shown actually comprise a series of one phase and two phase coexistence regions, with cubic phases distinguished by the structure or geometry of the bilayer surface, as determined from small-angle X-ray and neutron diffraction [36–38,71]. The position of these phases fits well with expectations, lying between the lamellar and inverted microemulsion regions of the ternary phase diagram.

(b) Anionic surfactants (Aerosol OT)

AOT is in many respects the antithesis of the cationic systems described above. Whereas DDAB has a strongly preferred interfacial curvature which determines the self-assembly structure, AOT changes its structure readily in response to most system changes. AOT forms microemulsions in combination with water and numerous oils, including but not limited to toluene, xylenes, cyclohexane, and the alkanes hexane through dodecane [14,55,57,72–76]. These are, throughout most of their region of stability, w/o structures, although they percolate to a conductive, bicontinuous structure at very high water content. This behaviour is typical of four and five component microemulsions (oil, water, surfactant, cosurfactant, salt), and is the reason why the behaviour of three component DDAB microemulsions was initially thought anomalous.

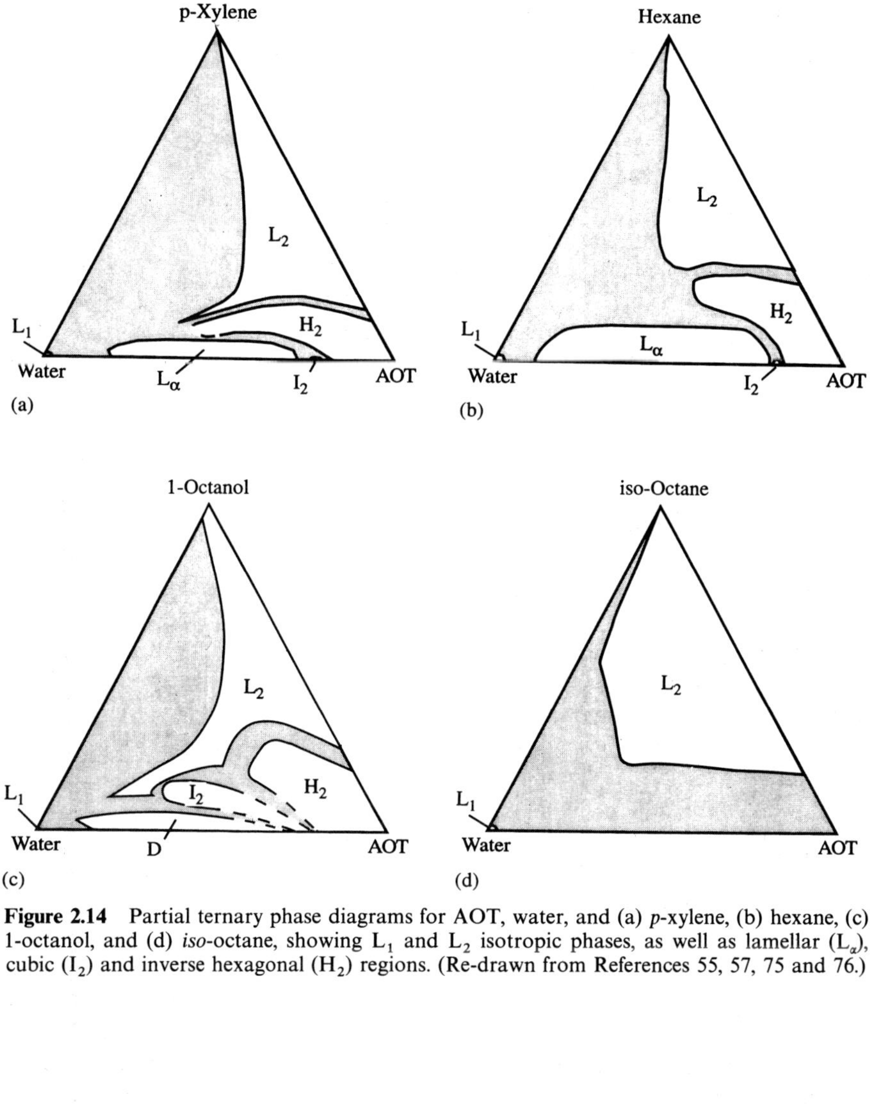

Figure 2.14 Partial ternary phase diagrams for AOT, water, and (a) *p*-xylene, (b) hexane, (c) 1-octanol, and (d) *iso*-octane, showing L_1 and L_2 isotropic phases, as well as lamellar (L_α), cubic (I_2) and inverse hexagonal (H_2) regions. (Re-drawn from References 55, 57, 75 and 76.)

Figure 2.14 shows isothermal ternary phase diagrams for AOT, water, and various oils, showing the position of the L_1 micellar and L_2 microemulsion phases. Unlike DDAB, AOT is soluble in these oils, and solubilizes water into the core of reverse micelles as it is added to the solution. The differences between the positions of the L_2 phases in this case are less dramatic than for DDAB-like systems, but the idea of oil penetration may still be observed through the maximum solubilization capacity, which increases going from *p*-xylene to iso-octane.

In AOT microemulsions, water droplet size increases with water/surfactant or surface/volume ratio. Several authors have verified this, arriving at empirical formulae for the radius of the water core of the form [15]:

$$r_C = 3\frac{v_w}{a_0}w$$

$$= 1.8w$$

where r_C is in Å, w is the molar ratio of water to surfactant and v_w is the molecular volume of water (30 Å^3) [72].

Note the absence of any oil specificity in these relationships. The oil is a true solvent and, at this level, a spectator which plays no role in determining droplet size. This attests to the flexibility of AOT at the oil/water interface. Recall that the mean curvature of an interface for an inverted sphere is $H = 2/r_C$. Thus the interfacial curvature of the AOT film varies continuously with w.

Oil specificity in AOT microemulsions does exist, but is a more subtle affair. That the droplets percolate to form a bicontinuous structure is the first clue to something more than a simple dispersion, and small-angle X-ray and neutron scattering shows that there is an attraction between microemulsion droplets [77]. The droplets are, at least, sticky spheres rather than hard spheres, and when they collide they do not simply bounce off one another. This attraction has also been inferred from other studies of AOT microemulsions, including light scattering and viscosity [78].

Percolation occurs when the concentration of droplets become high enough for the solution to become electrically conducting. This is usually accompanied by an increase in the diffusion coefficient of water as the droplet structure is transformed into a bicontinuous network. Percolation may therefore be interpreted either as an interdroplet attraction, or as a change in spontaneous interfacial curvature. Both effects manifest themselves as 'sticky' droplets.

Characterizing percolation is complicated by the fact that AOT microemulsions are more sensitive to temperature than DDAB systems. The range of stable compositions of L_2 phases changes significantly with temperature.

A common and simple method of characterizing percolation is therefore the *temperature percolation threshold* [79]. Here the conductivity is measured at a particular composition as a function of temperature as the sample is warmed, and the onset of percolation is measured as the point at which conductivity begins to suddenly increase.

The interfacial curvature energies in AOT systems are low compared with DAB systems, comparable with kT, and therefore easily affected by changes in temperature or solution conditions. This changing interfacial curvature is similar to the behaviour observed in aqueous solution.

The effect of temperature on the spontaneous curvature of the AOT-coated oil/water interface produces a striking contrast with non-ionic systems. Microemulsions of single-chained non-ionic surfactants have been

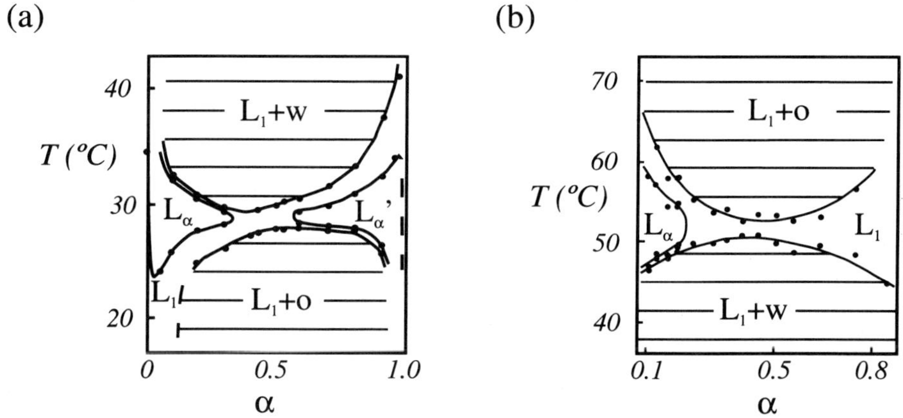

Figure 2.15 Typical slice through a surfactant/oil/water/temperature phase prism at constant surfactant content. The horizontal scale runs from surfactant in water (left, $\alpha = 0$) to surfactant in oil (right, $\alpha = 1$). (a) 7.0 wt% $C_{12}E_5$/1 : 1 weight mixture of cyclohexane and hexadecane/water [81]. (b) 5 wt% AOT/iso-octane/0.46 wt% NaCl solution [80].

widely studied by Kunieda and others, particularly the temperature dependence of their phase equilibria [14,74,80]. Typically the phase diagram viewed through a constant surfactant concentration slice looks like that shown in Figure 2.15(a). Here the abscissa is the fraction of oil in the oil + water mixture, α:

$$\alpha = \frac{\text{wt\% oil}}{\text{wt\% oil} + \text{wt\% water}}$$

This slice through the oil/water/surfactant/T phase prism is commonly studied as it systematically varies hydrophobic and hydrophilic volume fractions at (approximately) constant interfacial area, or constant surfactant content.

At low temperature such non-ionic systems form an oil-in-water microemulsion, which coexists with excess oil above approximately $\alpha = 0.1$. As temperature increases, non-ionic surfactants become less water soluble. It is equivalent to say that their hydrophile–lipophile balance (HLB) number increases, or that the spontaneous curvature goes from positive, through zero and becomes negative. The point of zero interfacial curvature may be referred to as the HLB temperature. Near the HLB temperature lamellar phases are observed, as shown in Figure 2.15(a), and at intermediate α a bicontinuous microemulsion is formed. As temperature is further increased the interfacial curvature becomes negative, and a water-in-oil microemulsion is formed, which coexists with excess water at $\alpha < 0.9$.

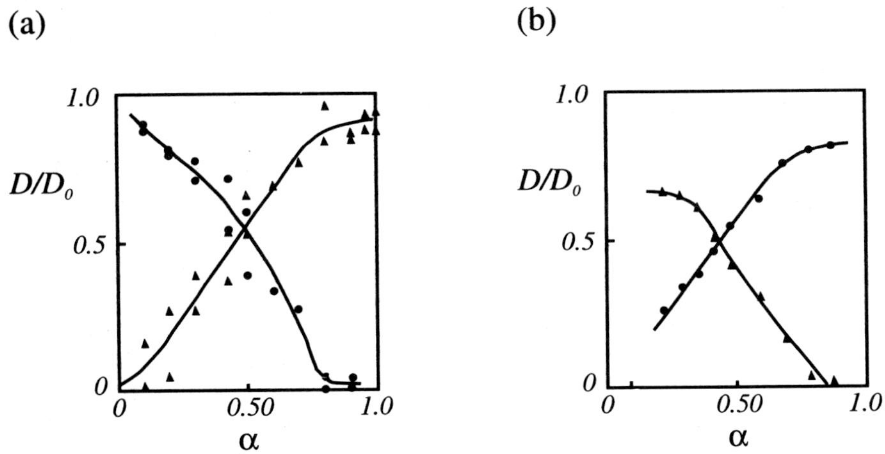

Figure 2.16 Self-diffusion coefficient of water (●) and oil (▲) along the single phase paths shown in Figure 2.15. The horizontal scale runs from surfactant in water (left, $\alpha = 0$) to surfactant in oil (right, $\alpha = 1$). D_0 is the diffusion coefficient of pure oil or water. (a) 7.0 wt% $C_{12}E_5$/1:1 weight mixture of cyclohexane and hexadecane/water [81]. (b) 5 wt% AOT/iso-octane/0.46 wt% NaCl solution [80].

This progression is reflected in self-diffusion coefficients measured through the single phase path shown in Figure 2.15(a). Water self-diffusion coefficients decrease as temperature is raised, while oil self-diffusion values increase. This signals the structural transformation from o/w through bicontinuous to w/o microemulsions as water breaks into droplets and oil becomes the continuous phase (see Figure 2.16(a)) [81].

Compare this with the behaviour of AOT, shown in Figure 2.15(b) and 2.16(b). AOT forms w/o microemulsions at low temperature, but on warming inverts itself to form o/w droplets via a bicontinuous intermediate structure. This may be seen in the self-diffusion coefficients of water and oil, which display the opposite behaviour to the non-ionic system.

Here the effect is due to a decrease in ion binding to the surface with increasing temperature, thereby increasing the equilibrium interfacial area per molecule. The spontaneous interfacial curvature changes from negative through zero to positive. Counterion effects are typically much smaller in anionic than in cationic systems—the differences between alkali metal ions are smaller than between halides. Nevertheless, here we have a striking example of a counterion effect setting curvature in an anionic surfactant system. In AOT, the HLB temperature is a sensitive function of the interfacial composition and can be tuned by changing counterion type and concentration, or by introducing other water-soluble additives to the system [79].

Magid *et al.* have compared the behaviour of AOT reverse micelles with a cationic analogue, BC8ASP, shown in Figure 2.3(b) [77]. Small-angle neutron scattering from reverse micelles of AOT and BC8ASP were compared, and it was found that the two surfactants behaved similarly, forming spherical aggregates which attracted each other weakly. Only slight differences were found between the aggregation numbers of AOT and BC8ASP in *iso*-octane. The micellization behaviour of BC8ASP has not been examined, but we would expect it to also parallel that of AOT. This result suggests that the geometrical constraints expressed through the tail structure and size of the polar group strongly control aggregation patterns.

The I_2 cubic phase of AOT can be swollen with the addition of a co-surfactant-like oil, 1-octanol [57]. This is shown in Figure 2.14. In this system, as with the DDAB systems shown in Figure 2.11, such phases form between the lamellar and L_2 microemulsion phases, indicating slightly negative interfacial curvature.

2.6 Triple-chained surfactants

2.6.1 *Aqueous solution of triple-chained surfactants*

$v/a_0 l_C$ considerations should preclude micellization by triple-chained surfactants. The existence of three tails strongly favours bilayer-based aggregates or inverted structures. However, as with AOT and SDBS, surfactant geometry cannot completely rule out aggregation into micelles in dilute solution.

Aggregation of triple-chained surfactants in water does take place at low concentrations, below that found for related single-chained surfactants. Measured CMCs are close to those of double-chained surfactants, around 1×10^{-5} M [9]. Some are reported to be nearer 1×10^{-4} M [82]. The effect of a variation in the chain length on the CMC is less than for double-chained analogues [9].

The type of aggregates formed above the CMC has been inferred from conductance and laser light scattering methods. The triple-chained trioctylmethylammonium chloride (TOMAC) forms small, tight aggregates [83], whereas much larger aggregates were reported for longer chained molecules [9].

Much of the interest in triple-chained amphiphiles centres on their ability to form bilayers. Identification of bilayers in vesicles and lamellae has been made by electron microscopy, after staining by uranyl acetate. The importance of a spacer unit in aiding the packing of these molecules into bilayers has been demonstrated by Kunitake *et al.* [9]. The three long tails become more compactly aligned in the bilayer because they are separated from the headgroup by ester or aromatic linkages. Some examples of these bilayer forming molecules are shown in Figure 2.2.

The modification of the shape of surfactant molecules by connector and spacer units allows bilayer membranes to be formed from single-, double- and triple-chained amphiphiles. The gel to liquid crystal phase transition of these membranes has been studied extensively by differential scanning calorimetry [12,13].

2.6.2 Reverse micelles and microemulsions

In the presence of oil and co-surfactant, aqueous triple-chained surfactants can form reversed micelle, creating a water in oil micromulsion. This behaviour is understandable given the already large volume occupied by three alkyl tails. The addition of co-surfactant or oil increases this volume, thus increasing the packing parameter, $v_{\mathrm{eff}}/a_0 l_{\mathrm{C}}$, to greater than 1. As with many microemulsions of double-chained surfactants, the curvature of the interface is negative.

Much work in this area has been on the application of reversed micelles to extract proteins and enzymes from aqueous solution [84–86]. The extraction system consists of an aqueous phase containing the protein to be removed. An organic microemulsion phase is composed of surfactant- and co-surfactant-coated reversed micelles in a continuous oil phase. Proteins are extracted into the aqueous microemulsion droplets from the bulk aqueous phase. There is, however, little in the way of systematic studies of phase equilibria or microemulsion structure in triple-chained surfactant systems.

References

1. Jungermann, E. (ed.) *Cationic surfactants.* Marcel Dekker, New York, 1970.
2. Ducker, W.A., Pashley, R.M. and Ninham B.W. *J. Colloid Interface Sci.*, 1989, **128**, 66.
3. Chen, V., Evans, D.F. and Ninham, B.W. *J. Phys. Chem.*, 1987, **91**, 1823.
4. Parsegian, V.A., Rand, R.P. and Fuller, N.L. *J. Phys. Chem.*, 1991, **95**, 4777.
5. Brady, J.E., Evans, D.F. *et al. J. Phys. Chem.*, 1986, **90**, 1853.
6. Kang, C. and Khan, A. *J. Colloid Interface Sci.*, 1993, **156**, 218.
7. Kunitake, T. and Okahata, Y. *et al. Chem. Lett.*, 1977, 387.
8. Gérin, M. and Fesco, J. *Analytica Chimica Acta*, 1978, **97**, 165.
9. Kunitake, T., Kimizuka, N. *et al. J. Am. Chem. Soc.*, 1984, **106**, 1978.
10. Kunitake, T., Okahata, Y. *et al. J. Am. Chem. Soc.*, 1980, **102**, 7877.
11. Kunitake, T., Higashi, N. *J. Am. Chem. Soc.*, 1985, **107**, 692.
12. Kunitake, T., Ando, R. and Ishikawa, Y. *Mem. Fac. Eng., Kyushu Univ.*, 1986, **46**, 245.
13. Kunitake, T., Ando, R. and Ishikawa, Y. *Mem. Fac. Eng., Kyushu Univ.*, 1986, **46**, 221.
14. Kunieda, H. and Shinoda, K. *J. Colloid Interface Sci.*, 1980, **75**, 601.
15. Eastoe, J., Fragneto, G. *et al. J. Chem. Soc. Faraday Trans.*, 1992, **88**, 461.
16. Wennerström, H., Khan, A. and Lindman, B. *Adv. Colloid Interface Sci.*, 1991, **34**, 433.
17. Gilchrist, C.A., Rogers, J. *et al. J. Colloid Interface Sci.*, 1967, **25**, 409.
18. Balmbra, R.R., Clunie, J.S. and Goodman, J.F. *Proc. Roy. Soc. London*, 1965, **286A**, 534.
19. Sein, A. and Engberts, J.B.F.N. *Langmuir*, 1995, **11**, 455.
20. Cheng, C.H. and Gulari, E. *J. Colloid Interface Sci.*, 1982, **90**, 410.

21. Dick, S.G. *MSc Thesis*, University of Melbourne, 1972.
22. Dick, S.G., Healy, T.W. and Fuerstenau, D.W. *J. Colloid Interface Sci.*, 1971, **37**, 595.
23. Gray, F.W., Gerecht, J.F. and Krems, I.J. *J. Org. Chem.*, 1955, **20**, 511.
24. Zhu, Y.-p., Masuyama, A. *et al. J. Am. Oil Chem. Soc.*, 1991, **68**, 539.
25. Zhu, Y.-p., Masuyama, A. *et al. J. Am. Oil Chem. Soc.*, 1992, **69**, 626.
26. Israelachvili, J.N., Mitchell, D.J. and Ninham, B.W. *J. Chem. Soc. Faraday Trans* **2**, 1976, **72**, 1525.
27. Gruen, D.W.R. *J. Phys. Chem.*, 1985, **89**, 146.
28. Tanford, C. *The hydrophobic effect*, 2nd Ed. Wiley Interscience, New York, 1980.
29. Mitchell, D.J., Waring, L. *et al. J. Chem. Soc., Faraday Trans 1*, 1983, **79**, 975.
30. Appell, J., Porte, G. and Poggi, Y. *J. Colloid Interface Sci.*, 1982, **87**, 492.
31. Rehage, H. and Hoffmann, H. *Faraday Disc. Chem. Soc.*, 1983, **76**, 363.
32. Laughlin, R.G. In Rubingh, D.N. and Holland, P.M. (eds), *Cationic Surfactants. Physical Chemistry.* Surfactant Science Series Vol. 37. Marcel Dekker, New York, 1991.
33. Hyde, S.T. *Colloids and Surfaces A*, 1995, **103**, 227.
34. Balmbra, R.R., Clunie, J.S. and Goodman, J.F. *Nature*, 1969, **222**, 1159.
35. Hyde, S.T. *J. Phys. Chem.*, 1989, **93**, 1458.
36. Ström, P. and Anderson, D.M., *Langmuir*, 1992, **8**, 691.
37. Barois, P. and Hyde, S.T. *Langmuir*, 1996, **6**, 1136.
38. Maddaford, P.J. and Toprakcioglu, C. *Langmuir*, 1993, **9**, 2868.
39. Balinov, B., Olsson, U. and Söderman, O. *J. Phys. Chem.*, 1991, **95**, 5931.
40. Zana, R. *J. Colloid Interface Sci.*, 1980, **78**, 330.
41. Lianos, P., Lang, J. and Zana, R. *J. Colloid Interface Sci.*, 1983, **91**, 276.
42. Warr, G.G., Magid, L.J. *et al. Langmuir*, 1988, **4**, 813.
43. Hoffmann, H. *Ber. Bunsenges. Phys. Chem.*, 1984, **88**, 1078.
44. Warr, G.G., Sen, R. *et al. J. Phys. Chem.*, 1988, **92**, 774.
45. Lusvardi, K.M., Full, A.P. and Kaler, E.W., *Langmuir*, 1995, **11**, 487.
46. Fontell, K., Ceglie, A. *et al. Acta Chimica Scand. Ser. A.* 1986, **A40**, 247.
47. Gazeau, D., Zemb, Th. and Dubois, M. *Prog. Colloid Polymer Sci.*, 1993, **93**, 123.
48. Madani, M. and Kaler, E. *Langmuir*, 1990, **6**, 125.
49. Franses, E., Puig, J.E. *et al. J. Phys. Chem.*, 1980, **84**, 1547.
50. Magid, L.J., Triolo, R. *et al. J. Phys. Chem.*, 1982, **86**, 164.
51. Das, S., Bhirud, R.G. *et al. J. Phys. Chem.*, 1992, **96**, 7454.
52. Ockelford, J., Timimi, B. A. *et al. J. Phys. Chem.*, 1993, **97**, 6767.
53. Caponetti, E., Triolo, R. *et al. J. Colloid Interface Sci.*, 1987, **116**, 200.
54. Binana-Limbelé, W., van Os, N.M. *et al. J. Colloid Interface Sci.*, 1991, **141**, 157.
55. Ekwall, P., Mandell, L. and Fontell, K. *J. Colloid Interface Sci.*, 1970, **33**, 215.
56. Sheu, E.Y., Chen, S.-H. and Huang, J.S. *J. Phys. Chem.*, 1987, **91**, 3306.
57. Gradzielski, M., Hoffmann, H. *et al. J. Colloid Interface Sci.*, 1995, **169**, 103.
58. Almgren, M., Grieser, F. and Thomas, J.K. *J. Amer. Chem. Soc.*, 1979, **101**, 279.
59. Hyde, S.T., Ninham, B.W. and Zemb, T. *J. Phys. Chem.*, 1989, **93**, 1464.
60. Chen, V., Warr, G.G. *et al. J. Phys. Chem.*, 1988, **92**, 768.
61. Barnes, I.S., Hyde, S.T. *et al. J. Phys. Chem.*, 1988, **92**, 2286.
62. Angel, L.R., Evans, D.F. and Ninham, B.W. *J. Phys. Chem.*, 1983, **87**, 538.
63. Chen, S.J., Evans, D.F. and Ninham, B.W. *J. Phys. Chem.*, 1984, **88**, 1631.
64. Chen, S.J., Evans, D.F. *et al. J. Phys. Chem.*, 1986, **90**, 842.
65. Gruen, D.W.R. and Haydon, D.A. *Pure Appl. Chem.*, 1980, **52**, 1229.
66. Gruen, D.W.R. and Haydon, D.A. *Biophys. J.*, 1981, **33**, 149.
67. Gruen, D.W.R. *Biochim. Biophys Acta.*, 1980, **595**, 161.
68. Barnes, I.S., Derian, P.J. *et al. J. Physique*, 1990, **51**, 2605.
69. Chen, V., Evans, D.F. and Ninham, B.W. *J. Phys. Chem.*, 1987, **91**, 1823.
70. Warr, G.G., Barnes, I.S. and Zemb, T.N. *Proc. 2nd World Surfactants Congress*, Paris, May 1988, 361.
71. Chen, C.-M. and Warr, G.G. In: Herb, C.A. and Prud'homme, R.K. (eds), *Structure and flow in surfactant solutions.* ACS Symposium Series no. 578, 1994, 306.
72. Robertus, C., Joosten, J.G.H. and Levine, Y.K. *J. Chem. Phys.*, 1990, **93**, 7293.
73. Hilfiker, R., Eicke, H.-F. *et al. Ber. Bunsenges. Phys. Chem.*, 1990, **94**, 677.
74. Chen, S.-H., Chang, S.-L. and Strey, R. *J. Chem. Phys.*, 1990, **93**, 1907.

75. La Mesa, C., Coppola, L. *et al. Langmuir*, 1992, **8**, 2616.
76. Tamamushi, B. and Watanabe, N. *Colloid Polym. Sci.*, 1980, **258**, 174.
77. Magid, L.J. *Colloids and Surfaces*, 1986, **19**, 129.
78. Smeets, J., Koper, G.J.M. *et al. Langmuir*, 1994, **10**, 1387 and references therein.
79. García-Río, L., Leis, J.R. *et al. Langmuir*, 1994, **10**, 1676.
80. Carnali, J.O., Ceglie, A. *et al. Langmuir*, 1986, **2**, 417.
81. Olsson, U., Nagai, K. and Wennerström, H. *J. Phys. Chem.*, 1988, **92**, 6675.
82. Fedchuk, T.M. and Tulyupa, F.M. *Colloid Journal of USSR*, 1988, **50**, 803.
83. Kunitake, T., Okahata, Y. *et al. J. Am. Chem. Soc.*, 1980, **102**, 7877.
84. Krei, G.A. and Hustedt, H. *Chem. Eng. Sci.*, 1992, **47**, 99.
85. Hilhorst, R., Fijneman, P. *et al. Pure and Appl. Chem.*, 1992, **64**, 1765.
86. Wolbert, R.B.G., Hilhorst, R. *et al. Eur. J. Biochem.*, 1989, **184**, 627.

3 Catanionic surfactants

A. KHAN and E. MARQUES

3.1 Introduction

In 1943, Scott, Tartar and Lingafelter [1] prepared, for the first time, the crystalline surfactant decyltrimethylammonium decylsulfonate. They designated the surfactant as 'double long-chain salt' to distinguish it from the usual 'single long-chain salt'. In 1987, Jokela, Jönsson and Khan [2] introduced the term 'catanionic' to designate this class of surfactant. The authors defined a catanionic surfactant as an equimolar mixture of two oppositely charged surfactants (the parent surfactants) from which the inorganic counterions are completely removed. The catanionic surfactant is thus uncharged, and one long alkyl chain organic ion acts as a counterion of the other. The organic counterions with short alkyl chain length (less than six carbon atoms), which do not form micellar aggregates with water, will not be discussed in this report. However, besides the catanionic surfactant systems, this report will discuss mixtures of cationic and anionic surfactants, either equimolar or non-equimolar, where the inorganic counterions are also present. These mixtures will be referred to as mixed cationic–anionic surfactant systems or, simply, as catanionic mixtures.

The catanionics are a new class of surfactants. Their aqueous mixtures exhibit many novel properties compared with those of the parent surfactants from which they are formed. Strong synergistic interactions, manifested as enhanced surface activity, enhanced adsorption and much lower critical aggregation concentrations, occur in dilute solutions of the catanionic surfactants. These solutions also often produce a precipitate with water, but the precipitate redissolves to form mixed micelles and, in some cases, vesicles, when an excess amount of one of the two parent surfactants is added (catanionic mixtures). Several liquid crystalline phases are also formed by the catanionics alone or by the mixed cationic–anionic surfactant systems.

Most surfactants used in practical applications actually consist of mixtures of surfactants, since these often perform better than the individual components. Rodlike micellar solutions have found use as drag-reduction agents in pipeline flow [3] and stable vesicle phases are of potential use as microreactors and agents for controlled drug release [4,5]. Dilute liquid crystals can be utilized in the formulation of liquid detergents [6]. The precipitation of mixed surfactants is useful in the quantitative analysis of surfactants [7,8]. In order to develop these and other applications, it is essential to understand how surfactants interact with each other and how these interactions influence the performance and phase equilibria of the surfactant systems. From a fundamental point of view, catanionic systems display many interesting features. The richness and complexity of their aqueous phase behavior permit qualitative discussions on the influence of electrostatics and molecular geometry, *inter alia*, on surfactant self-assembly. Moreover, the catanionic surfactant system has been found to be a superior model system for theoretical studies on the hydration force.

This is the first systematic report on these types of systems. The physico–chemical properties of several mixed cationic–anionic surfactant systems have been discussed in a book published recently [9]. Laughlin has also included a short report on the phase equilibria of a few binary catanionic surfactant systems [10]. It is the intention of the authors to provide the reader with a broad overview of the interesting properties of catanionic surfactant systems.

3.1.1 Nomenclature

Throughout this report, we have used abbreviations for many surfactants. The names and abbreviations for these surfactants are listed in Table 3.1.

3.2 Preparation of catanionic surfactants

Catanionic surfactants are prepared by mixing equimolar quantities of two oppositely charged surfactants. In such preparations, inorganic salts and ionic surfactants may be present as impurities. The aggregation behavior and phase equilibria of the catanionic surfactant system will be strongly affected by the presence of an ionic surfactant. On the other hand, the presence of a salt will affect the stability range of a given phase. A general description of the available methods for the preparation of catanionic surfactants is given below [2]. The suitability of one method over another largely depends on the chemical composition of the parent ionic surfactants.

3.2.1 Ion-exchange method

Both the cationic and the anionic surfactants are dissolved separately in a suitable solvent (water or mixed water–organic liquid). The cationic surfactant is converted to its hydroxide form and the anionic one to its acidic form, by eluting the surfactants through ion-exchange columns containing the appropriate ion-exchange resins. The two surfactant solutions are then carefully neutralized by titration. The solid catanionic surfactant is obtained

Table 3.1 Abbreviations for surfactants

Abbreviation	Name
Anionic surfactants	
Sulfates/sulfonate	
SOS	sodium octylsulfate
SDeS	sodium decylsulfate
SDS	sodium dodecylsulfate
LiDS	lithium dodecylsulfate
STS	sodium tetradecylsulfate
C_8EO_3S	octyl tri(ethylene oxide) sulfate
SAEOS	sodium alkyl ethylene oxide sulfate
SDBS	sodium dodecylbenzene sulfonate
HDBS	dodecylbenzene sulfonic acid
Carboxylates	
SO	sodium octanoate
PD	potassium dodecanoate
AOT	sodium bis-(2-ethyl-hexyl)sulfosuccinate
SLT	sodium *N*-lauroyl-*N*-methyltaurinate
Bile salts	
NaC	sodium cholate
NaDC	sodium deoxycholate
Perfluorinated surfactants	
LiFOS	lithium perfluorooctanesulfonate
SPFO	sodium perfluorooctanoate
Others	
VBAP	11-(*N*-4-vinylbenzamido)undecanyl *n*-hexadecanyl phosphate
Cationic surfactants	
Quaternary ammonium halides	
ATAB/C	alkyltrimethylammonium bromide/chloride
OTAB	octyltrimethylammonium bromide
DeTAB	decyltrimethylammonium bromide
DTAB/C	dodecyltrimethylammonium bromide/chloride
TTAB	tetradecyltrimethylammonium bromide
CTAB/C	cetyltrimethylammonium bromide/chloride
DDAB	didodecyldimethylammonium bromide
DODAB	dioctadecyldimethylammonium bromide
ODTAC	octadecyltrimethylammonium bromide
OTEAB	octyltriethylammonium bromide
Pyridinium halides	
C_8PyC	octylpyridinium chloride
DPyC	dodecylpyridinium chloride
TPyC	tetradecylpyridinium chloride
C_8EO_3Py	octyltri(ethylene oxide)pyridinium chloride
Perfluorinated surfactants	
DEFUMAC	diethanolheptadecafluoro undecanolmethylammonium chloride
Others	
DDAOH	*N*,*N*-dimethyl-*N*-dodecylamine oxide (protonated)

either by lyophilizing or by evaporating the solution under reduced pressure. The purity of the surfactant can be improved by multiple elutions of the parent surfactant solutions, and by recrystallizing the catanionic surfactant.

3.2.2 Precipitation method

The method can be applied in two different ways: (i) The silver salt of an anionic surfactant is precipitated from the aqueous solution of the anionic surfactant (counterion Na^+, K^+ or Li^+). The precipitate is separated, purified and dissolved in a suitable solvent, normally mixed water–organic liquid. The solution of silver salt is titrated with the solution of cationic surfactant (counterions Cl^- or Br^-) to the equivalence point. The silver halide precipitate formed in the titrant is separated from the catanionic surfactant solution. The white crystalline catanionic surfactant is then obtained and further purified as in section 3.2.1. The silver salt of short alkyl chain anionic surfactants may have considerable solubility in water, with the result of a low yield of catanionic surfactant formation.

(ii) Mixing of equivalent amounts of cationic and anionic surfactant solutions at a certain total surfactant concentration may yield a precipitate. Many mixed systems may form supersaturated solutions which may be stable for a long time (hours to weeks) prior to precipitation. By heating, the precipitate can be converted to a clear liquid, which under cooling slowly yields the crystalline precipitate of catanionic surfactant. The precipitate is filtered, carefully washed for removal of salt and adsorbed ionic surfactant and dried under vacuum.

3.2.3 Extraction method

The cationic and the anionic surfactants are dissolved separately in water. Equimolar amounts of the two surfactants are then mixed. By adding a suitable organic solvent, the liquid catanionic surfactant can be extracted from the aqueous phase to the organic phase. The organic phase is separated and the white crystalline solid of surfactant is obtained by evaporating the liquid under reduced pressure.

3.3 Phase equilibria

The study of phase equilibria in aqueous surfactant systems provides initial and valuable information on the general self-assembly properties of surfactants. Detailed investigation of the microstructure of individual phases or exploitation of particular aspects of the phase behavior can then be carried out. It is fruitful to begin with the study of binary single surfactant–water phase diagrams and then investigate the effect of adding one or more components to the system on the phase equilibria.

3.3.1 Binary systems

The phase equilibria for several catanionic surfactant–water systems have been investigated by Jokela *et al.* [2,11]. The four catanionic surfactants dodecylammonium dodecanoate (DAD), dodecylammonium dodecylsulfate (DADS), dodecyltrimethylammonium dodecylsulfate (DTADS) and dodecylethyldimethylammonium dodecylsulfate (DEADS) have been carefully synthesized. The phase diagrams for their binary aqueous mixtures are presented in Figure 3.1. Phase notations adopted in the work of Ekwall, Fontell and Mandell are used in this work [12,13].

Above the Krafft temperature there exists only one lamellar liquid crystalline phase, which is in equilibrium with excess water at low surfactant concentration. The phase behavior of these catanionic surfactant–water systems clearly resembles that of the lecithin–water system [14], in particular the observed swelling of the lamellar phase. For the catanionic systems the swelling results in a maximal water layer thickness of 6 Å for both DAD and DADS, 11 Å for DTADS and 14 Å for DEADS, whereas for the lecithin bilayer it is larger, approx. 19 Å at maximal water content. Electrostatic effects on the catanionic phase behavior can be neglected since: (i) there is negligible presence of excess ionic surfactant on the catanionic crystals (from conductivity measurements); (ii) differences in solubility between the two amphiphilic ions are insignificant; and (iii) electroneutrality is further ensured by use of two parent amphiphiles of equal chain length. If the parent surfactants are allowed to have different chain lengths, electrostatic effects on the swelling behavior may appear, as can be seen in Figure 3.2, where the surface charge density of the bilayer is shown to be dependent on the chain length of the anionic surfactant (A^-) for a given cationic surfactant (A^+) chain length, $n_{A^+} = 12$.

The bilayer swelling for the zwitterionic lecithin system can be explained on the basis of the hydration force, a short-range exponentially decaying repulsive force, the molecular origin of which is not yet fully understood [2]. The same force operates in the swelling of the catanionic bilayer and this is actually found to be a superior model system for the investigation of the nature of the hydration force. The advantage of using the catanionic bilayer is that the theoretical description of the bilayer–water interface is made easier [2]. In the lecithin bilayer, the free rotation of the choline headgroup makes the interface ill-defined, while there is good reason to believe that the charged headgroups in the catanionic bilayer lie at the same distance in the water–hydrocarbon interface, as can be seen in Figure 3.3. Another important difference is that the distance between the two opposite charges in the catanionic bilayer is not fixed.

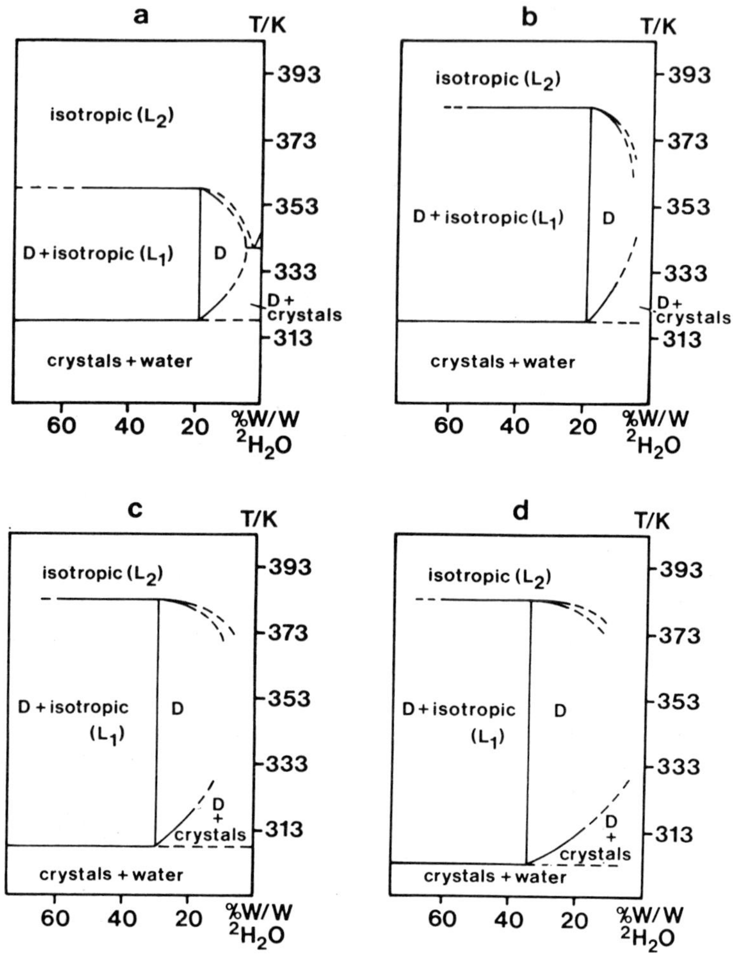

Figure 3.1 Experimental phase diagrams for the binary surfactant–water systems, (a) dodecylammonium dodecanoate, (b) dodecylammonium dodecyl sulfate, (c) dodecyltrimethylammonium dodecyl sulfate and (d) dodecylethyldimethylammonium dodecyl sulfate. Phase notations: L_1, L_2, normal- and reversed micellar solution phases; and D, lamellar liquid crystalline phase [2].

From the investigated phase diagrams the effect of the chemical composition and size of the polar headgroup on the interactions between the lamellar bilayers can be discussed [2]. The strength of the hydration force is found to be dependent on the polar headgroup size rather than on its chemical composition. These results can be interpreted in the light of the image–charge model suggested by Jönsson and Wennerström for the understanding of the hydration force on molecular level [15,16]. In later

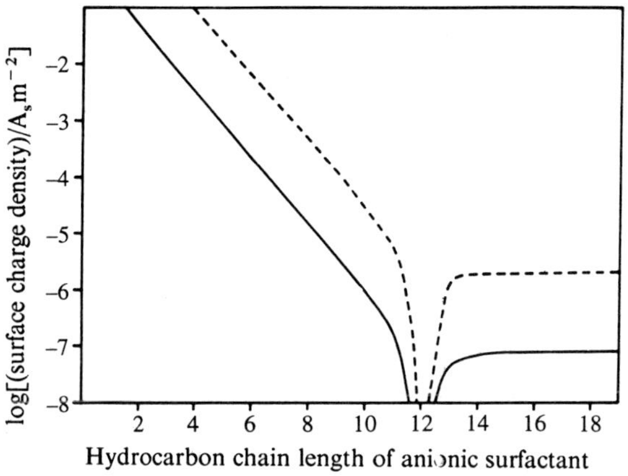

Figure 3.2 Surface charge density versus hydrocarbon chain length of the anionic surfactant, n_{A^-}, for a fixed length of the cationic surfactant, $n_{A^+} = 12$. Two values for the water layer thickness considered: solid line, 7 Å; dashed line, 200 Å. Data obtained from calculations [2].

work, the mechanism of the swelling is discussed on the basis of a protrusion force theory [17].

Other important observations arise by comparing the different binary diagrams, including that of the catanionic octylammonium octanoate (OAO)–water system [11]. There is an appreciable rise in the Krafft point (40°C) when the hydrocarbon chain length increases from 8 (in OAO) to 12 (in DAD). A higher swelling of the lamellar phase, D, occurs for the shortest chain surfactant, indicating that the hydration force is also dependent on the alkyl chain length. Above a certain temperature an isotropic solution (L_2) appears in all of the systems. The stability of the D phase relative to the L_2 phase is seen to be dependent on the chain length—higher stability for longer chains—and on the chemical composition of the headgroup rather

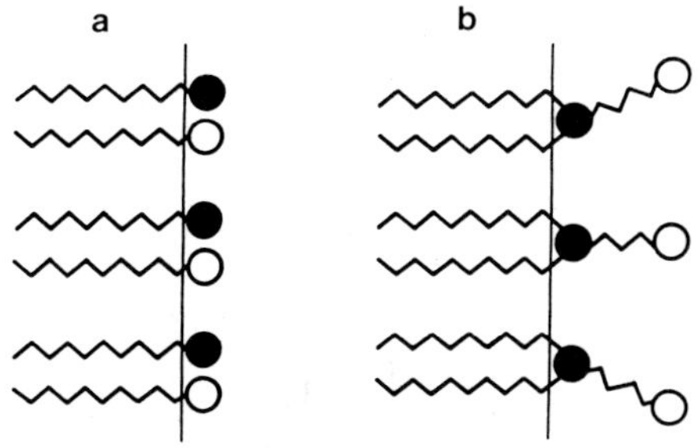

Figure 3.3 Orientation of the surfactant polar headgroup relative to the bilayer/water interface for (a) a catanionic surfactant and (b) a zwitterionic phospholipid [2].

than on its size, higher stability for the sulfate than for the carboxylate group. For the short chain liquid catanionic surfactant hexylammonium hexanoate (HAH) no liquid crystalline phases have been detected; instead, two solutions are found and they coexist in a wide two-phase region [18,19]. Self-diffusion data show that no appreciable amounts of micelle-like aggregates appear to be present in either of these solutions [19].

One can also obtain information on the phase behavior of catanionic surfactants from equimolar mixtures of cationic and anionic surfactants [20,21]. These equimolar mixtures contain, naturally, an amount of salt equivalent to the catanionic surfactant. The salt will act to screen the interactions between the oppositely charged headgroups of the catanionic surfactant. This may determine the existence range of a phase but not its existence *per se*. The effect of the molecular geometry of the parent surfactants on the self-assembly and phase behavior of the catanionic surfactant is apparent from Figure 3.4.

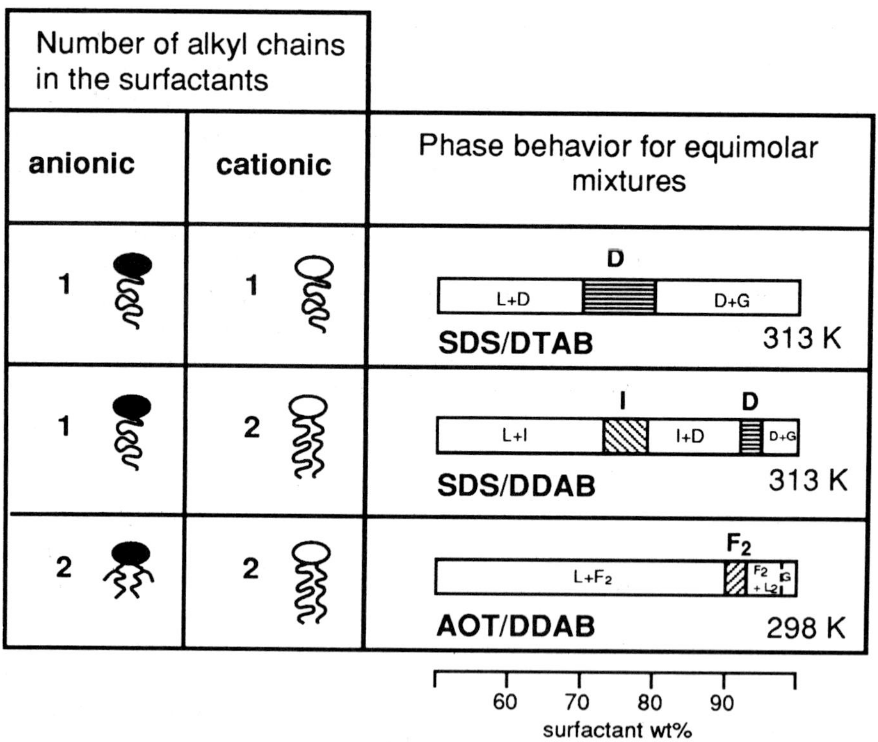

Figure 3.4 Effect of the number of alkyl chains of the parent surfactants on the aggregation behavior of the catanionic surfactant. Phase notations are: L, isotropic (almost pure water) solution; D, lamellar phase; I, reversed cubic phase; F_2, reversed hexagonal phase and G, hydrated surfactant crystals (redrawn from Reference 21).

It is well accepted that the preferred geometry for the surfactant aggregate can be predicted to some extent simply from its molecular geometry, by considering the so-called surfactant critical packing parameter (CPP) [22]

$$\mathrm{CPP} = V/a.l \tag{3.1}$$

where a is the headgroup area, and V and l are the apolar chain volume and length, respectively. For ionic surfactants, the headgroup area is largely dependent on electrostatics, with the result that the CPP is a function of surfactant and salt concentration at a given temperature. As a first approximation, the relation between the CPP and the preferred aggregate structure is as follows: for CPP = 0.33, spherical micelles; for CPP = 0.5, infinite cylinders; for CPP = 1, planar bilayers; for CPP > 1, inverted cylinders and micelles. For intermediate CPP values, different aggregate shapes or phase transitions may occur [23], e.g. if 0.33 < CPP < 0.5, rodlike micelles exist and if 0.5 < CPP < 1, vesicles occur.

The effective parameter of the catanionic surfactant is thus determined by two opposing effects: the electrostatic attraction between the oppositely charged headgroups and steric interactions between the tails. The parameter increases as the number of tails of the parent surfactants increases or, in terms of the flexible surface model [24] (complementary to the CPP model), the spontaneous curvature of a catanionic monolayer becomes increasingly negative (i.e. curvature towards water). Such an increase in the parameter is due to the fact that the tails cannot compress to compensate for headgroup interactions. This is clearly borne out from the phase diagrams in Figure 3.4. A lamellar phase (of zero average curvature) occurs for the single-chain–single-chain system, a cubic phase of reversed curvature occurs for the single-chain–double-chain system (in addition to a lamellar system of minute stability) and a reversed hexagonal phase (of high negative curvature) exists for the double-chain–double-chain system.

3.3.2 Ternary systems

Following the previous line of study, the effect of adding a third component, either an ionic surfactant [11] or an uncharged lipophile (long-chained alcohol) [25] to catanionic surfactant–water systems has also been investigated. The ternary phase diagrams for the systems OAO–sodium octanoate (SO)–water and DTADS–dodecanol–water are presented in Figure 3.5(a) and (b).

In both systems the swelling of the lamellar phases increases on addition of the third component, the effect being much more pronounced in the system with ionic surfactant. Only a small amount of charged amphiphile remarkably increases the water uptake of the bilayers, whereas a larger amount of uncharged lipophile causes a moderate swelling. The factor

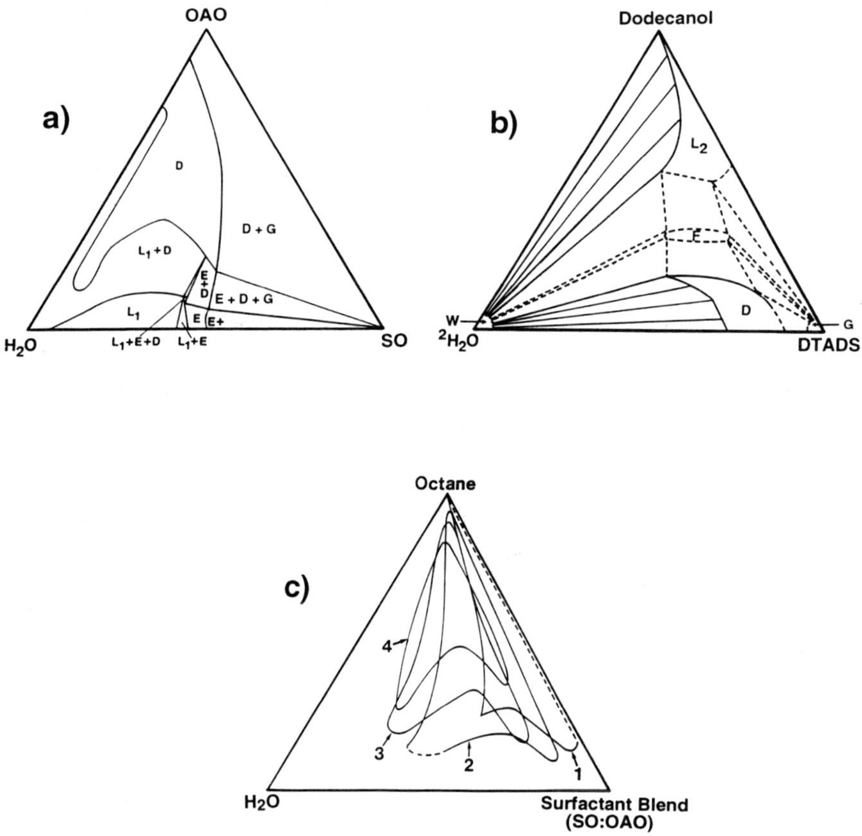

Figure 3.5 Ternary and pseudo-ternary phase diagrams for catanionic surfactant systems: (a) octylammonium octanoate (OAO)–sodium octanoate(SO)–water system at 298 K; (b) dodecyltrimethylammonium dodecylsulfate (DTADS)–dodecanol–water system at 333 K; and (c) microemulsion phase for the OAO–SO–water–octane system at 298 K, with increasing SO : OAO weight ratio: (1) 0, (2) 1 : 9, (3) 2 : 8 and (4) 3 : 7. Phase notations are the same as in Figures 3.1 and 3.4, and E, normal hexagonal liquid crystalline phase and W, almost pure water. Multiphase regions in (b) are shown and not labelled; the single microemulsion phase in (c) is enclosed by the boundary lines. (Redrawn from References 11, 25 and 27.)

responsible for the high swelling is, expectedly, the appearance of the long range repulsive double layer electrostatic force between the bilayers [11]. The swelling upon alcohol addition, which is not obvious, is interpreted within the mentioned image-charge model for the hydration force [25]. When the charge in the catanionic bilayer is diluted by the alcohol, the attractive component of the hydration force is strongly decreased, while the repulsive component is only slightly diminished, and thus the swelling can occur. Another marked difference between the two systems is the presence of reversed-type phases on the addition of dodecanol, these being a reversed hexagonal phase and, for higher alcohol concentration, a reversed micellar phase.

Microemulsions. The occurrence of the reversed micellar phase has been further exploited in an attempt to gain insight into the formation of microemulsions by catanionic surfactants. In water–catanionic surfactant–oil systems, the stability of the reversed micellar phase (L_2) is strongly dependent on the size of the polar headgroup [26,27]. Thus, for a large headgroup, such as that of DTADS, the L_2 phase is highly disfavored and absent in the ternary phase diagram for the DTADS–water–dodecane system, while it appears for the systems OAO–octane–water and DAD–dodecane–water [26]. The microemulsion phase of the OAO system is found to extend to the oil–water axis on addition of small amounts of ionic surfactant to the catanionic surfactant system (Figure 3.5(c)) [26,27]. This stabilization of the reversed micellar phase, at the expense of the pure water phase, can be explained in the same way as in the case of the lamellar phase described above. A decrease in the electrostatic interactions in the reversed micelle when the radius of the micelle increases (by water uptake) is the reason behind this increased stability. The stability of L_2 relative to lamellar phases on adding the ionic surfactant will decrease due to the lower electrostatic free energy of a surfactant in the lamellar sheet than in the reversed micelle [26].

In the hexylamine–hexanoic acid–water system at any mixing ratios, only an L_2 phase and an almost pure water phase occur, with an extensive two-phase region between them [18]. However, for the asymmetric system hexylamine–octanoic acid–water a wide lamellar liquid crystalline phase appears [28]. When hexane is added to the HAH–water system, isotropic liquid phases (not micellized) L_1 and L_2 appear, similar to the binary system [19]. On addition of heptane, the phase behavior resembles that of a water–oil–nonionic surfactant system; the microemulsion phase is in equilibrium with both almost pure water and almost pure oil in a large three-phase region [18].

3.3.3 Pseudo-ternary systems. When a system is composed of a cationic surfactant A^+X^-, an anionic surfactant A^-X^+ and water, with no common ion, it has four components according to the Gibbs phase rule [29]. This is due to the possibility of having the salts A^+A^- (catanionic surfactant) and X^+X^- (simple electrolyte), in addition to the cationic and anionic surfactant salts, bearing in mind electroneutrality. In principle, all four of these salts can form, and to have a correct graphical representation of the system all of them should be represented by a point in a phase diagram. Thus, the bidimensional triangular phase diagram is no longer the optimal representation for the portrayal of phase behavior, a third dimension is required. This can be achieved by using a pyramid [30] or a trigonal bipyramid [29]

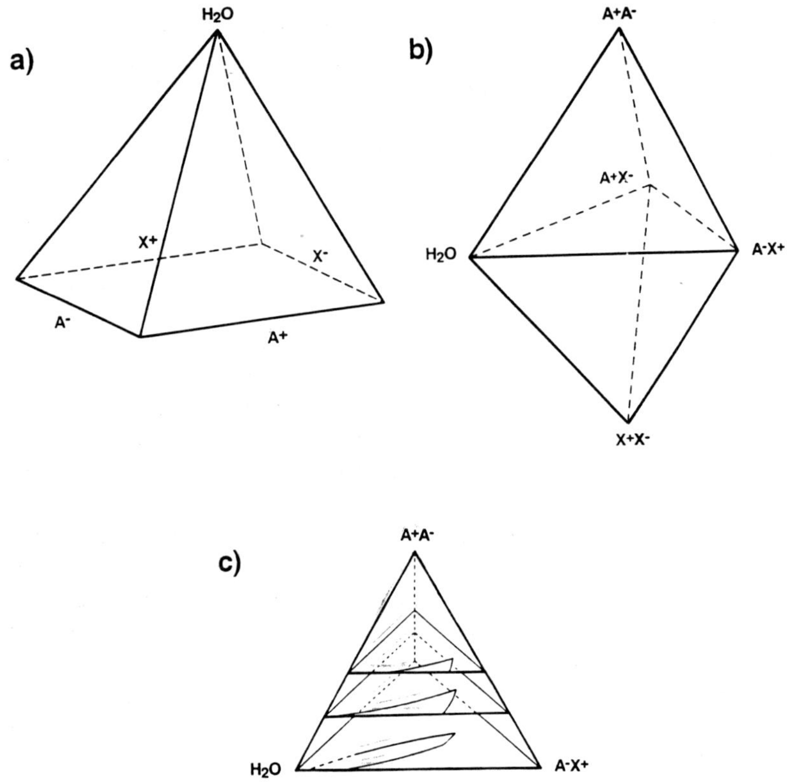

Figure 3.6 Phase diagram representations for the four component systems cationic surfactant(A^+X^-)–anionic surfactant(A^-X^+)–water: (a) pyramid and (b) trigonal bipyramid. For the location on these diagrams of a particular composition see References 29 and 30. (c) Model calculated extension in the bipyramid of the anionic-rich lamellar phase for different weight fractions of added A^+A^-: from bottom to top, 0 wt%, 25 wt% and 45 wt%. The cationic-rich lamellar phase, not shown, behaves symmetrically with respect to the equimolar line (line bisecting the water corner for each plane) [29].

representation, as seen in Figure 3.6(a) and (b); in either case, the ternary diagram constitutes a particular 2-D slice through the 3-D body.

When inspecting a triangular phase diagram for such four-component systems, one should then bear in mind that [20]: (i) one-phase regions in principle lie in cuts on 3-D one-phase bodies; (ii) thus, the same phase can be cut at distinct regions on the plane, but two such cuts, for example, are not distinct homogeneous phases; (iii) phase-separating samples may have constituent phases absent from the triangular plane; and (iv) the shape of three-phase regions is not limited to triangles. However, for practical purposes, triangular phase diagrams provide a satisfactory depiction of the phase behavior of catanionic mixtures.

Phase equilibria studies of catanionic mixtures have been reported for some time, but most of them usually addressed phenomena at high dilutions such as micellization and precipitation. Studies over the full range of composition, or at least for concentrated surfactant regions, have been less common and relatively recent [20,21,31–34]. The possibility of using thermodynamic model calculations to investigate the equilibrium between lamellar phases and salt crystals in concentrated systems has been presented by Jökela *et al.* [29]. Phase boundaries and tie lines are calculated by equating chemical potentials of each component in the two phases in equilibrium. The effect of adding excess catanionic surfactant A^+A^- on the stability of anionic- or cationic-rich lamellar phases relative to crystals is also discussed. As A^+A^- is added, the lamellar phases win over each surfactant crystal (they move towards higher surfactant concentration), but lose to the catanionic crystal (they depart from the equimolar line). These two effects can be seen in Figure 3.6(c), where only the anionic-rich lamellar phase is depicted for simple visualization. The same thermodynamic model has been used later to calculate phase boundaries between mixed micelles and catanionic crystals for the system SDS–DTAB–water in the dilute regime [35], as well as several micellar properties, with direct comparison with experimental results. The approach differs slightly in that direct minimization of free energy for each phase is used.

The complete phase diagrams for mixtures of the cationic double-chain surfactant DDAB with single-chain anionic SDS [20] and double-chain anionic AOT [36] have been investigated in detail by polarizing microscopy and water deuteron NMR (Figure 3.7).

On the basis of experimental evidence, the heterogeneous regions for the system DDAB–SDS–water are also presented. The most striking features when one inspects the two diagrams are the display of rich polymorphism of phase behavior and the appearance of phases, in particular liquid crystalline phases, that are not present in the two binary surfactant–water systems (at least not at the same temperature). In the DDAB–SDS system, two lamellar phases, D_{II} and D_{IV}, of limited stability, occur, as well as a cubic phase (I) of reversed structure. In the DDAB–AOT system, the phases are a wide cubic phase I_2 and a new reversed hexagonal phase F_2 for concentrated regions, and a sponge phase L_3 of limited stability at higher water content. A unique observation in the DDAB–AOT system is the coexistence of the two hexagonal phases F_1 and F_2 in a two-phase region [21].

Both diagrams can be qualitatively analysed on the basis of two effects: (i) long range electrostatic effects and (ii) molecular packing considerations. In both systems the phases formed by the anionic surfactant alone have a very limited capability of solubilizing the cationic surfactant (interestingly, the cubic phase I_1 shows a considerable degree of swelling on addition of the latter). On the other hand, a different behavior appears on the DDAB-

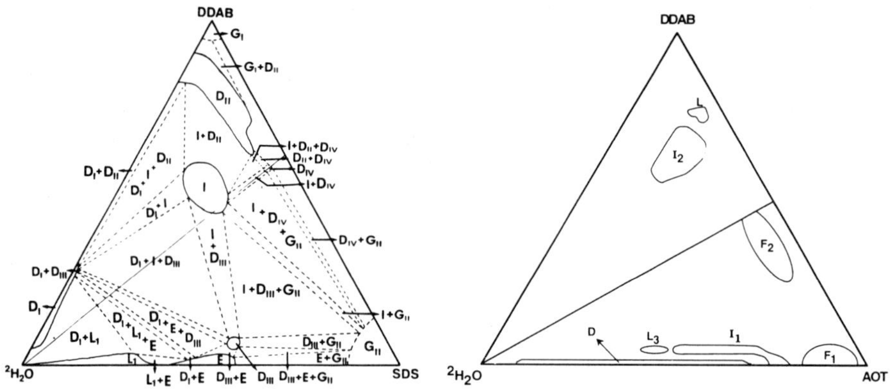

Figure 3.7 The pseudo-ternary phase diagrams for the systems (a) DDAB–SDS–water at 313 K (heterogeneous regions also depicted) and (b) DDAB–AOT–water at 298 K. Phase notations are as in Figures 3.1, 3.4 and 3.5, and L_3, sponge phase [20,36].

rich side: the water swollen lamellar phase D_I is destabilized by very small amounts of any of the anionic surfactants, while the water-poor (collapsed) lamellar phase D_{II} can incorporate a large amount of SDS, but no detectable amount of AOT. The first observation is an obvious consequence of double-layer electrostatic interaction effects. The latter is a clear indication of the geometric constraints imposed by the highly branched double tail AOT (as opposed to the single-tail SDS): the lamellae are immediately destabilized to phases of negative curvature, the cubic I_2 (and the isotropic liquid) and, ultimately, the reversed hexagonal phases. These studies indicate generally how a suitable blending of surfactants in water can dictate a particular aggregate structure and its stability region, which can be of interest for theoretical modelling and microstructure engineering.

Phase diagrams for other concentrated catanionic systems have been presented. The system (anionic) bile salt NaDC–CTAB–water [33] yields a novel cubic phase. The L_1 and E phases formed by CTAB incorporate large amounts of bile salt and, symmetrically, the bile salt micellar phase dissolves large amounts of surfactant. The same trends are observed for mixtures of bile salt with anionic (SDS and AOT) and non-ionic (Triton X-100) surfactants, giving an indication that no special interactions occur in the catanionic system [33]. Mixtures of the cationic perfluorosurfactant DEFUMAC with both anionic perfluorosurfactant LiFOS and anionic hydrocarbon surfactant LiDS have also been studied, at a total surfactant concentration of less than 20 wt% [34]. For the DEFUMAC–LiFOS system, a precipitate is formed at and around equimolarity, while for excess ionic surfactant, lamellar phases are detected. On the other hand, the DEFUMAC–LiDS system forms no precipitate and the phase behavior is entirely dominated by a lamellar phase. The microstructure for these lamellar phases has also been proposed [34]. Binding of sodium and water to magnetically aligned concentrated lamellar phases formed by CTAB–potassium dodecanoate (PD)–NaBr (with or without added decanol) in water has also been investigated and a three-site adsorption theory has been proposed [31]. In a later report, the phase diagram for surfactant–decanol–NaBr system, where the surfactant is either mixed PD–CTAB at varying mixing ratio or neat CTAB, is presented [32] and the transition between a nematic and a smectic lamellar phase is discussed on the basis of electrostatic and geometric effects.

3.4 Micellar phase

3.4.1 Critical micelle concentration (CMC)

For mixed surfactant systems, knowledge of the CMC of the aqueous mixture, its variation as a function of surfactant mixing ratios, as well as the composition of the mixed micelles at the CMC with respect to the two surfactants, is of both theoretical and practical importance. The CMC of the mixture is measurable by applying the methods that are commonly used for the single component surfactant systems. Surface tension and conductivity methods are extensively used to determine the CMC of the mixture experimentally.

The solution behavior of the mixed cationic and anionic surfactant systems deviates considerably from the ideal solution behavior which is normally applicable for the single component surfactant solutions. The regular solution approximation is employed to account for the observed behavior of the mixed solutions.

Within the regular solution approximation [37], the activity coefficient f_1 of solute 1 is related to the mole fraction x_2 of solute 2 in the binary solute mixture by

$$f_1 = \frac{a_1}{C_1} = e^{\beta x_2^2} \tag{3.2}$$

where a_1 and C_1 are, respectively, the activity and concentration of solute 1. The parameter β describes intermolecular interactions between solute 1 and 2 and for ideal behavior β is zero. It is related to the molar free energies of interaction between pure solutes (W_{11} and W_{22}) and between solute 1 and 2 (W_{12}) by

$$\beta = \frac{e^{(W_{11} + W_{22} - 2W_{12})}}{RT} \tag{3.3}$$

For surfactant mixtures, interactions occur when surfactant molecules are in close proximity, that is, in micelles and in the monolayer adsorbed at the air–water interface.

Within the phase separation approximation [38] for micelle formation, the CMC of the mixture (CMC_m), is related to the CMC of the pure components (CMC_1 and CMC_2) [39,40] as

$$\frac{1}{CMC_m} = \frac{\alpha_1}{f_1 CMC_1} + \frac{\alpha_2}{f_2 CMC_2} \tag{3.4}$$

where α_1 and α_2 are the overall mole fractions of components 1 and 2, respectively, in the surfactant mixture.

For regular solutions, the activity coefficients of molecules involved in micelles are related to the mole fraction of the two components in the micelles by [41]

$$\log(f_1) = \beta^m(1 - x_1)^2 \tag{3.5}$$

$$\log(f_2) = \beta^m x_1^2 \tag{3.6}$$

Using experimentally determined values of CMC_1 and CMC_2, for a given value of β, Equations (3.2) and (3.4) are solved iteratively for x_1, which allows the calculation of CMC_m as a function of α_1.

At higher surfactant concentration C, x_1 can be related as

$$x_1 = \frac{-(C - \Delta) + \sqrt{(C - \Delta)^2 + 4\alpha_1 C\Delta}}{2\Delta} \tag{3.7}$$

where $\Delta = f_2 CMC_m - f_1 CMC_m$. The monomer concentration can then be calculated by material balance. It has been shown that when the total surfactant concentration is high compared to the CMC_m, then the micelle composition becomes closer to the overall composition.

Some available experimental data on CMC *versus* composition of total surfactant can be fitted by adjusting the value of the single parameter β. Representative mixed systems are alkyltrimethylammonium bromide (ATAB) and chloride (ATAC)–sodium alkylbenzene sulfonate (SABS) [42–44] and ATAB–SAS [45,46]. These systems exhibit a drastic drop in the CMC values at different mixing ratios relative to the CMC of the pure components, and the CMC value for the mixed micelles is minimal as the bulk composition approaches equimolarity. Figure 3.8 illustrates the variation of the CMC measured for the system tetradecyltrimethylammonium bromide–sodium dodecylbenzene sulfonate [44] at different surfactant mixing ratios.

The measured CMC is found to be much lower than the CMC predicted for ideal mixing. Choosing $\beta = -24.1$, the calculated values from the prediction of the regular solution model are close to the experimental results. The figure also shows that the cationic–anionic interactions yield

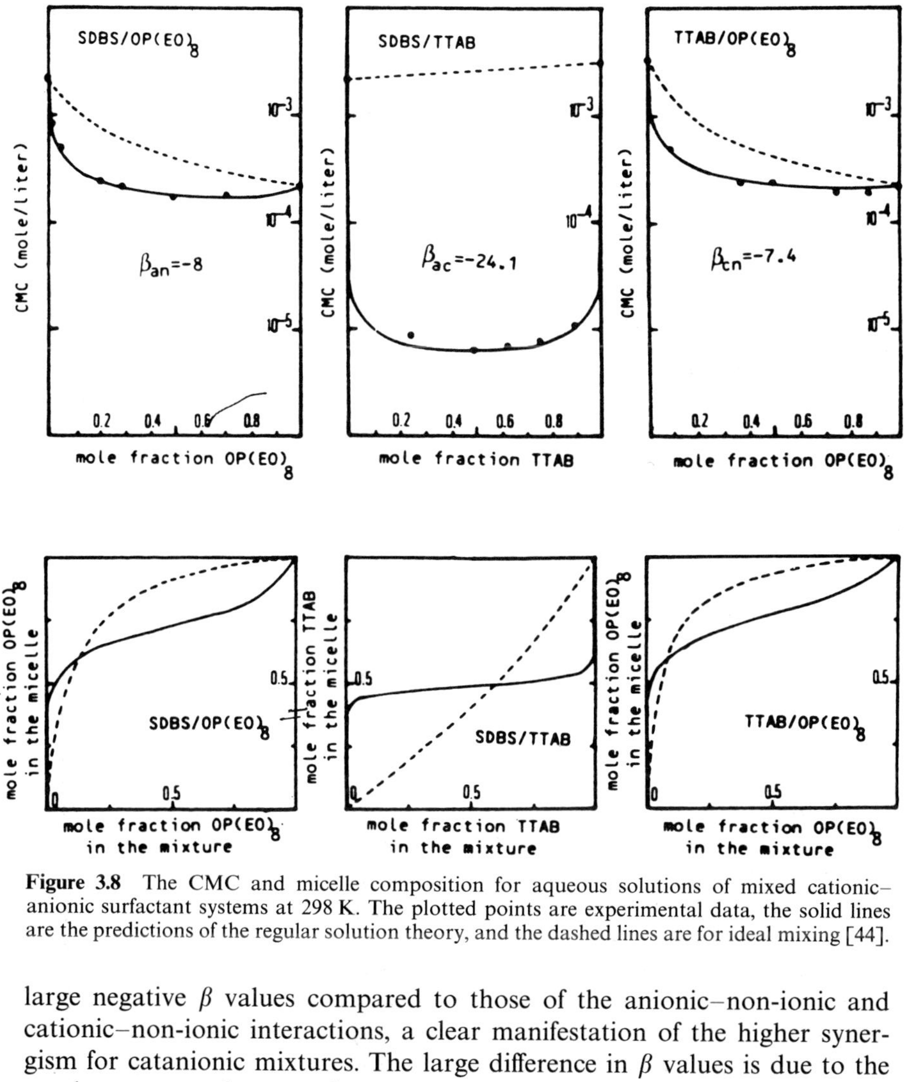

Figure 3.8 The CMC and micelle composition for aqueous solutions of mixed cationic–anionic surfactant systems at 298 K. The plotted points are experimental data, the solid lines are the predictions of the regular solution theory, and the dashed lines are for ideal mixing [44].

large negative β values compared to those of the anionic–non-ionic and cationic–non-ionic interactions, a clear manifestation of the higher synergism for catanionic mixtures. The large difference in β values is due to the much stronger electrostatic attractions between the oppositely charged headgroups in the catanionic mixtures. Very low CMC and large negative values for the interaction parameter β are also obtained for other mixed cationic–anionic surfactant systems [42,43,45,47–51]. CMC values of catanionic and cationic–anionic surfactants are given in Table 3.2. A few representative β values are shown in Table 3.3.

The micellar composition for the system [44] calculated at the CMC from the prediction of the regular solution theory is also shown in Figure 3.8. For comparison, the micelle composition calculated for ideal mixing is also included in the figure. The composition of the catanionic micelles is

Table 3.2 CMC values of aqueous solutions of catanionic surfactants and equimolar mixtures of cationic–anionic surfactants. The temperature is 298 K unless otherwise mentioned

Surfactant	CMC/mM	Reference
$C_6H_{13}N^+(CH_3)_3O^-SO_3H_{13}C_6$	110.0	62
$C_6H_{13}N^+(CH_3)_3O_3^-SH_{13}C_6$	220.0	62
$C_6H_{13}N^+(CH_3)_3O^-SO_3H_{17}C_8$	27.0	62
$C_8H_{17}N^+(CH_3)_3O^-SO_3H_{17}C_8$	7.5	62
$C_8H_{17}N^+(CH_3)_3O_3^-SH_{17}C_8$	20.2	1,62
$C_8H_{17}N^+(CH_3)_3O_3^-SH_{21}C_{10}$	5.8	61
$C_{14}H_{29}Py^+O_3^-SH_{15}C_7$	0.52	68
$C_{14}H_{29}Py^+O_3^-SH_{15}C_7$	0.63(323 K)	85
$C_{10}H_{21}N^+(CH_3)_3\ ^-OOCH_{15}C_7$	15.0	57
$C_{10}H_{21}N^+(CH_3)_3\ ^-OOCH_{17}C_8$	8.6	57
$C_{10}H_{21}N^+(CH_3)_3\ ^-OOCH_{19}C_9$	2.4	57
$C_{10}H_{21}N^+(CH_3)_3\ ^-OOCH_{21}C_{10}$	2.4	57
C_6PyB–SDS	4.6	51
C_8PyC–SC_8ED_3S	10.5	54
C_8EO_3PyC–SC_8S	3.16	54
TTAB–AEOS	0.04	87
CTAB–NaC	0.04	91,92
CTAB–NaDC	0.02	91
ODTAC–NaLMA	0.06(308 K)	88

found to maintain equimolarity over a wide concentration range, an observation that parallels the one made from the monolayer adsorption of mixed surfactant systems [52,53]. This behavior is very distinct from a system with ideal composition and can be attributed to the formation of a catanionic surfactant even in the presence of inorganic counterions. The equimolar composition in cationic and anionic surfactant is maintained

Table 3.3 The molar interaction parameter of the mixed micelles, β^m, and of the adsorbed mixed monolayer, β^s, for the aqueous mixtures of cationic and anionic surfactant

Surfactant pair	β^m	β^s	Temp/K	Reference
OTAB–SDeS	−10.4		298	46
DeTAB–SDeS	−11.3		298	46
OTAB–SDS	−10.8		298	46
DTAC–SDS	−25		298	79
TTAB–SDBS	−24.2		298	44
8ϕC16–OPEO8	−1.35		323	43
TTAB–OPEO8	−8.3		323	43
TTAB–8ϕC16	−15.2		323	43
C_8PyC–SC_8EO_3S	−4.1		298	54
C_8EO_3PyC–SC_8S	−8.4		298	54
ODTAC–NaLMA	−12.8		308	88
OTAB–SOS	−10.2	−14.2	298	150
DTAB–SDS	−25.5	−27.8	298	150
DDAO–$KC_{12}SO_3$	−21.6	−16.5	298	150
C_{12}NP–SDS	−15.7	−14.1	303	150

even upon addition of a non-ionic surfactant to the mixed micelles [44]. However, the micellar composition for the anionic–non-ionic and cationic–non-ionic surfactant mixtures does not maintain equimolarity as a function of mixture composition (Figure 3.8, bottom curves). For these systems, the micellar composition depends on the relative values of the CMC of the two components. It can be shown by combining Equations (3.4), (3.5) and (3.6) that equimolarity in the micelle is obtained for the same monomer composition.

The interaction is also very sensitive to the structure of the hydrophobic moiety of the surfactant and the length of the alkyl chain [43] (Figure 3.9). The figure shows that the absolute value of β increases when the anionic surfactant becomes more linear; this is attained either by changing the point of attachment of the benzene ring on the hexadecyl chain or by decreasing the alkyl chain length (here the benzene ring is attached to the centre of the alkyl chain). This is due to the fact that the electrostatic attraction increases as the anionic and cationic headgroups become closer.

Introduction of ethylene oxide groups, (EO), into either the cationic C_8Py or the anionic C_8S surfactant leads to weaker interactions due to the headgroup hindrance in the catanionic mixtures. However, the interaction between C_8EO_3Py and C_8S is stronger than that between C_8Py and C_8EO_3S ions. This effect is due to an enhanced electropositive character of the cationic surfactant, since the EO group becomes electropositive in aqueous solution [54].

In another study [46], CMC values determined over a wide range of mixing ratios of the two surfactants are used to obtain the composition of the mixed micelles by the Gibbs–Duhem equation within the realm of a pseudo-phase separation model [55] (Figure 3.10). The results shown in Figure 3.10 indicate that the composition of the two surfactants in the mixed micelles approaches equimolarity only for a long and symmetric pseudo-double tail catanionic surfactant ($DeTA^+DeS^-$). Moreover, the equimolar micelle composition is maintained over a wide range of mixing ratios. The results calculated by using the regular solution model show large deviations in micellar compositions within the chosen concentration range. The discrepancy of results between the two models is explained on the basis of two effects. First, in the regular solution model, the micellar phase is treated as a regular solution, but for the catanionic surfactant mixtures the micelles are more tightly packed (due to strong electrostatic interactions) than in other types of micelles. Therefore, the behavior of catanionic micelles may deviate from regular solution predictions. Secondly, in this model, the contribution of the inorganic counterions is neglected, although the concentration of these ions may vary significantly in mixed micelles from that in pure micelles.

Formation of a soluble ion-pair for long alkyl chain surfactants has also been observed at concentrations below the CMC for many cationic–anionic

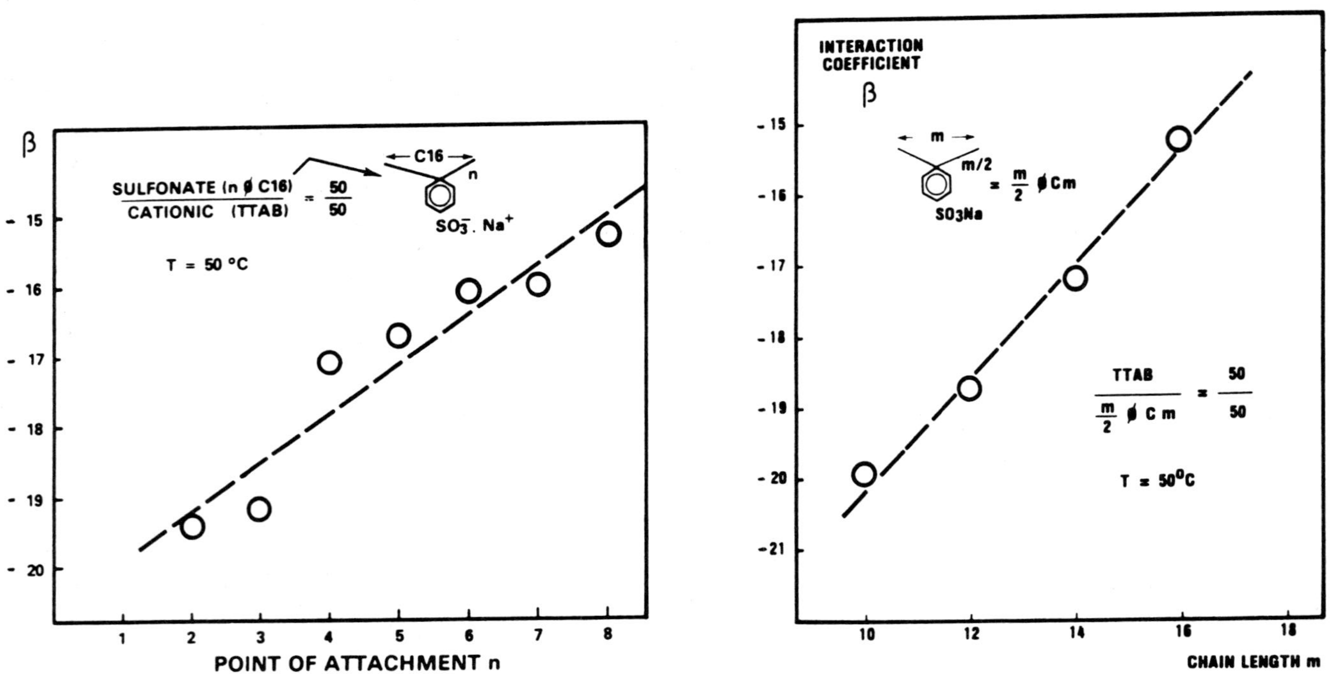

Figure 3.9 Effect of the polar headgroup branching and the alkyl chain length on the molar interaction parameter for aqueous mixtures of mixed cationic–anionic surfactant systems at 323 K [43].

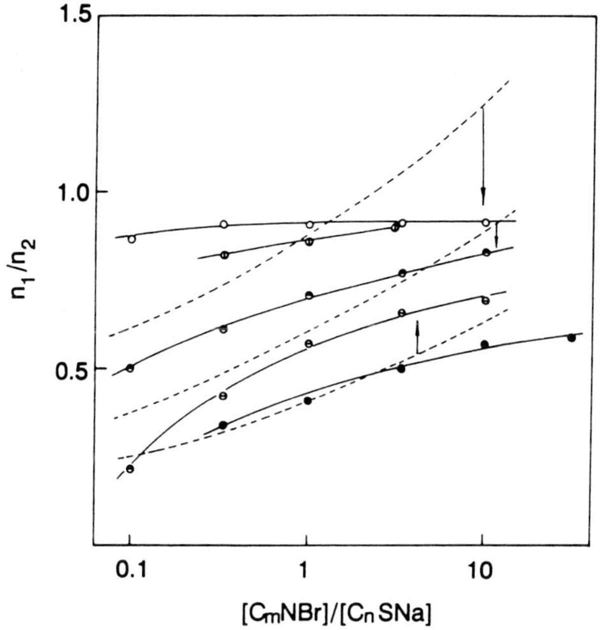

Figure 3.10 Molar ratio of cationic to anionic surfactant in the micellar phase, n_1/n_2, as a function of total surfactant concentration at different mixing ratios at CMC for the systems alkyltriethylammonium (C_mN^+) bromide-sodium alkyl sulfate (C_nS^-)–water, at 298 K. ○, $m = 10$, $n = 10$; ⦶, $m = 8$, $n = 8$; ◓, $m = 8$, $n = 10$; ⊖, $m = 8$, $n = 12$; and ●, $m = 7$, $n = 12$. The solid lines are from the prediction of the Gibbs-Duhem equation and the dashed lines are from the regular solution model [46].

surfactant mixed solutions [56–60]. The ion-pair is a charged complex held together by both hydrophobic and electrostatic attractive interactions and the association constant for many systems has been determined.

Thus, in summary, synergism in the mixed micelles is manifested as the reduction of the CMC values and maximal syngergism is attained at equimolarity, where the CMC reaches a minimum value. As expected, the synergism is the highest for the catanionic mixture as compared to that of the ionic–non-ionic surfactant mixture.

3.4.2 *Micellar properties*

Due to the strong electrostatic attraction of oppositely charged headgroups and the incorporation of pseudo-double chains in the mixed micelles, the CPP (Equation 3.1) is expected to attain larger values for the mixed surfactants in water compared to the single surfactant–water system at identical concentrations. Thus, a micellar growth leading to non-spherical aggregates, such as rod- or disc-like micelles, is expected to occur in these systems. The CPP will increase as the composition of the two surfactants approaches equimolarity and at equimolar (catanionic) mixtures, the CPP is expected to attain its maximum value (maximum micellar growth). Several techniques, such as light scattering (static and dynamic), fluorescence quenching, NMR, cryo-TEM, etc., are used to characterize important micellar properties like micellar size, shape, aggregation number and surface charge.

Catanionic surfactants of the type alkylquaternaryammonium alkylsulfate or sulfonate and alkylpyridinium alkylsulfate or sulfonate are frequently used to investigate the micellization processes. Catanionic surfactants with two symmetric alkyl chains with $C_n = C_m > 8$ (where C_n and C_m represent the alkyl chain length) or with asymmetric alkyl chains — one long alkyl chain, C_n or $C_m > 10$ and one short alkyl chain, C_n or $C_m > 8$ — are practically insoluble in water. At concentrations just above the maximal concentration where only monomer is present, these systems form a precipitate, a turbid mixture, an opalescent solution or a supersaturated metastable solution [1,61–65]. Large colloidal particles are shown to exist in metastable solutions. The particles are almost uncharged and composed of an equimolar ratio of the two surfactants. Addition of simple inorganic salts, such as NaCl and NaBr, does not have any significant effect on the composition of the aggregates [62,65,66].

The situation is, however, different for water soluble catanionic surfactants. These surfactants yield stable micellar solution phases, since the formation of mixed micelles from chains of different lengths is energetically favored to that from chains of equal length [67]. Octyltrimethylammonium octanesulfonate (C_8–C_8) [1] and decanesulfonate (C_8–C_{10}) [1,61], for example, form stable micellar solutions above the Krafft point. A comparison between calculated and observed molecular masses for micelles of the two systems rules out the formation of spherical micelles. Small spherical micelles formed for the system hexadecylpyridinium dodecylsulfate [65] experience an initial mild micellar growth with increased temperature and then a reduction in size between 20 and 65°C. The globular micelles formed for the system tetradecylpyridinium heptylsulfonate [68] at low concentrations undergo a rod-to-disc type transition on increasing concentration. The micellar size is calculated [69] by incorporating the chain of C_7 component ions into the core of the mixed micelles. At higher concentrations and temperature, the system also forms anisotropic liquid crystals [68]. A large micellar solution phase is obtained for the system dodecyltrimethylammonium nonanoate in which large rod-like mixed micelles are detected [70a], but no micellar aggregates are detected for the short chain surfactant hexylammonium hexanoate [18,19].

The precipitate formed by the long chain catanionics above the monomer concentration can be redissolved on adding an excess amount of one of the two parent ionic surfactants and the system forms stable cationic- and anionic-rich micellar solution phases on either side of the equimolar composition. Since the CMC of the mixed systems is extremely low, the monomers will have negligible effect on micellization at concentrations much higher than the CMC. The composition of the micelles will correspond to that of the overall composition and their properties will be determined by the mixing ratios of the two surfactants. The surface of the cationic- and anionic-rich micelles becomes, respectively, positively and negatively charged and that for equimolar mixtures is uncharged. There is also a considerable decrease of 'fluidity' of the micellar surface for the equimolar mixture [71].

In these systems, the counterion of the 'minority' surfactant exists free in the bulk solution and the 'minority' surfactant ions incorporate into the mixed micelles. On the other hand, the degree of counterion binding of the 'majority' surfactant increases with the increase of its concentration in the mixture, as shown by measuring the Na^+ and Br^- ion binding of the DTAB–SDS system [72]. The composition-dependent counterion binding is also measured in micellar solutions for the systems DTAC–SDS and TPC–STS and no detectable Na^+ and Cl^- ion binding is observed for the equimolar composition [73]. Theoretical calculations based on a cell model [35,74] correctly predict the observed counterion binding for the DTAC–SDS system. For large hydrophobic organic counterions, the model [75–77] also predicts, as found experimentally, a complete incorporation of both counterions to form the mixed micelles; this results in a decrease of surface charge density leading to the formation of larger aggregates. The absence of the salt effect on the equimolar mixtures and the negligible amounts of counterion binding to the 'minority' surfactant at different mixing ratios demonstrate the strong association between the two large organic ions. The association, however, becomes weakened if the alkyl chain of one of the surfactants is made shorter.

Addition of SDS to DTAC and of DTAC to SDS at high total surfactant concentration results in a significant increase of aggregation number which is, however, more pronounced with the SDS-rich micellar solution [35,78]. The increase of the aggregation number is more dramatic as the mixed micelle composition approaches equimolarity, indicating a spherical to non-spherical micelle shape change (Figure 3.11).

The large increase of the aggregation number in the SDS-rich side can be understood in terms of a decrease in the *gauche/trans* conformation ratio in the tails of the surfactants. This is a result of the increased crowding of the methylene chains accompanying the shape change. Simultaneously, there is an increase in SDS headgroup ordering in SDS-rich mixed micelles and a dominance of interaction with DTA^+ ions in DTAC-rich mixed micelles [79]. The formation of relatively larger, non-spherical mixed micelles exhibiting a synergistic increase in oil solubilization capacity is also shown in Figure 3.11 [79].

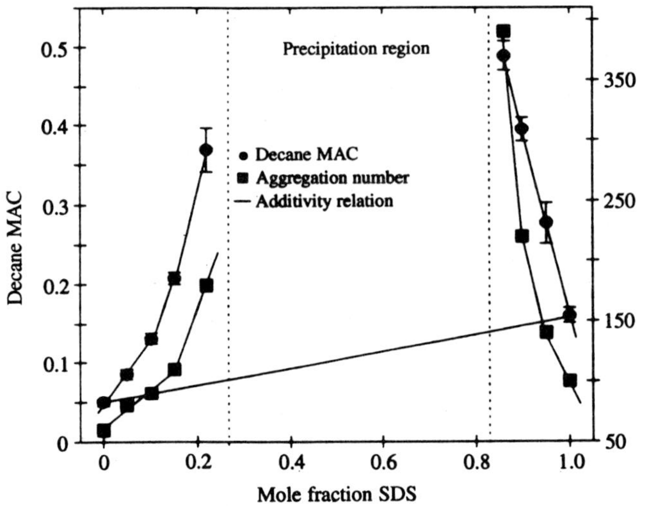

Figure 3.11 Variation of the mixed micelle aggregation number with the SDS mole fraction for the system DTAC–SDS–H_2O at 298 K. The total surfactant concentration is 0.3 M. Also plotted, MAC—moles of decane solubilized per mole of micelle [78,79].

A similar increase in aggregation number is also measured for cationic-rich, anionic-rich and for both cationic- and anionic-rich micellar solutions for the mixed systems CTAB–sodium alkyl sulfonate [80]. ATAC(B)–SDS [81] and OTEAB–sodium alkyl sulfate [82]. Both the decrease in temperature and the addition of salt also increase the aggregation number for the ionic-rich solutions of these systems, but for the equimolar mixtures an addition of salt has no detectable effect on the aggregation number. The aggregation number N also varies with the length of the alkyl chain of the sulfonate surfactant for the mixtures sodium alkyl sulfonate–CTAB, in the order $N(C_{16}) > N(C_{14}) > N(C_8)$ [80] indicating the importance of both electrostatic and hydrophobic effects in the micellization process. For the OTAB–SDS systems [83–85], an increase in aggregation number is measured for the SDS-rich solution. The change in the aggregation number with total surfactant concentration and mixing ratio can be explained in terms of a change in the micellar composition. At high total concentration, the micellar composition is nearly equal to the bulk composition. As the bulk concentration decreases, the fraction of the 'minority' surfactant in the mixed micelles increases [84]. Some systems also exhibit viscoelastic properties at high dilution [82–85].

Only one type of mixed micelle is formed for the system OTAM–SPFO [49], in contrast to the reported demixing of fluorinated ionic and hydrogenated non-ionic surfactants. Yet the exchange rate between the monomer state and the aggregate state for the system is found to be slow compared to the exchange rate between monomers and normal-size micelles [86]. The increase in the mean lifetime is possibly caused by the existence of larger and more ordered aggregates.

Most of the studied cationic–anionic surfactant mixtures are insoluble or slightly soluble with water. Therefore their practical use, in areas where high concentrations of surfactant mixtures are needed, is very limited. There is a need to search for soluble mixed surfactant systems. At present very few systems, such as mixed TTAB–SAEOS and TTAB–APE (organic alkoxy phosphate esters) [87], are found to produce clear solution phases over a wide concentration range at equimolar composition. They show high synergistic properties (Table 3.2). Like non-ionic surfactants, cloud points are observed in these systems (Figure 3.12).

The cloud point phenomenon has also been observed for the mixture STAC–sodium-*N*-lauroyl-*N*-methyl-β-alanine at equimolar concentration [88]. The mixed system α,ω-type (bolaform) cationic surfactant (*N*,*N*′-1,12-dodecanediyl-bis(triethylammonium bromide))–SDS [89] is soluble with water forming a clear solution. The system forms mixed micelles, whereas the α,ω-type cationic surfactant alone does not form micelles with water. The micellar size is largest at 1 : 1 molar ratio and almost equivalent to that of an unipolar surfactant with an alkyl chain half as long.

Bile salts, biologically important molecules, are anionic surfactants which self-associate in water forming small micelles [90]. They form mixed micelles with cationic surfactants [91–94] and soluble ion-pairs at concentrations

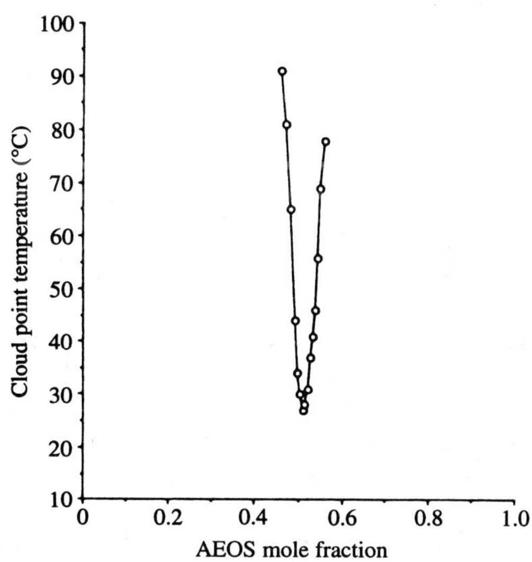

Figure 3.12 Cloud point temperature for solutions of the sodium poly(oxyethylene)sulfate (AEOS)–tetradecyltrimethylammonium bromide–water system. Total surfactant concentration is 0.05 M [87].

below the CMC [91,95]. The formation of ion-pairs is accounted for by the modified form of the Onsager equation [96] and it yields large association constants. A large reduction of the CMC for the mixed micelles is also measured (Table 3.2). Bile salts such as sodium deoxycholate (NaDC) form an isotropic solution phase with CTAB and, at equimolar ratio of the two surfactants, two liquids coexist at low total surfactant concentration. Addition of NaDC to a constant concentration of CTAB or CTAC leads to a dramatic increase of the micellar size (rod-type aggregates), whereas the mixed micelles for the system sodium cholate–CTAB are small and almost spherical in shape. The size of the micelles is almost independent of sample composition. The striking difference of the effects between the trihydroxy and dihydroxy bile salts in forming the mixed micelles can be found in the structure of bile salts [94,97,98]. The antibacterial agent, chlorhexidine digluconate (CG) is a divalent anionic surfactant. It also forms mixed micelles with ATAB surfactants. Both the CMC and the aggregation number are decreased on adding CG to micelles of the cationic surfactants [99].

3.5 Vesicle phase

3.5.1 *The stability of vesicles*

In recent years, studies on catanionic mixtures have made an important contribution to the fundamental issue of vesicle formation and stability in aqueous surfactant systems. Vesicles are closed, usually spherical, bilayer shells and the interest in them ranges from basic sciences, such as biosciences and physical chemistry, to application aspects in specialized chemistry, pharmacology (drug delivery) and medicine [4,5,100]. The establishment of vesicles as a thermodynamically stable phase is an on-going theoretical debate. From a theoretical point of view, the stabilization of a vesicle phase is seen as a competition between entropy (translational entropy or entropy of mixing) and curvature elastic energy [22,101]. It is clear that lipid vesicle solutions commonly obtained by sonication, extrusion or other non-spontaneous methods exist in a metastable state, i.e. are only kinetically stable [102]. Upon ageing, they evolve back to their equilibrium state, either a single lamellar phase or, more often, a two-phase equilibrium, usually μm size multilamellar droplets in excess solution. In other systems, vesicles are simply considered as very dilute lamellar dispersions, stabilized by translational entropy rather than by curvature energy.

The search for thermodynamically stable vesicles (also designated here as equilibrium vesicles) has nevertheless attracted considerable attention from experimentalists and theoreticians. By thermodynamically stable vesicles we mean those having the following characteristic features: (i) vesicles generated spontaneously (simple mixing of components) and reversibly; (ii) vesicles stable in time (no size/shape changes, no phase separation); (iii) vesicles in which shape, size distributions and stability are independent of preparation methods; and (iv) vesicles in equilibrium with other phases. Regular reports of formation of equilibrium vesicles in aqueous mixtures of lipids (fatty acids, phospholipids, fatty acids–phospholipids, bile salts–phospholipids) have appeared in the past two decades (see compiled references in References 119,125). More recently, in binary double-chain surfactant–water [103–108] and single-chain surfactant–water [109] systems and in systems composed of surfactant–alcohol–water or brine [110–113] equilibrium vesicles have been claimed to exist. Experimental studies on vesiculation in catanionic mixtures have immensely contributed to this debate and models by Safran and coworkers [114,115] have provided theoretical support. Although work on these mixtures has been done previously by Hargreaves and Deamer [116], the interest in the field is relatively recent.

3.5.2 Phase equilibria and properties of catanionic vesicles

Kaler *et al.* in 1989 [117] have presented evidence for equilibrium vesicle formation in the catanionic system CTAT–SDBS and, thereafter, a number of other experimental studies have been reported. The availability of elucidating techniques for such highly dilute surfactant solutions creates experimental difficulties. The methods used usually involve direct visualization of the aggregates by microscopy, optical (video-enhanced) and transmission electron microscopy (TEM), either with the freeze-fracture (FFTEM) or with the cryogenic technique (cryo-TEM), supported by more quantitative techniques such as light scattering (both static and dynamic), fluorescence (in particular, time-resolved fluorescence quenching, TRFQ) and NMR (chemical shifts, molecular self-diffusion and relaxation measurements).

These reports either directly address equilibrium vesicle formation (no non-spontaneous methods used) or consist of a more classical nature in the sense that sonicated catanionic (1:1) vesicles are used. Equilibrium vesicles have been claimed to occur in the following catanionic mixtures: (i) sodium alkyl benzene sulfonates–alkyltrimethylammonium tosylate/halides [117–120]; (ii) sodium alkylsulfates–quaternary alkyl ammonium halides [20,35,121] or catanionic siloxane halide surfactants [122]; (iii) sodium alkyl carboxylates (different chain lengths)–quaternary alkyl ammonium bromides (different headgroup sizes) [123]; (iv) dodecylpyridinium chloride–sodium acylated aminoacid surfactants [124–126] and (v) oppositely charged single-chained perfluorinated surfactants [127,128]. Detailed phase diagrams have been studied for systems (i) and (ii). Studies with sonicated catanionic vesicles comprise mixtures of perfluorocarbon [129] or hydrocarbon–fluorocarbon surfactants [129] and sodium alkyl carboxylates–quaternary alkyl ammonium bromides [123,130].

Single-phase regions of vesicles are usually found to exist with excess of either cationic or anionic surfactant, within a limited range of surfactant concentration and mixing ratios. The appropriate two- and three-phase regions are in many cases also detected, but in others multiphase samples cannot be clearly resolved. This can be seen in Figure 3.13, where two representative phase diagrams are depicted [118,121]. Direct imaging by electron microscopy of the vesicles is also presented.

As the A^+/A^- molar ratio or the total surfactant concentration is varied, different aggregate shape transitions and phase equilibria can be found: (i) micelles/vesicles; (ii) vesicles/liquid crystals; (iii) vesicles/precipitate; and (iv) monomers/vesicles. A schematic diagram depicting the equilibria between the catanionic vesicles and the other phases is shown in Figure 3.14.

The micelle-to-vesicle transition, which is of great theoretical interest, has been found experimentally to occur in two ways. If there is limited micellar growth, the transition is gradual and via a single thermodynamic phase of coexisting micelles and vesicles. These micelles can grow into rods or even disks as it is apparent from cryo-TEM observations for the system SDS–DDAB, where disk-like micelles are clearly observed [121] (see Figure 3.13(b)). If there is a substantial micellar growth, a phase separation can actually occur prior to the vesicle region, indicating a first-order phase transition. This has been seen to occur in the system SDS–DTAB [35], where long rod-like micelles phase separate with vesicles, and in other investigated systems at equimolar composition [123], where ultracentrifugation leads to clear phase separation.

In most systems, the vesicles are also in equilibrium with lamellar phases. This is not unexpected and it may in fact be the factor responsible for the limited extension of the vesicle region in the direction of increased total surfactant concentration. When the volume fraction of vesicles of a given size increases up to a point of dense packing, either the vesicles shrink or a phase transition occurs to accommodate further surfactant. In this case, phase separating micron-size multilamellar droplets (also called spherulites or onions) may form. They usually show flow birefringence and, at higher concentrations, static birefringence. Different regimes of bilayers (vesicles–onions–flat lamellae) have been proposed for some pseudo-ternary surfactant systems [110–113], in connection with theoretical models [101]. Liquid crystalline phases other than lamellar have been reported to exist in equilibrium with vesicles in multiphase samples, although their existence as single phase regions has not been determined: an hexagonal phase for the DTAB–SDS system [35] and a cubic phase for the SDS–siloxane surfactant system [122]. Vesicles are also found in two phase regions with equimolar precipitates (catanionic surfactant), as observed in several systems [35,118,121].

These vesicles are usually spherical and very polydisperse (as detected from TEM and light scattering), with mean diameters varying markedly from system to system, from small unilamellar vesicles in some systems, through large unilamellar vesicles to multilamellar vesicles in others. With sulfate and sulfonate headgroups vesicles only form with excess of ionic surfactant, while equimolarity leads to precipitation [35,117–121]. With the carboxylate headgroup (where protonation depends on pH) 1:1 mixtures display coexistence of micelles and vesicles, with no precipitation [123]. On the other hand, the catanionic sonicated vesicles (salt absent) are short-lived (hours) and a precipitate is readily formed [130].

It has been shown that the larger the size of the headgroup of one of the surfactants, the more difficult it is to form vesicles [123]. Longer sonication time and higher surfactant concentration are required in the order methyl < ethyl < propyl < butyl as substituent groups in a quaternary ammonium headgroup; in the case of methyl substitution, vesicles are formed even without sonication. The same trend has been seen with *N*-methylated *N*-acylated aminoacid surfactants–DPyC mixtures [125,126], if only the N or the α-C are methylated, vesicles coexist with micelles, but when both the N and α-C atoms are methylated only micelles are formed. Thus, it seems that specific interactions between the paired surfactants are important, since an increase in steric hindrance decreases vesicle stability. Headgroup interactions are also seen to change upon vesiculation in systems where two distinct headgroup stereochemical configurations for one of the surfactants are possible [126].

Vesicle formation as a function of surfactant alkyl chain length has also been investigated for the systems C_nCOONa–C_mNMBr and it is found that it occurs, either spontaneously or by sonication, only if the total number of C atoms is higher than 19, $n + m \geqslant 19$ [123]. An asymmetry in chain length between the two surfactants is also seen to affect vesicle stability [120]. The vesicle region in the phase diagram is larger in the short-chain surfactant rich side than in the long-chain rich side. This is observed for the systems DTAB, TTAB or CTAB with SDBS and the effect is even more pronounced for the CTAB–SOS mixture, where an unusually extended region of SOS-rich vesicles is found, but CTAB-rich vesicles are absent. According to Kaler *et al.* the bilayers rich in short-chain surfactant should have higher curvature, leading to smaller vesicles and therefore extension of the vesicle region toward higher surfactant concentration.

As might be expected, in systems where one of the surfactants has an ionizable headgroup, pH changes may affect vesiculation at fixed composition and temperature. Such changes in systems where the anionic headgroup is the carboxylate ion have actually been seen to affect vesicle stability [123]. An interesting case is presented by the system DDAO–SLT [124], where DDA^+O^- is a surfactant which is zwitterionic at high pH but cationic, DDA^+OH, after protonation of the negatively charged O atom. It is

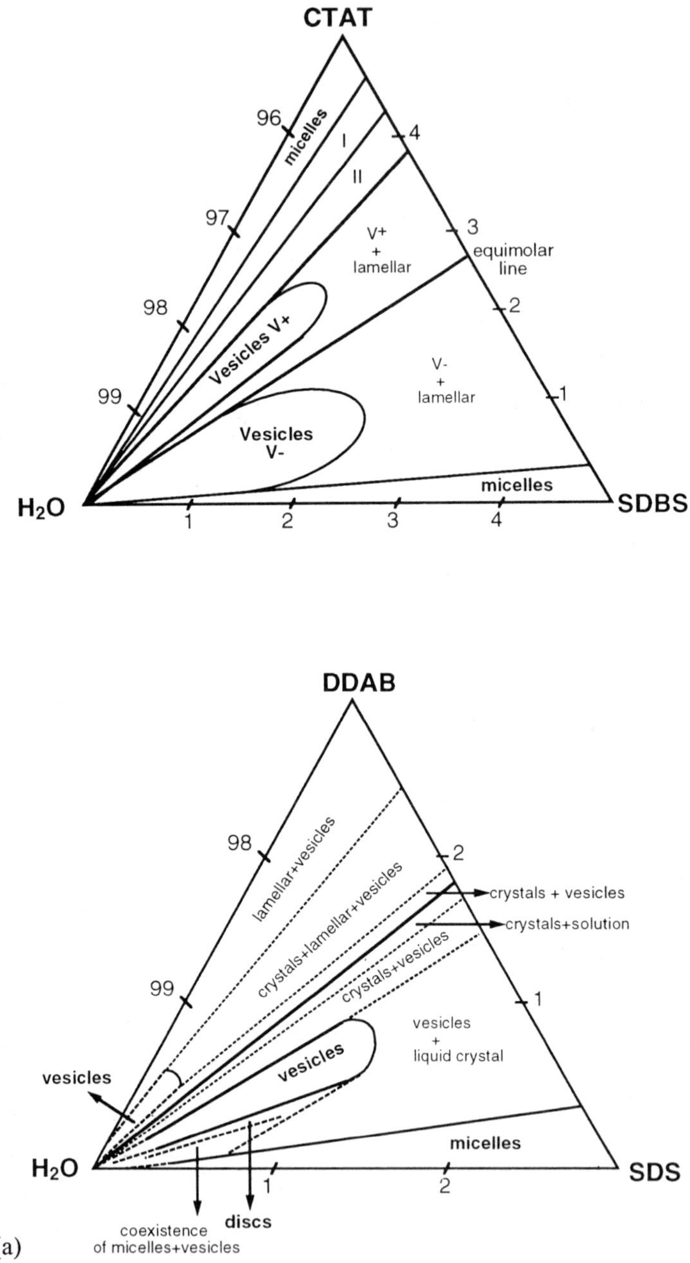

(a)

Figure 3.13 (a) Phase diagrams of the dilute region for the systems CTAT–SDBS–water at 298 K (redrawn from Reference 118) and DDAB–SDS–water at 298 K [121]. (b) TEM imaging of the vesicles detected in these systems: SDBS-rich vesicles (top left) and CTAT-rich vesicles (bottom left), as observed by FFTEM in the CTAT–SDBS–water system, bar = 100 nm. SDS-rich vesicles (top right) and disc-like aggregates in coexistence with vesicles (bottom right), as observed by cryo-TEM in the DDAB–SDS–water system, bar = 200 nm.

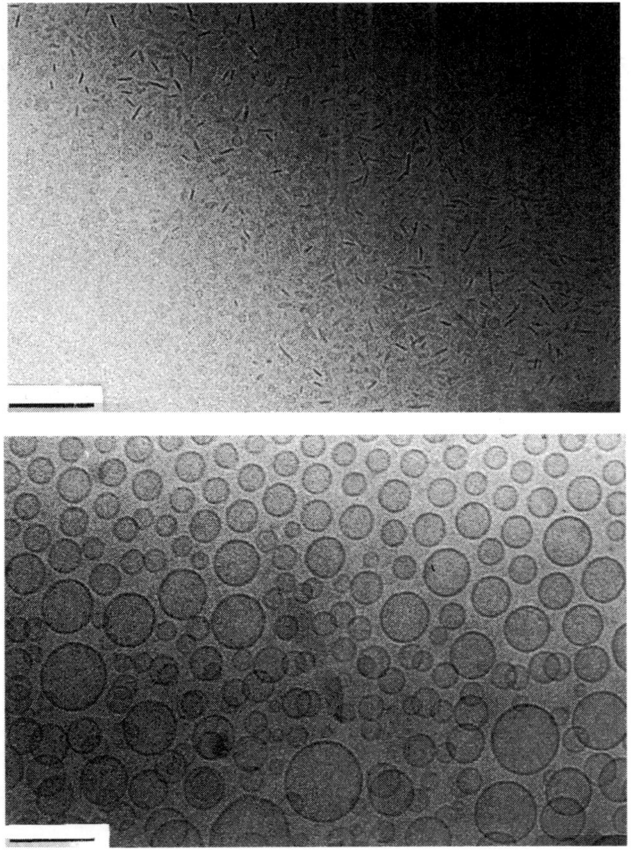
(b)

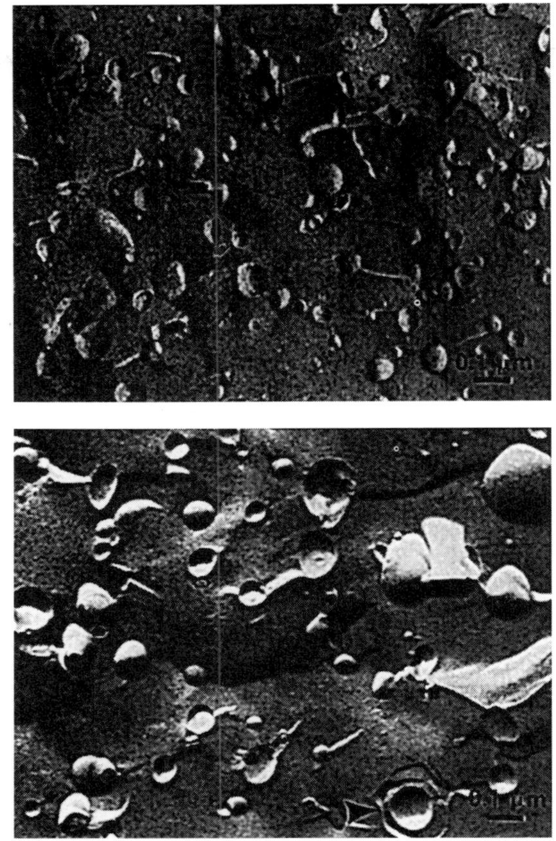

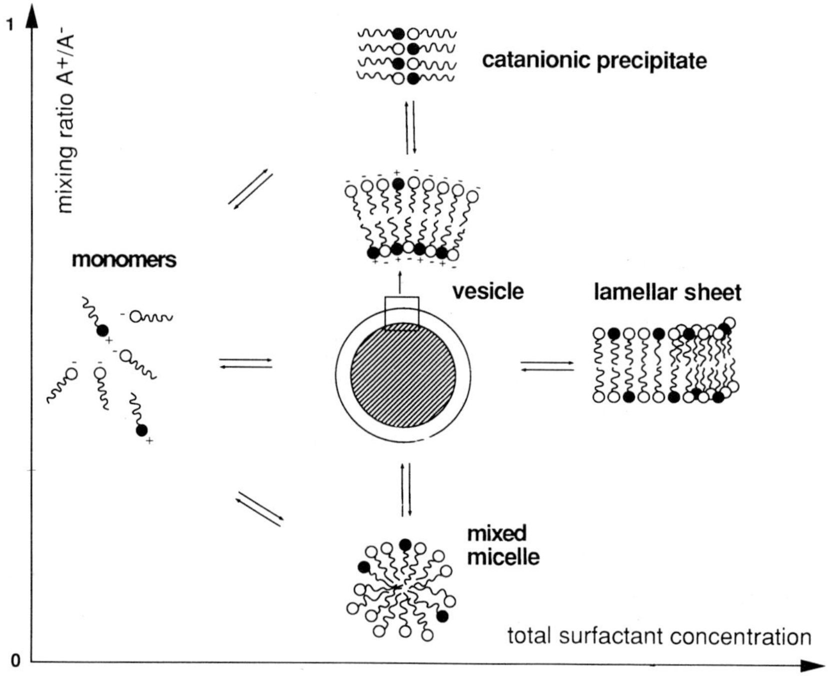

Figure 3.14 The different phase equilibria for the catanionic vesicles as a function of mixing ratio and total surfactant concentration. A detailed view of the possible structure for the vesicle bilayer is also presented (see text).

observed that the vesiculation in this system depends not only on the mixing composition and total surfactant concentration (as expected) but also on the protonation degree of DDAO. Upon pH decrease, leading to the formation of the cationic species and, thus, to the catanionic pair $DDAO^+HLT^-$, vesicle formation is induced from an initial micellar solution. The dependence of the phase behavior on the degree of neutralization and concentration for a given composition and temperature is shown in Figure 3.15. The possibility of a reversible micelle-to-vesicle transition simply by forward and backward pH titration is also mentioned. This system provides a good example of reversibility of vesicle formation in a catanionic mixture.

As stated previously, the use of vesicles for practical purposes is of great interest. A possible advantage of catanionic vesicles is their intrinsic stability; systems as described above where a micelle-to-vesicle transition is easily tunable by pH changes may be of potential use. Work carried out recently [131,132] provides a good example of application possibilities. Magnetic nanoparticles of Fe_2O_3 and/or Fe_3O_4 have been synthesized inside the pool of spontaneously formed vesicles of HDBS–CTAB. This is achieved basi-

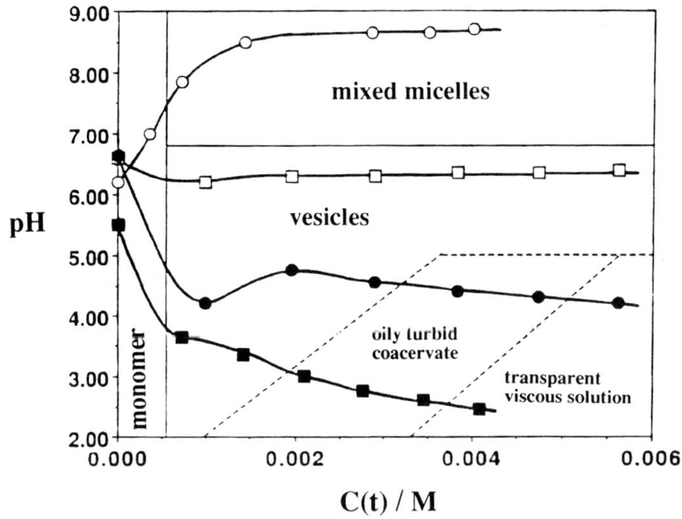

Figure 3.15 Titration curves for the system DDAO–SLT with a mixing ratio 3 : 1 at different degrees of neutralization r: $r = 0.0$ (empty circles); $r = 0.25$ (full circles); $r = 0.5$ (empty squares) and $r > 1.0$ (full squares). The parameter r is defined as $r = [\text{added HCl}]/[C_t]$, where C_t is the total surfactant concentration. NaCl is also present; the temperature is 298 K (redrawn from Reference 124).

cally by treating with NaOH a solution of vesicles in which Fe^{2+} ions have already been internalized.

From the experimental studies, it becomes evident that the presence of a catanionic ion-pair is essential for equilibrium vesiculation in these systems. Safran *et al.* [114,115] have presented a model which accounts for the energetic stabilization of vesicles by considering specific interactions between oppositely charged surfactants and a bilayer composition asymmetry. A single surfactant monolayer assumes a certain spontaneous curvature with respect to the polar/non-polar interface, which is the result of the optimization of free energy in its several terms—hydrophobic, surface and packing free energy terms [23] and is essentially dependent on the geometric shape of the molecule [22]. In a vesicle, the outer and inner monolayers have curvatures opposite in sign; thus, for a single component vesicle a curvature frustration occurs and, as a result, the vesicle is not energetically stable and reverts to the flat membrane state. However, the coexistence of one of the parent surfactants with the catanionic pair may lead to vesicle stabilization, if non-ideal mixing across the bilayer is assumed and curvature frustration concomitantly relieved. The outer layer could be enriched in the excess surfactant, whereas the inner layer would have essentially the catanionic pair of suitable geometry for the required spontaneous negative curvature. This demixing across the bilayer, although involving an entropic penalty, should compensate the curvature free energy cost associated with the vesicle formation.

As has been discussed before, the effect of the headgroup interactions and alkyl chain volume on the preferred aggregate curvature has been clearly seen from phase diagram studies. This strengthens the notion of the ability of catanionic pairs to adopt negative curvatures compatible with vesicle inner layers. So far, it has not been shown experimentally that such bilayer composition asymmetry exists. Interestingly, the existence of ion-pair segregation domains in sonicated vesicles of the catanionic system DODAB–VBAP [133] has been reported on the basis of absorption, emission and NMR spectroscopy data. Two models for the organization of these pairs in the membrane are presented, but curiously the model for preferential catanionic pair segregation in the inner layer, consistent with curvature arguments, is dismissed. Another recent model for equilibrium vesicles has been proposed, based on an asymmetry in charge density between the two membrane monolayers [134]. Naturally, the stability of vesicles has also to be understood in terms of vesicle–vesicle (i.e. inter-bilayer) interactions [4,118]; these interactions are mainly of four kinds: van der Waals, electrostatic, hydration and steric interactions.

3.6 Precipitation

Many mixed cationic–anionic surfactants spontaneously produce well-defined crystalline precipitates at very dilute solutions. For some systems, especially those containing alkyl carboxylate [63] as one of the surfactants, a longer time is needed for the formation of a precipitate. On the other hand, the highly symmetric straight chain pairs DTAB–SDS [35], DAC–SDS [56,135–138] and DPyC–SDS [139,140] can pack quite well into a crystalline lattice. As a result, the heterogeneous precipitate region dominates the phase behavior in the dilute part of the phase diagram, at the expense of other phases. These results are quite different from the phase behavior reported for the mixtures of CTAT with a highly branched dodecyl chain of SDBS [117,118]. Owing to packing constraints, this system precipitates only at equimolar mixtures and cannot yield any precipitation at non-equimolarity. The precipitate is neutral, composed of an equimolar ratio of the two surfactants over a wide range of mixing ratios, and possesses the characteristic crystal structure.

Boundary lines between the heterogeneous region containing the precipitate and the single isotropic solution phase have been determined for several mixed cationic–anionic surfactant systems [35,56,135,136,139,140]. The phase diagrams for two representative systems determined at very high water contents are shown in Figure 3.16.

(a)

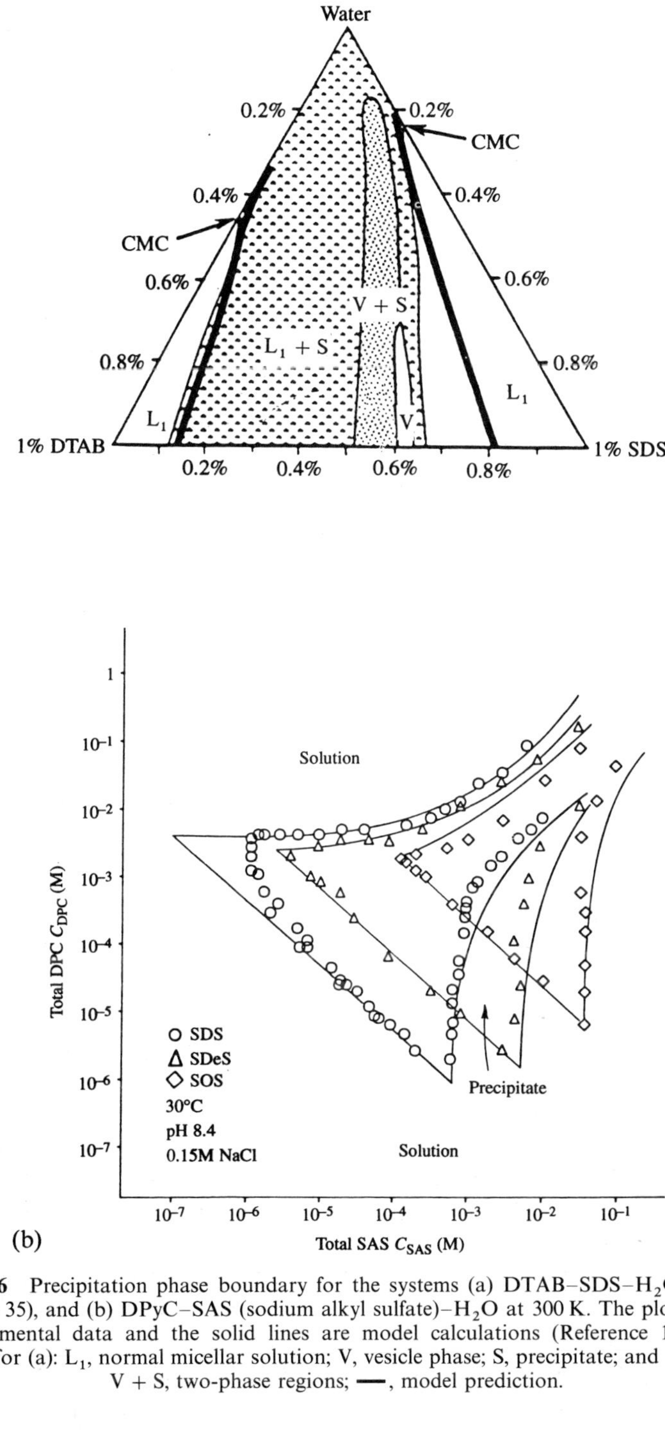

(b)

Figure 3.16 Precipitation phase boundary for the systems (a) DTAB–SDS–H_2O at 298 K (Reference 35), and (b) DPyC–SAS (sodium alkyl sulfate)–H_2O at 300 K. The plotted points are experimental data and the solid lines are model calculations (Reference 139). Phase notations for (a): L_1, normal micellar solution; V, vesicle phase; S, precipitate; and L_1 + S and V + S, two-phase regions; —, model prediction.

The main difference between the two phase diagrams is that for the DTAB–SDS system a vesicle phase is identified, whereas this phase is absent in the DPyC–SDS system. Very recently, it has been shown mainly by cryo-TEM, FFTEM and NMR techniques that the existence of vesicle phases is common for mixed oppositely charged surfactant systems. The techniques used to determine the phase diagram shown in Figure 3.16(b) are not amenable to the detection of vesicles even if they exist in the DPyC–SDS system.

The molar enthalpy change of precipitation is almost an order of magnitude larger than that of mixed micellization indicating that precipitation is primarily enthalpy driven, while micellization is primarily entropically driven [139]. The results in Figure 3.16(b) and Reference 56 show that an increase in the alkyl chain length increases the tendency for both precipitation and micellization. Addition of a non-ionic surfactant, $NP(EO)_8$ results in a lower tendency for precipitation in DPyC–SDS mixtures [141].

The thermodynamic cell model [142,143] which has been used extensively to predict the phase behavior and micellar properties of surfactant systems, is extended to account for the observed phase equilibrium between a micellar phase and a crystalline precipitate [35]. The theoretical phase diagram calculated from the prediction of the model is also shown in Figure 3.16(a). There is a good agreement between the observed and the model calculated phase diagram for the DTAB-rich side, but for the SDS-rich side the phase boundary predicted by the model begins to deviate with increasing SDS concentration. At around this concentration, micellar growth becomes significant. The underestimation of the stability range of the micellar phase relative to the precipitate phase is due to the presence of nonspherical micelles, since the model is developed for monodisperse spherical micelles.

The regular solution theory, within the limits of the pseudo-phase separation approximation, is also applied to account for the observed stability range of the precipitate region (Figure 3.16(b)). For the system in Figure 3.16(b), it is assumed that the precipitate is in equilibrium with surfactant monomers only at concentrations where no micelles are present. At concentrations above the CMC, the precipitate is in equilibrium with micelles only. A solubility product is determined for the former equilibrium and the regular solution theory as described earlier is used to calculate the monomer concentration of each surfactant in the bulk phase, as well as the composition of the micelles in terms of the two surfactants from the monomer–micelle equilibrium. The precipitation boundary is then calculated from the prediction of the model. Though good agreement is obtained for the phase boundary in the DPyC-rich side, the model calculations deviate considerably at high concentrations in the anionic-rich part of the phase diagram. It has been realized in later studies [141] that the experimentally determined phase boundary around this concentration range was not accurate due to the presence of crystals of non-visible size. However, it is reported that the phase boundary can be determined accurately if the system contains an excess amount of salt. A good agreement is then obtained between the experimentally determined and the theoretically calculated phase boundary. The effect of added non-ionic surfactant, alkyl chain length and temperature on the stability range of the precipitate region can also be well predicted from the model calculations [56,139–141].

3.7 Adsorption

It is well known that the surface activity of mixed surfactant solutions is often greater than would be expected in the absence of the mutual influence of the two surfactants. Such synergistic effects are much larger in catanionic mixtures than in ionic–non-ionic surfactant mixtures, while solutions of non-ionic–non-ionic surfactants exhibit negligible extent of synergism. The strong synergism in mixed cationic–anionic systems shows as an enhanced adsorption of the electroneutral catanionic molecules at the solution–air interface, as compared to the adsorption of its parent surfactant ions. A very strong cationic–anionic synergism is apparent in Figure 3.17(a). This effect shows up more obviously (Figure 3.17(b)) in the equilibrium surface pressure change that is obtained by mixing solutions of DTAB and SDS for various mixing ratios at a constant total surfactant concentration. No significant change in the surface pressure is observed even with solutions of the pure catanionic, DTA^+DS^- (no inorganic counterions) [52,144].

For example, a 1 : 1 molar solution yields a surface pressure which is higher by about $40\,mN\,m^{-1}$ than that of either pure component. Similar observations are made from surface tension data for the systems DPyC–SDS, DTAC–SDS [145,146], alkyltrimethyl (and ethyl) ammonium bromide–sodium alkyl sulfate [48,147] and *N*-dodecyl-β-alanine-sodium alkyl sulfate [148]. The Gibbs adsorption equation is used to determine the surface excesses, Γ, and the composition of the adsorbed film from the measured surface tension data. The finding of an equimolar composition of the adsorbed layer for a wide range of bulk mixing ratios is in good agreement with the results from tracer electrophoresis [48,148]. The occurrence of an equimolar composition for mixed micelles at the CMC for a wide range of mixing ratios of the two surfactants has already been mentioned.

The regular solution theory has been extended to account for the adsorption of the components of binary surfactant mixtures at the aqueous solution–air interface [149–154]. Equations similar to those for mixed micelles are derived in order to obtain the composition of the adsorbed layer

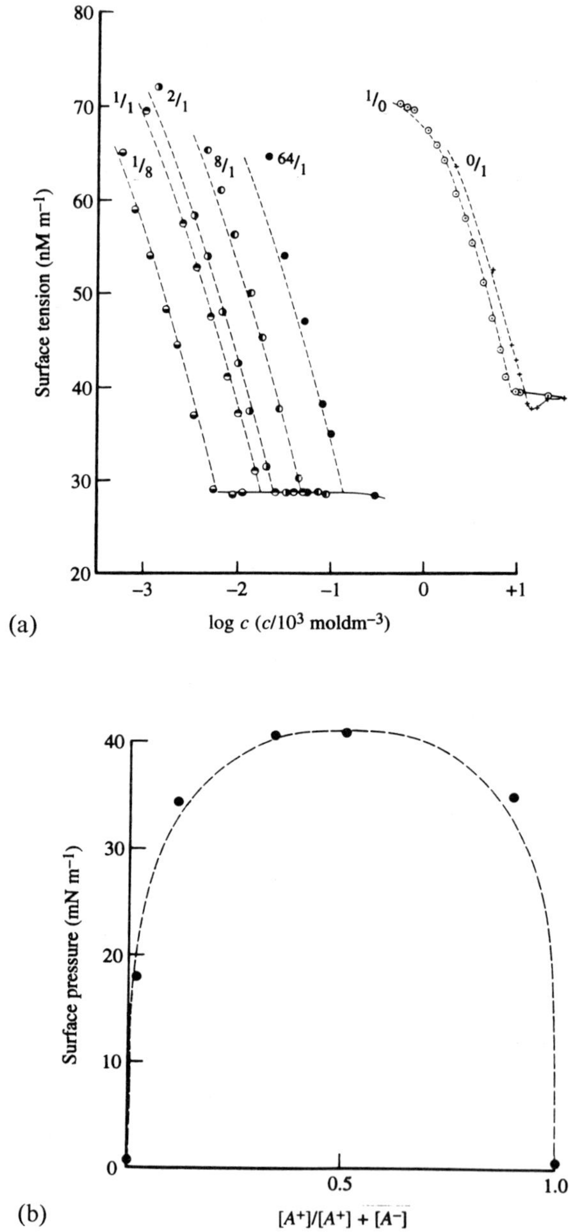

Figure 3.17 (a) Surface tension versus total surfactant concentration for different mixing ratios, [SDS]/[DTAB] = 0/1, 64/1, 1/0, etc.; (b) surface pressure versus solution composition at constant total surfactant concentration, 3×10^{-5} M, for the system DTAB–SDS–H_2O at 298 K. Redrawn from Reference 52.

and to establish conditions under which synergism in effectiveness of surface tension reduction will exist at the CMC in binary mixtures of surfactants. When this type of synergism exists, the mole fraction of either surfactant at the solution–air interface equals its mole fraction in the mixed micelles. The important conditions for synergism are: (1) $\beta^s < 0$ and (2) $\beta^s - \beta^m < 0$, where β^s and β^m are the experimentally determined molecular interaction parameters for mixed monolayer and mixed micelle formation, respectively. The limited data available (Table 3.3) indicate that synergism in terms of effectiveness of surface tension reduction occurs only in cationic–anionic surfactant mixtures. For a given value of surface tension of the mixed solution, the magnitude of β^s is higher for systems with symmetric than with asymmetric alkyl chains.

For mixed systems with asymmetric alkyl chains, e.g. DTAB–SOS and OTAB–SDS [147], an asymmetric composition of the adsorbed monolayer is obtained, in particular, at non-equivalent bulk compositions, due to an enhanced adsorption of the C_{12} component compared to C_8. For similarly charged surfactant mixtures [70b], the enhanced adsorption can be explained in terms of a favorable free energy of transfer of the longer chain surfactant from the aqueous environment. However, for oppositely charged surfactant mixtures, an additional opposing term, the electrostatic energy term, opposes any departure from electroneutrality, so that the equilibrium surface composition is determined by the two opposing effects. The result of these two effects can be monitored by adding salt to the mixed systems. In the presence of salt, the surface excess of the longer chain species (dominant free energy of transfer term) is increased, leading to the formation of an asymmetric surface composition. For the symmetric alkyl chains (dominant electrostatic free energy term), the salt effect is negligible and the adsorbed layer will retain equimolar composition [155–160].

Individual cationic surfactants can be adsorbed strongly onto negatively charged silica surfaces, but no significant adsorption of anionic surfactants is detected from their respective binary solutions. However, in the mixed cationic–anionic systems, the adsorbed amounts of both the cationic and the anionic surfactant ions are enhanced and the excess adsorption of the cationic ions is exactly equal to the adsorption of the anionic ions [161,162]. This observation suggests that the excess adsorption of the two oppositely charged ions may occur in the form of ion-pairs [162]. From measurements of the ζ potential and molecular model calculations for mixed surfactant solutions, it is suggested that the cationic and anionic surfactants are adsorbed on solid interfaces in the form of bilayers [163] and multilayers [164]. Composition-dependent adsorption of cationic and anionic surfactant ions is observed on cellulose fibres and the addition of non-ionic surfactants is found to impair the adsorption of the ionic species [165].

3.8 Concluding remarks

In this chapter we have presented an overview of the novel and important physico–chemical properties of catanionic surfactants and catanionic mixtures. The rich display of phase behavior of the concentrated systems has been discussed. It has been shown how the surfactant film curvature can be varied with change of surfactant molecular geometry, concentration and mixing ratios. The lamellar and microemulsion phases occurring in these systems have been studied, both experimentally and theoretically and interpreted in terms of hydration and electrostatic forces. Further development of the thermodynamic cell model, which has successfully accounted for the self-assembly of ionic surfactants, may increase our understanding of catanionic systems.

For the dilute systems, important bulk and interfacial solution properties and precipitation phenomena have been described. Of these, the synergistic properties, manifested as a large decrease of the CMC and enhanced surface activity and adsorption, have been extensively studied experimentally. Theoretical models for the quantitative description of these properties are available but a fine-tuning of the predictions is required. A search for soluble catanionic surfactants is also of interest to further exploit this synergistic behavior for application purposes. The equilibrium vesicles recently discovered in these mixtures are found to have much potential in terms of applications, besides being interesting from a fundamental point of view. Until now, a limited number of systems has been investigated.

Acknowledgment

John Daicic is gratefully acknowledged for his comments on the manuscript. We are thankful to Bengt Jönsson and Björn Lindman for valuable discussions during the preparation of the article. This work is financed by the Swedish Research Council for Engineering Sciences (TFR). E.M. acknowledges financial support from Junta Nacional de Investigação Científica e Tecnológica (JNICT), Portugal.

References

1. Scott, A. B., Tartar, H.V. and Lingafelter, E.C. *J. Am. Chem. Soc.*, 1943, **65**, 698.
2. Jokela, P., Jönsson, B. and Khan, A. *J. Phys. Chem.*, 1987, **91**, 3291.
3. Harwigsson, I., Khan, A. and Hellsten, M. *Tenside. Deterg.*, 1993, **30**, 174.
4. Lasic, D.D. *Liposomes: from physics to applications*. Elsevier, Amsterdam, 1993.
5. Ostro, M.J. *Liposomes: from biophysics to therapy*. Marcel Dekker, New York, 1987.
6. v.d. Pas, J.C. *Tenside Surf. Det.*, 1991, **28**, 158.

7. Cropton, R.W.G. and Joy, A.S. *Analyst*, 1963, **88**, 516.
8. Hummel, D. *Identification and analysis of surface active agents.* John Wiley and Sons, Inc., London, 1962.
9. Ogino, K. and Masahiko, A. (eds) *Mixed surfactant systems.* Marcel Dekker, New York, 1993.
10. Laughlin, R.G.*The aqueous phase behavior of surfactants.* Academic Press Inc., San Diego, 1994.
11. Jokela, P., Jönsson, B. *et al. Langmuir*, 1988, **4**, 187.
12. Ekwall, P. *Adv. Liq. Cryst.*, 1975, **1**, 1.
13. Mandell, L., Ekwall, P. and Fontell, K. *Acta Polytech. Scand. Met. Ser.*, 1968, **74**, I.
14. Sadaghiani, A., Khan, A. and Lindman, B. *J. Colloid Interface Sci.*, 1989, **132**, 352.
15. Jönsson, B. and Wennerström, H. *J. Chem. Soc., Faraday Trans. 2*, 1983, **79**, 19.
16. Jönsson, B. and Wennerström, H. *Chem. Scr.*, 1985, **25**, 117.
17. Nilsson, U., Jönsson, B. and Wennerström, H. *J. Phys. Chem.*, 1993, **97**, 5654.
18. Backlund, S., Karlsson, S. and Sjöblom, J. *J. Dispers. Sci. Technol.*, 1994, **15**, 561.
19. Khan, A. and Mendonca, C. *J. Colloid Interface Sci.*, 1995, **169**, 60.
20. Marques, E., Khan, A. *et al. J. Phys. Chem.*, 1993, **97**, 4729.
21. Khan, A., Regev, O. *et al. Progr. Colloid Polym. Sci.* 1994, **97**, 146.
22. Israelachvili, J.N. *Intermolecular and surface forces, 2nd edition*, Academic Press, San Diego, 1992.
23. Evans, D.F. and Wennerström, H. *The colloidal domain—where physics, chemistry, biology and technology meet*, VCH, New York, 1994.
24. Helfrich, W. *Z. Naturforsch.*, 1973, **28c**, 693.
25. Jokela, P. and Jönsson, B. *J. Phys. Chem.*, 1988, **92**, 1923.
26. Jönsson, B., Jokela, P. *et al. Langmuir*, 1991, **7**, 889.
27. Sadaghiani, A.S. and Khan, A. *J. Colloid Interface Sci.*, 1991, **146**, 69.
28. Cox, J.M. and Friberg, S.E. *J. Amer. Oil Chem. Soc.*, 1981, **58**, 743.
29. Jokela, P., Jönsson, B. and Wennerström, H. *Progr Colloid Polym. Sci*, 1985, **70**, 17.
30. Thalberg, K., Lindman, B. and Karlström, G. *J. Phys. Chem.*, 1991, **95**, 6004.
31. Hecker, L., Reeves, L. W. and Tracey, A.S. *Mol. Cryst. Liq. Cryst.*, 1979, **53**, 77.
32. Boonbrahm, P. and Saupe, A. *Mol. Cryst. Liq. Cryst.*, 1984, **109**, 223.
33. Mesa, C.L., Khan, A. *et al. J. Colloid Interface Sci.*, 1985, **103**, 373.
34. Tamori, K., Kihara, K. *et al. Colloid Polym. Sci.*, 1992, **270**, 885.
35. Herrington, K. L., Kaler, E.W. *et al. J. Phys. Chem.*, 1993, **97**, 13792.
36. Caria, A. and Khan, A. (manuscript submitted) 1995.
37. Hildebrand, J.H., Prausnitz, J.M. and Scott, R.L. *Regular and related solutions.* van Nostrand Reinhold, Princeton, 1970.
38. Wennerström, H. and Lindman, B. *Phys. Rev.*, 1979, **52**, 1.
39. Clint, J.H. *J. Chem. Soc., Faraday Trans. 1*, 1975, **71**, 1327.
40. Lange, H. and Beck, K.H. *Kolloid-Z.u. Z. Polymere*, 1973, **251**, 424.
41. Rubingh, D.N. In: Mittal, K.L. (ed.) *Solution chemistry of surfactants.* Plenum, New York, 1979, vol. 1, p. 337.
42. Treiner, C., Nortz, M. and Vaution, C. *Langmuir*, 1990, **6**, 1211.
43. Bourrel, M., Bernarde, D., and Graciaa, A. *Tenside Deterg.*, 1984, **21**, 311.
44. Graciaa, A., Ghoulam, M.B. *et al. J. Phys. Chem.*, 1989, **93**, 4167.
45. Treiner, C., Khodja, A.A., and Fromon, M. *Langmuir*, 1987, **3**, 729.
46. Yu, Z.-J. and Zhao, G.-X. *J. Colloid Interface Sci.*, 1993, **156**, 325.
47. Holland, P.M. and Rubingh, D.N. *J. Phys. Chem.*, 1983, **87**, 1984.
48. Corkill, M. and Goodman, J. *Proc. Royal Soc. Ser A*, 1963, **273**, 24.
49. Guo, W., Guzman, E.K. *et al. Langmuir*, 1992, **8**, 2368.
50. Funasaki, N. and Hada, S. *J. Phys. Chem.*, 1982, **86**, 2504.
51. Yu, Z.-J. and Zhao, G.-X. *J. Colloid Interface Sci.*, 1989, **130**, 414.
52. Lucassen-Reynders, E.H., Lucassen, J. and Gilles, D.J. *J. Colloid Interface Sci.*, 1981, **81**, 150.
53. Rodakiewicz-Novak, J. *J. Colloid Interface Sci.*, 1982, **85**, 856.
54. Li, X.-G. and Zhao, G.-X. *Colloids Surfaces*, 1992, **64**, 185.
55. Herman, C.B. In: Silva, A.F. (ed.) *Trends in interfacial electrochemistry.* Reidel, Dordrecht, 1986.

56. Filipovic-Vincekovic, N., Bujan, M. *et al. Colloid Polym. Sci.*, 1995, **273**, 182.
57. Oakenful, D.G. and Fenwick, D.E. *Aust. J. Chem.*, 1973, **26**, 2649.
58. Mitsuishi, M. and Hashizume, M. *Bull. Chem. Soc. Japan*, 1973, **46**, 1946.
59. Smith, D.L., Cox, M.F. *et al. J. Am. Oil Chem. Soc.*, 1989, **66**, 718.
60. Tomlison, E., Davis, S.S. and Mukhayer, G.I. In: Mittal, K.L. (ed.) *Solution chemistry of surfactants*. Plenum, New York, 1979, vol. 1, p. 3.
61. Anacker, E.W. *J. Colloid Interface Sci.*, 1953, **8**, 402.
62. Corkill, J.M., Goodman, J.F. *et al. Trans. Faraday Soc.*, 1966, **62**, 994.
63. Scowen, R.V. and Leja, L. *Can J. Chem.*, 1967, **45**, 2821.
64. Hoffmann, H., Rehage, H. *et al. Colloid Polym. Sci.*, 1982, **260**, 1042.
65. Yu, Z.-J. and Zhao, G.-X. *J. Colloid Interface Sci.*, 1989, **130**, 414.
66. Rivara-Minten, E., Baglioni, P., and Kevan, L. *J. Phys. Chem.*, 1988, **92**, 2613.
67. Marszall, L. *Langmuir*, 1990, **6**, 347.
68. Hoffmann, H., Kalus, J. and Schwandner, B. *Ber. Bunsenges. Phys. Chem.*, 1987, **91**, 99.
69. Tanford, C. *The hydrophobic effect*. J. Wiley and Sons, New York, 1973.
70. (a) Regev, O. and Khan, A. (in press). (b) Aniansson, G. *J. Phys. Chem.*, 1951, **55**, 1286.
71. Mukhayer, G.I. *J. Colloid Interface Sci.*, 1975, **1975**, 224.
72. Kamenka, N., Chorro, M. *et al. Colloids Surfaces*, 1992, **67**, 213.
73. Filipovic-Vincekovic, N. and Skritic, D. *Colloid Polym. Sci.*, 1988, **266**, 954.
74. Jönsson, B., Wennerström, H. and Halle, B. *J. Phys. Chem.*, 1980, **84**, 2179.
75. Jansson, M. and Stilbs, P. *J. Phys. Chem.*, 1987, **91**, 113.
76. Jansson, M. and Jönsson, B. *J. Phys. Chem.*, 1989, **93**, 1451.
77. Jansson, M. and Stilbs, P. *J. Phys. Chem.*, 1986, **89**, 4868.
78. Malliaris, A., Binana-Limbele, W., and Zana, R. *J. Colloid Interface Sci.*, 1986, **110**, 114.
79. Scheuing, D.R. and Weers, J.G. *Langmuir*, 1990, **6**, 665.
80. Li, G.-Z., Li, F. *et al. Colloids Surfaces*, 1993, **76**, 257.
81. Almgren, M. and Swarup, S. *J. Phys. Chem.*, 1983, **87**, 876.
82. Yu, Z.-J. and Zhao, G.-X. *J. Colloid Interface Sci.*, 1989, **130**, 421.
83. Kato, T., Iwai, M., and Seimya, T. *J. Colloid Interface Sci.*, 1989, **130**, 439.
84. Kato, T., Takeuchi, H. and Seimya, T. *J. Colloid Interface Sci.*, 1990, **140**, 253.
85. Barker, C.A., Saul, D. *et al. J. Chem. Soc. Faraday Trans.*, 1974, **5**, 154.
86. Aniansson, E.A., Wall, S.N. *et al. J. Phys. Chem.*, 1976, **80**, 905.
87. Mehreteab, A. and Loprest, F.J. *J. Colloid Interface Sci.*, 1988, **125**, 602.
88. Nakama, Y., Harusawa, F. and Murotani, I. *J. Am. Oil Chem. Soc.*, 1990, **67**, 717.
89. Ishikawa, M., Matsumura, K. *et al. J. Colloid Interface Sci.*, 1991, **141**, 10.
90. Nair, P.P. and Kritchevsky, D. (eds). *The bile acids*. Plenum Press, New York, 1971.
91. Jana, P.K. and Moulik, S.P. *J. Phys. Chem.*, 1991, **95**, 9525.
92. Barry, B.W. and Gray, G.M.T. *J. Colloid Interface Sci.*, 1975, **52**, 327.
93. Wu, K. and McGown, L.B. *J. Phys. Chem.*, 1994, **98**, 1185.
94. Vethamuthu, M.S., Almgren, M., *et al. Langmuir*, 1992, **8**, 2396.
95. Oakenful, D. and Fenwick, D.E. *Aust. J. Chem.*, 1977, **1977**, 335.
96. Robinson, R.A. and Stokes, R.H. *Electrolyte solutions*. Butterworths, London, 1955.
97. Mazer, N.A., Benedek, G.B. and Carey, M.C. *Biochemistry*, 1980, **19**, 601.
98. Carey, M.C. *Hepatology*, 1984, **4**, 138S.
99. Attwood, D. and Patel, H.K. *J. Colloid Interface Sci.*, 1989, **129**, 222.
100. Fendler, J.H. *Membrane mimetic chemistry*. John Wiley & Sons, New York, 1983.
101. Simons, B.D. and Cates, M.E. *J. Phys. II France*, 1992, **2**, 1439.
102. Lasic, D.D. *J. Colloid Interface Sci.*, 1990, **140**, 302.
103. Brady, J.E., Evans, D.F. *et al. J. Am. Chem. Soc.*, 1984, **106**, 4279.
104. Brady, J.E., Evans, D.F. *et al.*, *J. Phys. Chem.*, 1986, **90**, 1853.
105. Evans, D.F. and Ninham, B.W. *J. Phys. Chem.*, 1986, **90**, 226.
106. Ninham, B.W., Evans, D.F. and Wel, G.J. *J. Phys. Chem.*, 1983, **87**, 5020.
107. Talmon, Y., Evans, D.F. and Ninham, B.W. *Science*, 1983, **221**, 1047.
108. Regev, O., Kang, C. and Khan, A. *J. Phys. Chem.*, 1994, **98**, 6619.
109. Cantú, L., Corti, M. *et al.*, *Europhys. Lett.*, 1990, **13**, 561.
110. Hervé, P., Roux, D. *et al. J. Phys. II France*, 1993, **3**, 1255.
111. Hoffmann, H., Thunig, C. and Munkert, U. *Langmuir*, 1992, **8**, 2629.

112. Hoffmann, H., Munkert, U. *et al. J. Colloid Interface Sci.*, 1994, **163**, 217.
113. Hoffmann, H., Thunig, C. *et al. Langmuir*, 1994, **10**, 3972.
114. Safran, S.A., Pincus, P. and Andelman, D. *Science*, 1990, **248**, 354.
115. Safran, S.A., MacKintosh, F.C. *et al., Progr. Colloid Polym. Sci.*, 1991, **84**, 3.
116. Hargreaves, W.R. and Deamer, D.W. *Biochemistry*, 1978, **17**, 2804.
117. Kaler, E.W., Murthy, A.K. *et al. Science*, 1989, **245**, 1371.
118. Kaler, E.W., Herrington, K.L. *et al. J. Phys. Chem.*, 1992, **96**, 6698.
119. Kaler, E.W., Herrington, K.L. *et al.* In: Chen, S.-H., Huang, J.S. and Tartaglia, P. (eds) *Structure and dynamics of strongly interacting colloids and supramolecular aggregates in solution.* Kluwer Academic, Dordrecht, 1992, p. 571.
120. Kaler, E., Herrington, K.L. and Zasadzinski, J.A.N. In: Sirota, E.B., Weitz *et al.* (eds) *Mat. Res. Soc. Symp. Proc., Complex Fluids.* Material Research Society, Pittsburgh, 1992, vol. 248, p. 3.
121. Marques, E., Regev, O. *et al.* (unpublished).
122. Lin, Z., He, M. *et al. J. Phys. Chem.*, 1993, **97**, 3571.
123. Huang, J.-B. and Zhao, G.-X. *Colloid Polym. Sci.*, 1995, **273**, 156.
124. Watzke, H.J. *Progr. Colloid Polym. Sci.*, 1993, **93**, 15.
125. Ambühl, M., Bangerter, F. *et al. Langmuir*, 1993, **9**, 36.
126. Ambühl, M., Bangerter, E. *et al. Prog. Colloid Polym. Sci.*, 1993, **93**, 183.
127. Szönyi, S. and Watzke, H.J. *Prog. Colloid Polym. Sci.*, 1993, **93**, 364.
128. Szönyi, S., Cambon, A. *et al.* In: Lipowsky, R., Richter, D. and Kremer, K. (eds.) *The structure and conformation of amphiphilic membranes. Springer Proc. in Physics.* Springer-Verlag, Berlin, 1992, vol. 66, p. 198.
129. Esumi, K. *Colloids Surfaces*, 1994, **84**, 49.
130. Fukuda, H., Kawata, K. *et al. J. Am. Chem. Soc.*, 1990, **112**, 1635.
131. Yaacob, I.I., Nunes, A.C., *et al. J. Colloid Interface Sci.*, 1994, **168**, 289.
132. Yaacob, I.I., Nunes, A.C. and Bose, A. *J. Colloid Interface Sci.*, 1995, **171**, 73.
133. Lerebours, B., Watzke, H.J. and Fendler, J.F. *J. Phys. Chem.*, 1990, **94**, 1632.
134. Kumaran, V. *J. Chem. Phys.*, 1994, **99**, 5490.
135. Tomasic, V., Filipovic-Vincekovic, N. *et al. Colloid Polym. Sci.*, 1991, **269**, 1289.
136. Filipovic-Vincekovic, N., Skritic, D. and Tomasic, V.-X. *Ber. Bunsenges. Phys. Chem.*, 1991, **95**, 1646.
137. Skrtic, D., Filipovic-Vincekovic, N. and Tomasic, V. *J. Colloid Interface Sci.*, 1992, **151**, 288.
138. Skrtic, D., Babic-Ivancic, V. *et al. Ber. Bunsenges. Phys. Chem.*, 1993, **97**, 121.
139. Amante, J.C., Scamehorn, J.F. and Harwell, J.F. *J. Colloid Interface Sci.*, 1991, **144**, 243.
140. Stellner, K.L., Amante, J.C., *et al. J. Colloid Interface Sci.*, 1988, **123**, 186.
141. Shiau, B.-J., Harwell, J.H. and Scamehorn, J.F. *J. Colloid Interface Sci.*, 1994, **167**, 332.
142. Jokela, P., Jönsson, B. and Wennerström, H. *Prog. Colloid Polym. Sci.*, 1985, **70**, 17.
143. Gunnarsson, G., Jönsson, B. and Wennerström, H. *J. Phys. Chem.*, 1980, **84**, 3114.
144. Lucassen-Reynders, E.H. *Kolloid-Z. u. Z. Polymere*, 1972, **250**, 356.
145. Lange, H. and Schwuger, V. M. J. *Kolloid-Z. u. Z. Polymere*, 1971, **243**, 120.
146. Schwuger, V.M.J. *Kolloid-Z. u. Polymere*, 1971, **243**, 129.
147. Corkhill, J.M., Goodman, J.F. *et al. Trans. Faraday Soc.*, 1967, **63**, 247.
148. Nakamura, A. and Muramatsu, M. *J. Colloid Interface Sci.*, 1977, **62**, 165.
149. Hua, X.Y. and Rosen, M.J. *J. Colloid Interface Sci.*, 1982, **90**, 212.
150. Zhu, B.Y. and Rosen, M.J. *J. Colloid Interface Sci.*, 1984, **99**, 435.
151. Rosen, M.J. and Hua, X.Y. *J. Colloid Interface Sci.*, 1982, **86**, 164.
152. Rosen, M.J. and Zhu, Y. *J. Colloid Interface Sci.*, 1984, **99**, 427.
153. Rosen, M.J. and Murphy, D.S. *J. Colloid Interface Sci.*, 1986, **110**, 224.
154. Rosen, M.J. and Gu, B. *Colloids Surfaces*, 1987, **23**, 119.
155. Góralczyk, D. *J. Colloid Interface Sci.*, 1994, **167**, 172.
156. Góralczyk, D. *Colloid Polym. Sci*, 1994, **272**, 204.
157. Góralczyk, D. *Tenside Deterg.*, 1992, **29**, 39.
158. Góralczyk, D. *Colloids Surfaces*, 1992, **66**, 241.
159. Góralczyk, D. *J. Colloid Interface Sci.*, 1980, **77**, 68.
160. Rodakiewicz-Nowak, J. *J. Colloid Interface Sci.*, 1981, **84**, 533.

161. Muller, H. and Krempl, E. *Tenside*, 1968, **5**, 333.
162. Huang, Z., Yan, Z. and Gu, T. *Colloids Surfaces*, 1989, **36**, 353.
163. Capovilla, L., Labbé, P. and Reverdy, G. *Langmuir*, 1991, **7**, 2000.
164. Meguro, K., Adachi, T. *et al. Langmuir*, 1988, **4**, 1160.
165. Hellsten, K.M.E., Klingberg, A.W. and Karlsson, B.T.G. *J. Am. Oil Chem. Soc.*, 1989, **66**, 1381.

4 Bolaform and dimeric (gemini) surfactants

R. ZANA

4.1 Bolaform surfactants

4.1.1 Definition

The word 'bolaform' was first used associated to 'electrolytes' to refer to organic molecules made up of two charged groups connected by a linear polymethylene chain as shown in Figure 4.1 (a 'bola' is a long cord with heavy balls at each end used by the gauchos in the Argentinian pampa to capture cattle) [1]. When the polymethylene chain is sufficiently long, the bolaform electrolyte becomes a bolaform amphiphile (bolaamphiphile) or surfactant. Indeed, it is then surface-active and capable of self-associating in water to form micelles. Bolaform surfactants come with a wide variety of ionic head groups and also with non-ionic head groups as shown by the few examples listed in Table 4.1 [2–6]. However, as shown below bolaforms are, in general, much less good surfactants than conventional ionic surfactants (made up of one charged group connected to one alkyl chain). Nevertheless, bolaforms often display biological activity [7,8] and some drugs are in fact bolaform amphiphiles, such as surfactant 4 in Table 4.1. Also, some special bolaforms are capable of giving rise to organized assemblies of peculiar structure.

4.1.2 Synthesis and purification

Bolaform surfactants have been synthesized and purified in many different

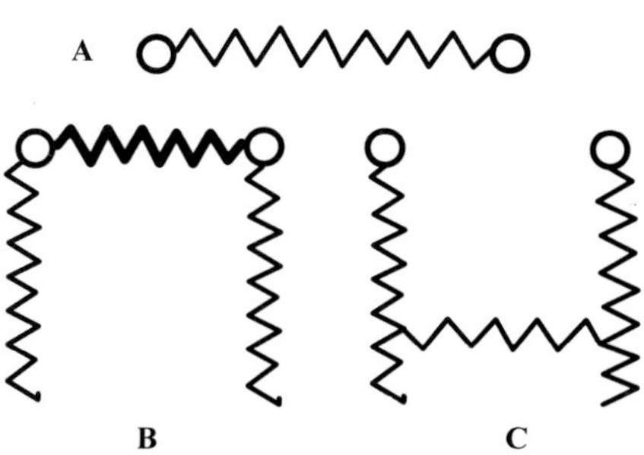

Figure 4.1 Schematic representation of a bolaform surfactant (A); and a dimeric surfactant (B). Surfactant C where the spacer connects the two amphiphilic moieties far from the head groups is in fact a branched bolaform surfactant.

Table 4.1 Examples of bolaform surfactants

Surfactant	Chemical Formula[a]	CMC (mM)
1. Hexadecanediyl-1,16-bis-(trimethylammonium bromide)	$(CH_3)_3N^+(CH_2)_{16}N^+(CH_3)_3 2Br^-$	42[b] (0.9[b])
2. Potassium hexadecanedioate	$^-O_2C(CH_2)_{14}CO_2^- 2K^+$	45[c] (3[b])
3. Dodecanediyl-1,12-bis(tetrakis-(oxyethylene)monoether)	$HO_4(C_2H_4O)(CH_2)_{12}(OC_2H_4)_4OH$	≈5.3[b] (<0.1[b])
4. Demecarium bromide	$(CH_3)_3N^+\Phi OCON(CH_3)(CH_2)_{10}$-$N(CH_3)COO\Phi N^+(CH_3)_3 2Br^-$	8.8[b]

[a]Φ refers to a phenylene group, substituted in *meta* in surfactant 4; [b]Values at 25°C; [c]Values at 40°C.

ways. The surfactants in Table 4.1 are taken as representative examples. Surfactant 1 and its homologues were obtained by reacting trimethylamine with the corresponding α,ω-dibromoalkane in dry ethanol under reflux for 48 h and purifying the raw surfactant by several recrystallizations from an ethylacetate–ethanol mixture [2]. Surfactant 2 was prepared by stoichiometric neutralization of the corresponding diacid by alcoholic KOH in a water +30% ethanol mixture at boiling point. The surfactant precipitated upon cooling [4]. Surfactant 3 was obtained by a Williamson ether synthesis and purified by passage through a separation column with mixtures of acetone and benzene, followed by recrystallization from diethylether–acetone mixtures [5].

4.1.3 *Behavior at interfaces*

Bolaforms adsorb at the air–water interface but are much less efficient than the corresponding conventional surfactants in lowering the surface tension of water. This is illustrated in Figure 4.2 which shows that at a given surfactant concentration the measured surface tension remains larger with the bolaform. The values of the surface area per surfactant, obtained from such plots, suggest that most bolaforms adopt a wicket-like (inverted U shaped) or looped conformation at the air–water interface, with the polar groups anchored at the interface and the alkanediyl chain forming a more or less wide loop in the air [7,9]. Some bolaforms have been reported to lie flat in the interface [3].

4.1.4 *Micellization in aqueous solution*

(a) Critical micelle concentration (CMC), micelle ionization degree and thermodynamics of micellization. Bolaform surfactants are much more hydrophilic than the corresponding conventional surfactants, owing to their additional polar group. This results in much larger values of the critical micelle concentration (CMC), as seen in Table 4.1. The difference depends

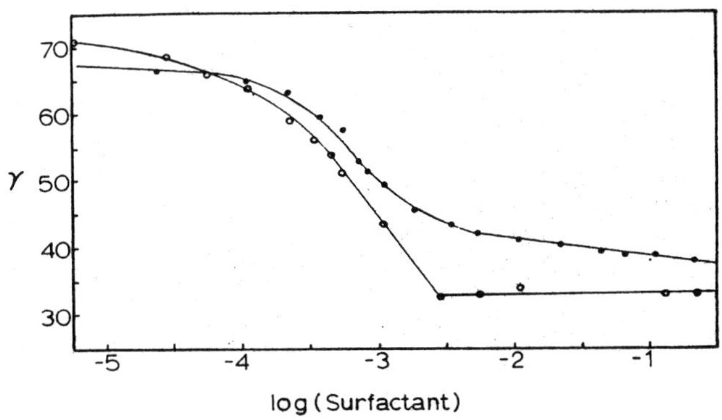

Figure 4.2 Variation of the surface tension γ (in mN/m) with the surfactant concentration at 24°C for the bolaform $(C_4H_9)_3N^+C_{12}H_{24}N^+(C_4H_9)_32Br^-$ (●) and the conventional surfactant $C_{12}H_{25}N^+(C_4H_9)_3Br^-$(O). The Gibbs equation leads to values of 1.46 and 0.73 nm^2 for the surface areas per surfactant for the bolaform and the corresponding conventional surfactant. The factor of 2 between these values suggests a wicket-like conformation of the bolaform at the air–water interface. *Reproduced from Reference 7, with permission of the American Chemical Society.*

on the nature (hydrophile–lipophile balance) of the head group and can be very large, the more so the more hydrophilic the head group. For the same reason, the minimum carbon number of the bolaform alkanediyl chain for which micelle formation can be unambiguously characterized is larger than that of the alkyl chain of the corresponding conventional surfactant. Thus, the CMC of hexylammonium chloride is about 0.95 M [10] whereas the bolaform dodecarediyl-1,12-bis(trimethylammonium bromide) forms only rather small aggregates at above 0.6 M [11]. The bolaform decanediyl-1,10-bis(trimethylammonium bromide) does not form micelles even at concentrations as high as 1 M [8]. Likewise, while the CMC of potassium hexanoate is about 1 M [12], there is no self-association in solutions of potassium decanedioate [4] but micelles form in solutions of sodium dodecanedioate at around 0.25–0.4 M [13]. Roughly speaking, the CMC of a bolaform with an alkanediyl chain containing *m* carbon atoms is near that of the conventional surfactant with the same polar head and an alkyl chain with *m*/2 carbon atoms.

It is interesting to compare the dependence of the CMC on the carbon number *m* for bolaform and conventional surfactants. The variations of lnCMC with the carbon number are generally linear and the free energy of transfer of a methylene group from the aqueous phase to the micellar pseudo-phase, $\Delta G_t^0(CH_2)$, can be determined from the slopes of the plots of lnCMC versus *m*, on the basis of the equations 4.1 for conventional surfactants [14] and 4.2 for bolaforms [15].

$$\Delta G_t^0 = RT(1 + \beta)/\ln CMC \quad (4.1)$$

$$\Delta G_t^0 = RT(1 + 2\beta)\ln CMC \quad (4.2)$$

In these equations ΔG_t^0 is the free energy of transfer of one surfactant from water to the micellar pseudo-phase and β the fraction of charges of micellized surfactant neutralized by bound counterions, a quantity experimentally accessible from electrical conductivity or potentiometric measurements. Figure 4.3 shows the variation of ΔG_t^0 for a series of conventional surfactants, the alkyltrimethylammonium bromides, and for two series of bolaform surfactants. The values of ΔG_t^0 have been calculated from the reported values of the CMC and micelle ionization degree [2,11,16–19]. The values of $\Delta G_t^0(CH_2)$ for the bisquaternary ammonium bolaforms and for their corresponding conventional surfactants, the alkyltrimethylammonium bromides, are found to be fairly close: −3.0 and −3.45 kJ/mol, respectively. The results for the second series of bolaform surfactant, the alkanediyl-α,ω-bis(carboxylatepropylenesulfonate, sodium salt) or $(CH_2)_s[CO_2(CH_2)_3SO_3^- Na^+]_2$ yield surprisingly low values of $\Delta G_t^0(CH_2)$. However, they were obtained at 40°C and the proximity of the Krafft temperature may have affected the measurements [17].

The fractional charge $\alpha = 1 - \beta$, of a bolaform micelle is generally larger than that of the corresponding conventional surfactant. Thus, for instance, the values of α are 0.50 for hexadecanediyl-1,16-bis(trimethylammonium bromide) [18] and 0.20 hexadecanetrimethylammonium bromide [16].

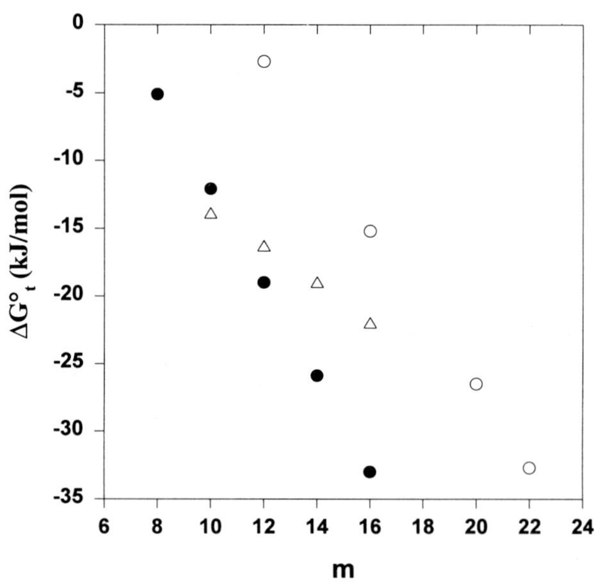

Figure 4.3 Variation of ΔG_t^0 with the carbon number *m* of the alkane or alkanediyl group for the alkyltrimethylammonium bromides (●, from the data in Reference 16), the alkanediyl-α,ω-bis(sodium carboxylatepropylenesulfonate) (△, from the data in Reference 17), and the alkanediyl-α,ω-bis(trimethylammonium bromide) (○, from the data in References 18 and 19).

Calorimetric measurements revealed that the enthalpy of micellization of decanediyl-1,10-bis(triethanolammonium carboxylate) is close to zero at room temperature [13], a result similar to that found for conventional ionic surfactants. The thermodynamics of micellization of alkanediyl-α,ω-bis-(sodium sulfate) has been investigated [20].

(b) Conformation of the bolaform alkanediyl chain. Controversies exist in the literature about this question for bolaforms in aqueous solution in both the molecularly dispersed state and micellized state. Electrical conductivity studies of dilute (molecularly dispersed) aqueous solutions of various types of dicationic bolaforms provided evidence for some association of one halide ion to the doubly charged cation [21]. The values of the intercharge distance, calculated from the bolaform ion limiting equivalent conductivity, were found to depend much less than expected on the alkanediyl chain carbon number [22] and it was suggested that in the ion-pair formed by a bolaform ion and an anion, the alkanediyl chain may be curled [21]. However, these calculations have been strongly criticized [23].

In the micellized state, the fast exchange of surfactant observed to occur between micelles and bulk phase in micellar solutions of the bolaform docosanediyl-1,22-bis(trimethylammonium bromide) led to the conclusion that the micellized bolaform ions are probably folded, as at the air–water interface [18]. However, a later NMR study of micellar solutions of eicosanediyl-1,20-bis(triethylammonium bromide) concluded that the eicosanediyl group adopted a predominantly stretched conformation in the bolaform micelles [24]. An attempt was made to reconcile these seemingly contradictory results on the basis of the 'surfactant block model' for micelles [25]. The issue is apparently still unsettled.

(c) Micelle size. The size of bolaform micelles has been characterized by the micelle aggregation number N, the micelle molecular weight M_w, or the micelle hydrodynamic radius R_H, obtained from fluorescence quenching [5,18,19,26], light scattering [6,9,19,27] and quasi-elastic light scattering [27], respectively. In all instances, as to be expected, the aggregation number of bolaform micelles has been found to be smaller than that of the corresponding conventional surfactant micelles. For instance, the aggregation number of micelles of docosanediyl-1,22-bis(trimethylammonium bromide), $(CH_3)_3N^+(C_{22}H_{44})N^+(CH_3)_3 2Br^-$, ranged between 15 and 25 [18,19] whereas the N value for $C_{22}H_{45}N^+(CH_3)_3$, Br^- should be well above 100 [28]. However, the aggregation number of undecyltrimethylammonium bromide is about 45. Thus, the volumes of the hydrophobic cores of the micelles of this surfactant and of docosanediyl-1,22-bis(trimethylammonium bromide) should be about the same, on the basis of these N values. In general, the aggregation number of a bolaform has a value somewhat

lower than half that of the conventional surfactant having an alkyl chain carbon number half that of the bolaform alkanediyl chain [27].

Other results show that the aggregation number of bolaform micelles increases with the ionic strength of the solution [27], decreases when the size of the head group is increased (bisquaternary ammonium [19] and non-ionic [5] bolaforms) and increases with the carbon number of the alkanediyl group (bisquaternary ammonium [18,19] and bis sodium sulfate [26,29] bolaforms. (Note that the value of 90 reported for the aggregation number of dodecanediyl-1,12-bis(sodium sulfate) [9] was later corrected to 5 to 10 [29]). All these results are very similar to those found for conventional surfactants.

(d) Mixed micellization. Bolaform surfactants have been shown to comicellize with conventional surfactants. Thus, mixed micellization between alkyltrimethylammonium bromides and alkanediyl-α,ω-bis(trimethylammonium bromide) has been investigated [30]. At low concentration, close to the CMC of the mixture, comicellization occurs only if the carbon number of the alkyltrimethylammonium bromide is only slightly smaller than half the bolaform carbon number. Thus, the hexadecyltrimethylammonium bromide comicellizes with docosanediyl-1,22-bis(trimethylammonium bromide) but does not comicellize with the shorter homologues of this bolaform. Nevertheless, the results showed that mixed micellization is enhanced at higher total surfactant concentration. Because of the tendency of bolaforms to form small micelles, the aggregation numbers of the mixed micelles are smaller than that of the conventional surfactant [30]. A similar result was reported for the mass of the mixed micelles of alkyltrimethylammonium bromides and of the bolaform drug, chlorhexidine digluconate, which forms micelles of aggregation number of about 4 [31].

In contrast, the mixtures of sodium dodecylsulfate (SDS) and of dodecanediyl-1,12-bis(triethylammonium bromide) were shown to have very low CMCs (synergism) and the mixed micelles were larger than those of SDS. The maximum micellar molecular weight was obtained at the stoichiometric composition corresponding to charge neutralization [32]. Again, all of the above results are similar to those reported for binary mixtures of conventional surfactants.

4.1.5 Phase behavior

There have been relatively few studies of the phase behavior of bolaforms. The study of the thermotropism of the pure tetradecanediyl and eicosanediyl-α,ω-bis(X^+-carbosylate), where X^+ is an alkali metal ion, showed the existence of two lamellar phases, one where the alkyl chains and head groups are in a crystalline, ordered state and another where they are in a disordered liquid state [33].

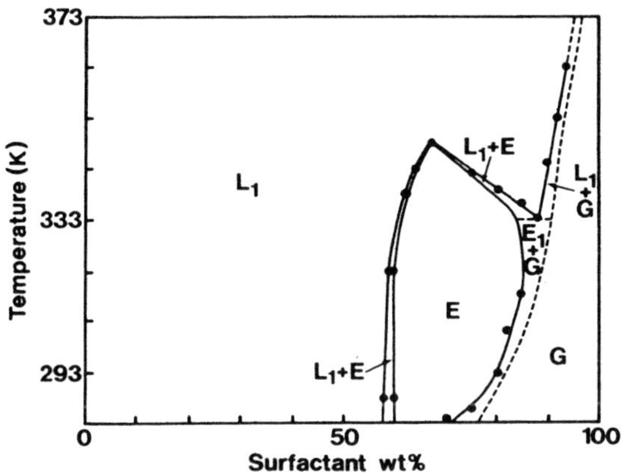

Figure 4.4 Phase diagram of the binary mixture water/eicosanediyl-1,20-bis(triethylammonium bromide). (L1) micellar solution; (E) hexagonal liquid crystalline phase; (G) solid surfactant. *Reproduced from Reference 36, with permission of Academic Press.*

The lyotropic behavior of tetradecanediyl-1,14-bis(potassium carboxylate)/water mixtures has been thoroughly investigated [34,35]. Evidence was reported for a lamellar phase [34,35] and, possibly, a cubic phase [35]. In contrast, only a hexagonal phase was observed with the eicosanediyl-1,20-bis(triethylammonium bromide)/water mixtures [36], as shown in Figure 4.4. The large difference in head group size between these two bolaforms is responsible for their different phase behavior, the larger cationic head group favoring the more curved hexagonal structure compared to the lamellar structure. Note that the realm of the hexagonal phase tends to be reduced at higher temperature.

4.1.6 Peculiar bolaforms

The results reviewed above concerned bolaforms giving rise to spherical or close to spherical micelles. However, the theory of the self-association of bolaforms has been put forward [37] and the calculations show that bolaforms can give rise to a variety of organized assemblies.

Peculiar bolaforms, where two chains connect the two polar groups [38–42], contain an amino acid polar group and an amide group [43], or contain rigid paraquat head groups [44], have been synthesized. Only the bisparaquat bolaform is soluble in water where it forms micelles of diameter below 10 nm. The other bolaforms give rise to various structures including ultra-thin monolayer membranes, vesicles and tubules. Notice that the cell wall of thermophilic and acidophilic archaebacteria, which grow best at low pH and high temperature, are made up of bolaform lipids [38,40].

4.1.7 Conclusions

The results reviewed show that bolaforms behave in many respects as conventional surfactants. However, they are much less efficient than conventional surfactants as far as useful properties are concerned. Being much more hydrophilic they form micelles only at high concentrations. These micelles have much smaller sizes and solubilizing capacity [19,36] than the micelles of the corresponding conventional surfactants which reduces the interest of bolaforms in the usual uses of surfactants. Nevertheless, bolaforms appear to have interesting biological properties. Also the bolaforms discussed in section 4.1.6 may prove useful.

4.2 Dimeric (gemini) surfactants

4.2.1 Definition

Dimeric surfactants are made up of two identical amphiphilic moieties connected at the level of the head groups or very close to the head groups by a spacer group, as shown in Figure 4.1 [45]. They are also referred to as gemini [46] or bipolar [38] surfactants. The expression 'siamese surfactants' has been proposed [47] because like Siamese twins they contain two identical and connected parts. Various types of spacer groups, flexible ones such as polymethylene [45], short polyoxyethylene [48], heteroatomic [49,50], and rigid ones such as xylylene or vinylene di-*p*-phenylene (stilbene derivatives) [46], have been used. Some examples of dimeric surfactants are listed in Table 4.2 [46, 50–52]. Note that if the spacer group is a polymethylene chain which connects the two amphiphilic moieties far from the head groups, close to the end of the alkyl chains, the dimeric surfactant is then simply a bolaform surfactant with a branched alkanediyl spacer. This is clearly seen in Figure 4.1.

Most reported physico–chemical results on dimeric surfactants concern bisquaternary ammonium surfactants of the alkanediyl-α,ω-bis-

Table 4.2 Examples of dimeric surfactants

Chemical Formulae	Reference	CMC (mM)[a]
1. C_3H_6-1,3-bis[$(CH_3)_2N^+C_{12}H_{25}Br^-$]	51	1.1 (15)
2. C_6H_4-1,4-bis[$CH_2(CH_3)_2N^+C_{12}H_{25}Br^-$]	46	1.0 (15)
3. [–$SCH_2CH_2NHCOCH_2N^+(CH_3)_2C_{12}H_{25}Cl^-$]$_2$	50	0.041 (<1)
4. O[–$CH_2CH(OCH_2CO_2^-Na^+)CH_2OC_{10}H_{21}$]$_2$	52	0.084
5. [–$O(CH_2CH_2O)_3$–][–$CH_2CH(OCH_2CO_2^-Na^+)CH_2OC_{10}H_{21}$]$_2$	52	0.37

[a]The values in brackets are the CMC for the corresponding monomeric surfactants, when available.

(dimethylalkylammonium halide) type which are referred to as *m-s-m*,$2X^-$ surfactants below, *s* and *m* being the carbon numbers of the alkanediyl spacer and of the alkyl chain of the amphiphilic moieties, respectively, and X^- the halide counterion. These surfactants are formally the dimers of the quaternary ammonium surfactants with two unequal chains $C_mH_{2m+1}(C_{s/2}H_{s+1})N^+(CH_3)_2X^-$, thus the name dimeric surfactants.

As is shown below, the characteristics of dimeric surfactants are extremely promising. Thus, these surfactants appear to be superior to their monomeric counterparts in all respects, giving them 'the potential for being the next generation of surfactants, the surfactants of the nineties' [53]. Besides, their aqueous solutions have interesting structural features and rheological properties. This explains the current interest in these surfactants both in industrial and academic spheres.

4.2.2 *Synthesis and purification*

The synthesis of the dimeric surfactants will not be reviewed except for surfactant 1 in Table 4.2. This surfactant and its homologues were obtained via the single step bisquaternization reaction of dodecyldimethylamine with 1,3-dibromopropane (molar ratio 2.1 : 1) in dry ethanol, under reflux for 48 h [51]. However, the synthesis of the other surfactants in Table 4.2 involved two or more reaction steps.

The real difficulty when dealing with dimeric surfactants lies in the purification of the raw surfactants. Indeed the last reaction step which leads to the dimeric surfactant usually involves a reaction at the two ends of an intermediate compound. This reaction rarely reaches completion and leads to the formation of a one-chain surfactant in addition to the dimeric surfactant. The separation of these two compounds is difficult to achieve by successive recrystallizations, as the two surfactants co-crystallize, and it usually involves chromatography. This is specially true for compounds with a short spacer. Thus surfactants 2–4 in Table 4.2 were purified by chromatography [50,52].

4.2.3 *Behavior at interfaces*

There have been many measurements of the surface tension of aqueous solutions of dimeric surfactants, investigating the effect of the length of the spacer and of the side alkyl chain [46,48–50,52,54–61]. The dimeric surfactants are always more efficient in lowering the surface tension of water than the corresponding conventional (monomeric) surfactants because, as is shown in the next paragraph, they have much lower CMC values. The surface tension isotherms have been used to obtain the values of the surface area, *a*, per surfactant molecule, using the Gibbs equation [62]. In all instances the values are large (generally above $10\,nm^2$) and indicate that when the spacer group is flexible and long enough, i.e. contains 10 or more carbon atoms, the dimeric surfactants probably adopt the same wicket-like or looped conformation at the air–water interface [48–50,52,56,58–61] as bolaform surfactants. In dimeric surfactants with short spacer groups, the spacer may be lying flat at the interface. This is suggested by the rapid increase of the surface area per dimeric surfactant with the spacer carbon number [60], as well as by some results concerning the structure of the lyotropic mesophases evidenced in water–dimeric surfactant mixtures [63].

The situation appears to be very different when rigid spacers are used. Thus, Menger and Littau concluded on the basis of the measured values of *a* that, when using rigid aromatic spacers (xylylene and stilbenylene), the whole dimeric surfactant (spacer and side alkyl chains) lies flat in the air–water interface [46]. Also peculiar ageing effects were observed in the measurements of surface tension.

Figure 4.5 shows the variation of *a* with the spacer carbon number *s* for the alkanediyl-α,ω-bis(dodecyldimethylammonium bromide) or 12-*s*-12,2Br$^-$ dimeric surfactants [60]. A pronounced maximum is observed for $s = 10–12$. Fragmentary results reported for quaternary ammonium bolaform surfactants also reveal the presence of a maximum in the plot of *a* versus the carbon number of the alkanediyl group of these surfactants [7]. The interpretation given to the plot in Figure 4.5 is that at low *s*, below 10, the spacer is in contact with water and lies flat at the interface. For

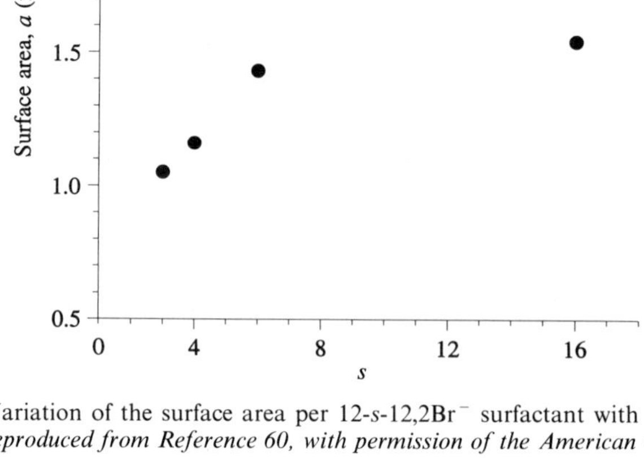

Figure 4.5 Variation of the surface area per 12-*s*-12,2Br$^-$ surfactant with the spacer carbon number *s*. *Reproduced from Reference 60, with permission of the American Chemical Society.*

$s > 10$–12, the spacer becomes too hydrophobic to remain in contact with water and moves to the air side of the interface, adopting a wicket-like or looped conformation, thus giving a decrease of a [7,60]. A complementary or alternative interpretation of these results is based on the fact that at low s values the spacer is rigid. The dimeric surfactant would then lie flat at the interface, just as for dimeric surfactant with rigid aromatic spacers. At higher s values the spacer becomes flexible and the alkyl chains can then orient differently, making an increasing angle with the interface, resulting in a rapid decrease of a. Note that the maximum in the plot of Figure 4.5 has been accounted for theoretically and the attractive and repulsive interactions of the surfactant molecules and the conformational entropy of the spacer found to be the dominant factors in determining the dependence of a on s [64]. The theory gave a good account of the increasing part of the plot but the decreasing part was not well accounted for. The difference may well arise from the effect just discussed and which was not taken into account in the calculations.

4.2.4 *Micelle formation*

(a) Critical micelle concentration, micelle ionization degree and thermodynamics of micellization. The formation of micelles in aqueous solution of dimeric surfactants has been much investigated. In all instances, where the comparison is possible, the CMCs of dimeric surfactants have been found to be at least one order of magnitude smaller than for the corresponding monomeric surfactants. Thus, the CMC value of 12-3-12,2Br^-(propanediyl-1,3-bis(dodecyldimethylammonium bromide)) is 1 mM whereas that of the corresponding monomeric surfactant, DTAB (dodecyltrimethylammonium bromide), is 15 mM [51]. This, of course, is very interesting for potential applications. The most likely explanation for these low CMC values is that, in solutions of dimeric surfactants, two alkyl chains at a time are transferred from water to the micellar phase, resulting in nearly double the free energy change of transfer, and thus, in a much lower value of the CMC. Indeed, for an oligomeric surfactant made up of i identical amphiphilic moieties each comprising an alkyl chain and one uni-univalent head group, the free energy of transfer per alkyl chain of surfactant, ΔG_t^0 is given by [15]:

$$\Delta G_t^0 = [(1/i) + \beta]RT\ln CMC \tag{4.3}$$

where all symbols have the same meaning as in equation 1 and where the CMC is expressed in mole per liter or mole fraction of alkyl chain. This equation yields values of −18.4 and −20.5 kJ/mol of dodecyl chain for DTAB and the dimeric surfactant 12-3-12,2Br^-, respectively. These two values differ by only a small amount. Inserting in equation 4.3 the values of the CMC of the bisquaternary ammonium surfactants with hexanediyl spacer [47,51] and using for β the same values as for the corresponding monomeric surfactants (see below) the free energy of transfer per CH_2 group of dimeric surfactant, $\Delta G_t^0(CH_2)$, was found to be very close to that for the corresponding monomeric surfactant, that is -3.2 ± 0.3 kJ/mol.

Figures 4.6 show the variation of the CMC of a series of homologous m-s-m,$2Br^-$ dimeric surfactants with the carbon number of the alkyl chains, m, at $s = 6$ and with the spacer carbon number, s, at $m = 12$. The variation of the CMC with m is linear in the semilogarithmic representation used, as for conventional (monomeric) and bolaform surfactants. On the contrary, the plot of CMC versus s shows a maximum at around $s = 5$, irrespective of the value of m between 10 and 16 [51,57]. The peculiar behavior of the plot as $s < 5$ was attributed to a change of the spacer conformation as s is progressively increased, to its resulting effect on the hydration of the head groups [57] and on the orientation of the alkyl chains [51]. The linear decrease of the CMC upon increasing m at $s \geqslant 10$ in the semilogarithmic representation used has been attributed to the progressive penetration of the spacer in the micelle hydrophobic core [51].

For rigid spacers it has been observed that the plot of CMC versus m goes through a marked minimum and increases for $m > 16$ [46]. This *very peculiar* behavior has been attributed to the self-coiling of the alkyl chains and/or the formation of submicellar aggregates.

Figure 4.6a also shows the variation of the micelle ionization degree, α, with the spacer carbon number for the 12-s-12,$2Br^-$ surfactant series [51]. The sigmoidal increase of α with s is similar to that found with the series of the corresponding monomeric surfactants, $C_{12}H_{25}(C_{s/2}H_{s+1})N^+$-$(CH_3)_2Br^-$, upon increasing $s/2$.

The CMC and enthalpies of micellization of the dimeric surfactants of general formula $[Cl^-C_mH_{2m+1}COOCH_2N^+(CH_3)_2CH_2]_2$ for $m = 8$ to 16, have been measured and the free energy, entropy and enthalpy of micellization thus obtained [65]. However, an incorrect equation was used to calculate the free energy of micellization from the value of the CMC and the reported thermodynamic quantities are affected by errors difficult to evaluate.

Some authors have reported the existence of two CMCs for the solutions of some dimeric surfactants [47,54,55]. The most recent study [47] showed that the true CMC probably corresponds to the larger of the two values. The first break may reveal the occurrence of some ion-pairing between the counterion and the dimeric surfactant ion [47,59], as postulated for bolaform ions [21], but this explanation is still speculative.

(b) Micelle size. There have been few determinations of size or aggregation number of micelles of dimeric surfactants. The most extensive results

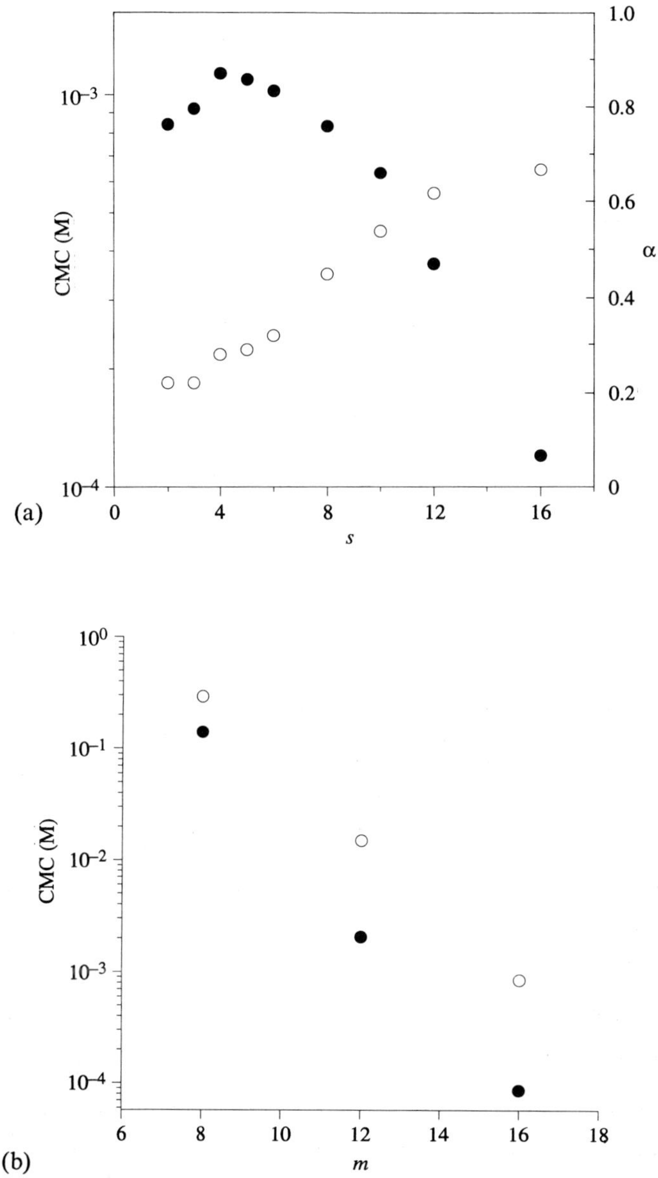

Figure 4.6 (a) Variation of the CMC (●) and of the micelle ionization degree α (○) with the spacer carbon number s, for the 12-s-12,2Br$^-$ surfactants. *Reproduced from Reference 51, with permission of the American Chemical Society.* (b) Variation of the CMC for the alkyltrimethylammonium bromides (○) and for the m-6-m,2Br$^-$ (●) surfactants with the carbon number m of the alkyl chain.

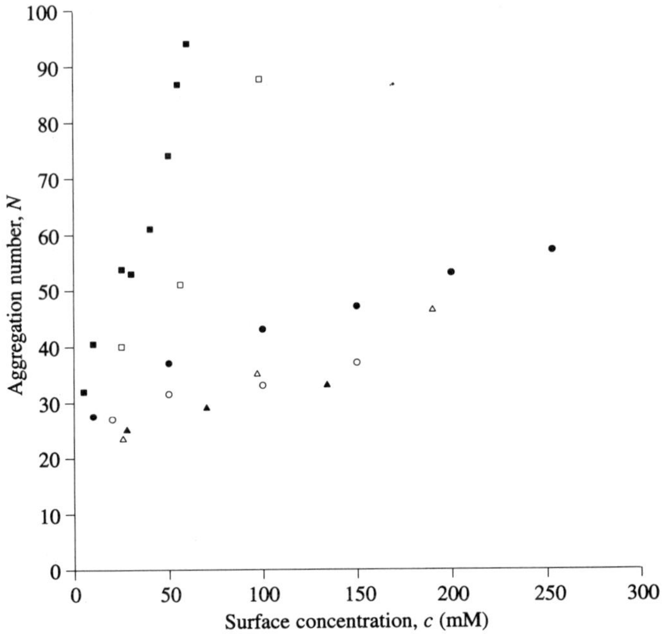

Figure 4.7 Variation of the aggregation number of the 12-*s*-12,2Br$^-$ surfactant micelles with the surfactant concentration for (■) 12-3-12,2Br$^-$, (□) 12-4-12,2Br$^-$, (•) 12-5-12,2Br$^-$, (○) 12-6-12,2Br$^-$, (▲) 12-8-12,2Br$^-$, and (△) 12-10-12,2Br$^-$. *Reproduced from Reference 67, with permission of the American Chemical Society.*

concern the 12-*s*-12,2Br$^-$ surfactant series [66–68]. Figure 4.7 shows the variations of *N* with the surfactant concentration, *C*, for surfactants of increasing spacer carbon number. At low *C*, the various plots tend to values of *N* of 25 to 30, consistent with spherical or spheroidal micelles (indeed, the aggregation number of the minimum spherical micelle of a surfactant with a dodecyl alkyl chain is of about 60) [16]. In all instances *N* increases with *C* and the increase becomes steeper as *s* decreases. For long spacers ($s = 6$ to 10) the increase of *N* is moderate to small and the micelles probably retain the same shape. For short spacers ($s = 3$ to 5), the values of *N* at high *C* become large and clearly inconsistent with a spheroidal micelle shape. It has been shown that the micelles are then polydisperse [67]. The 12-2-12,2Br$^-$ homologue has been found recently to give rise to micelles of still larger aggregation numbers [68]. The substitution of the counterion Br^- by Cl^- results in a significant decrease of the micelle aggregation number, as in the case of conventional monomeric surfactants [67,68].

The size of the micelles of the dichained–diglucamide head group 'sugar' surfactant, $(C_6H_{13})_2C[(CH_2NHCO)(CHOH)_4CH_2OH)]_2$ has been investigated by neutron scattering [69]. At concentration as low as 16 mM this surfactant already forms micelles 14 nm long and of radius 1.2 nm.

(c) Microstructure of solutions of dimeric surfactants. The cryo-transmission electron microscopy which permits direct visualization of the particles present in the system under investigation has been very successfully used to investigate the microstructure of solutions of dimeric surfactants of the *m*-*s*-*m*,$2Br^-$ type, with $m = 12$ and 16 [45,66,67]. The micrographs in Figure 4.8 show typical cryogenic transmission electron micrographs for the homologues of the 12-*s*-12,$2Br^-$ series with $s = 2$, 3, 4 and 8. Giant and entangled thread-like micelles are clearly seen in the 12-2-12,$2Br^-$ micrograph. The presence of such micelles explains the viscoelasticity of the solutions of this surfactant, at concentration as low as 2%. The 7% 12-3-12,$2Br^-$ solution also shows thread-like micelles, although much shorter, but not the 2% solution of this surfactant. The 12-4-12,$2Br^-$, 12-8-12,$2Br^-$, 12-10-12,$2Br^-$ and 12-12-12,$2Br^-$ solutions only show the presence of densely packed spherical or spheroidal micelles (Figures 4.8c and 4.8d and Reference 67). These trends are in agreement with those for the aggregation numbers just discussed. For still longer spacers ($s = 16$ and 20), spherical vesicles, often double-walled and some with defects, were observed particularly with the 12-20-20,$2Br^-$ surfactant [67]. Tubular vesicles were also present. Thus, with increasing spacer carbon number the 12-*s*-12,$2Br^-$ series shows the very unusual sequence of structures: thread-like micelles → spheroidal micelles → vesicles.

A number of bola-amphiphile lipids having the same chemical structure as dimeric surfactants and possessing a long spacer group (more than 16 atoms) have been synthesized and mixed with other lipids to produce vesicles [70–73]. Further investigations indicated that a significant fraction of the dimeric lipids in these mixed vesicles adopt a conformation where one head group is located at the vesicle outer layer and the other at the vesicle inner layer, as shown in Figure 4.9. These compounds have been referred to as bilayer bridging lipids. The stability of such mixed vesicles is enhanced whereas their permeability is lowered.

With the 16-*s*-16,$2Br^-$ series, unilamellar vesicles, some showing defects, were found to coexist with bilayer membrane fragments and thread-like micelles in solutions of 16-3-16,$2Br^-$, whereas only thread-like micelles were present in solutions of 16-4-16,$2Br^-$ and spherical micelles in solutions of 16-6-16,$2Br^-$ [67].

Although the above results refer to the dimeric bisquaternary ammonium surfactants, they are probably of more general character. Thus, preliminary experiments showed that solutions of dimeric surfactants with short spacers and mixed fluorinated/hydrogenated alkyl chains were viscoelastic, probably revealing the presence of thread-like micelles [74]. In view of these results, it was concluded [67] that dimeric surfactants with short spacers

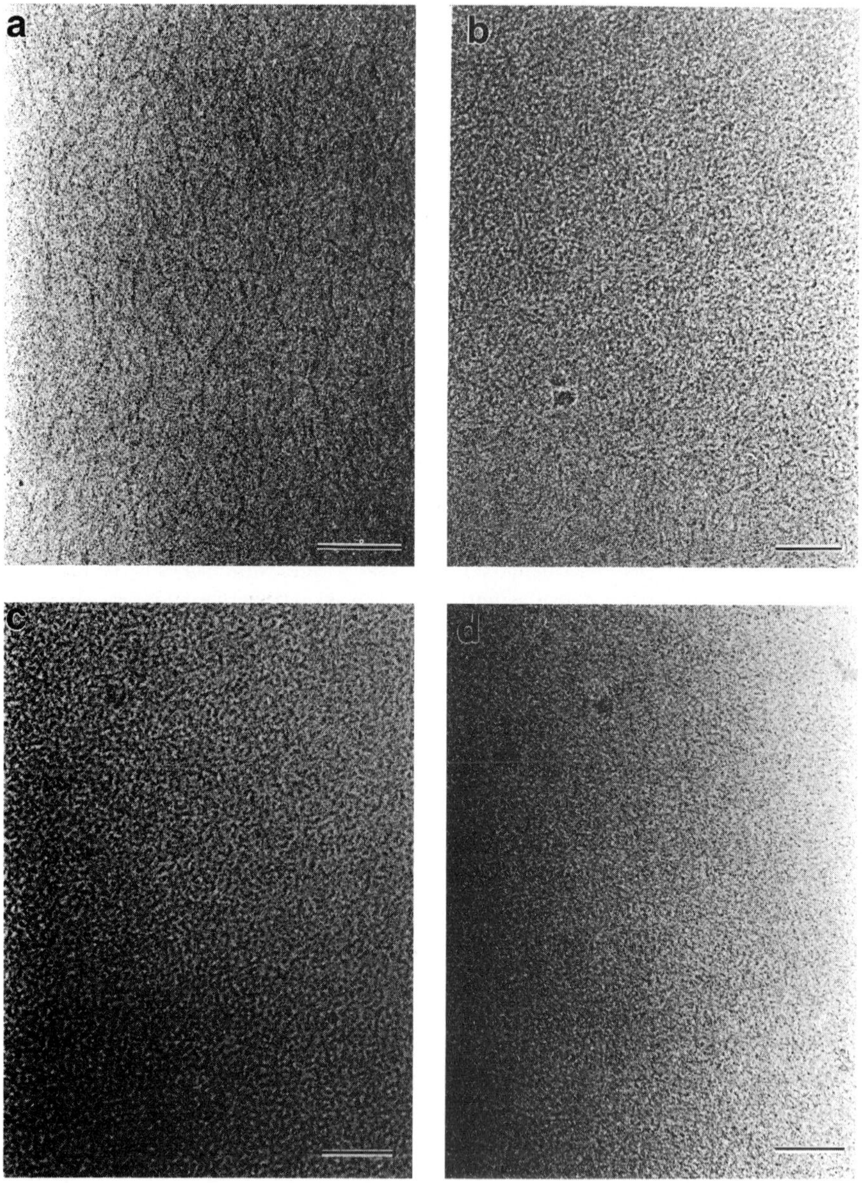

Figure 4.8 Cryo-transmission electron micrographs of vitrified solutions of 12-*s*-12,2Br$^-$ dimeric surfactants: (a) $s = 2$, 1.3%; (b) $s = 3$, 7%; (c) $s = 4$, 5%; (d) $s = 8$, 5%. Bars = 100 nm. *Reproduced from Reference 67, with permission of the American Chemical Society.*

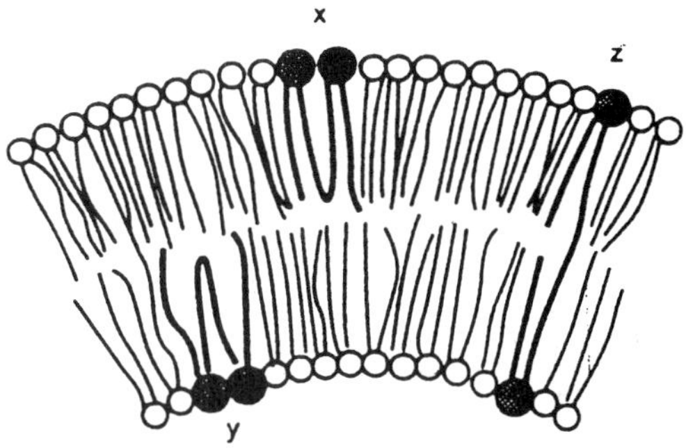

Figure 4.9 Schematic representation of possible arrangements of a dimeric lipid (black heads and bold chains) in a host bilayer of conventional lipid molecules (light head groups and thin chains): (x) and (y) show lipids with wicket or u-shaped conformation in the outer and inner layer; (z) shows a membrane-spanning dimeric lipid, with the spacer in an extended conformation. *Reproduced from Reference 71, with permission of the American Chemical Society.*

tend to form organized assemblies less curved than the corresponding monomeric surfactants and, therefore, that the 'dimerization' of surfactants offers a new way to control the architecture of surfactant assemblies and the rheology of their solutions. This is of much importance in the field of applications. Some aspects of the rheology of solutions of dimeric surfactants are described below.

(d) Dynamics of micelles dimeric surfactants. The kinetics of the exchange of dimeric surfactants between its micelles and the intermicellar solution has been investigated by the ultrasonic relaxation method for two short chain dimeric surfactants: 8-3-8,2Br^- and 8-6-8,2Br^- [47]. The results showed that dimeric surfactants behave very much like their monomeric counterparts: the incorporation of a dimeric surfactant into its micelles is controlled mainly by diffusion, whereas its exit rate is very much dependent on the hydrophobicity of the dimeric surfactant ion. No direct study of the kinetics of the micelle formation–breakdown processes has been reported yet for dimeric surfactants. It is hoped that such studies will be performed in the near future as they are of importance for a better understanding of the rheological behavior of the solutions of dimeric surfactants (see section 4.2.5).

(e) Mixed micellization. Mixed micellization has been observed between anionic dimeric surfactants and one zwitterionic and several polyoxyethylenated non-ionic conventional surfactants [48,75]. The results reveal synergism in mixed micelle formation with the zwitterionic surfactant but not with the non-ionic surfactants used. However, synergism in surface tension reduction effectiveness was reported. The dynamic surface tension of dimeric surfactant/conventional surfactant mixtures is under investigation

[76]. The first results indicate that the formation of the surface layer is diffusion-controlled.

Mixed micellization has been observed between the dimeric surfactant 12-2-12,2Br$^-$ and the corresponding monomeric surfactant DTAB by using cryo-transmission electron microscopy [45]. The micrographs showed a progressive breakdown of the thread-like 12-2-12,2Br$^-$ micelles and their transformation into mixed spherical micelles, upon addition of DTAB to the dimeric surfactant solution. A small angle neutron scattering of the mixtures showed that the DTAB was distributed somewhat uniformly all over the mixed micelles and not preferentially localized in the hemispherical endcaps of the thread-like micelles, within the experimental accuracy [77].

It has also been observed that the addition of DTAB to a vesicular solution of 12-20-12,2Br$^-$ dimeric surfactant results in the progressive transformation of the 12-20-12,2Br$^-$ vesicles into spherical mixed micelles of this surfactant and of DTAB. No intermediate structures, such as bilayer membrane fragments and giant thread-like micelles, usually observed during this transformation, were seen in this instance [78].

4.2.5 Rheology of the 12-2-12,2Br$^-$ solutions

This dimeric surfactant is the only one whose rheological behavior in solution has been investigated extensively as a function of concentration and temperature, in the absence of added salt, in order to gain some insight into the growth of the thread-like micelles [79]. Figure 4.10 shows a typical variation of the zero-shear viscosity, η, with the surfactant volume fraction, Φ. The clear crossover at around $\Phi = 0.015$, separates the dilute regime where the micelles grow very slowly with Φ and a regime where micelle growth is very rapid, in agreement with the theoretical predictions. The decrease of η at high Φ has been interpreted in terms of a decrease of micelle length associated to a decrease of the energy for forming a hemispherical endcap. Molecular dynamics simulations [80] of solutions of dimeric surfactants suggested an alternative explanation in terms of the formation of branched micelles (tree-like micelles) with sliding crosslinks, as discussed theoretically [81]. The simulations suffer from the high surfactant concentrations required for keeping the time of calculation within reasonable range.

The analysis of the rheological data was based on the theory of Cates for living polymers. Indeed, the thread-like micelles can be considered as polymers with fluctuating length owing to the micelle formation–breakdown processes [82,83]. This made it possible to estimate the time for micelle breakdown (time over which a micelle breaks in two smaller micelles). The values decrease much upon increasing surfactant volume fraction and temperature, between 10 and 0.10 s [79]. More studies are needed to understand better the rheological behavior of dimeric surfactants.

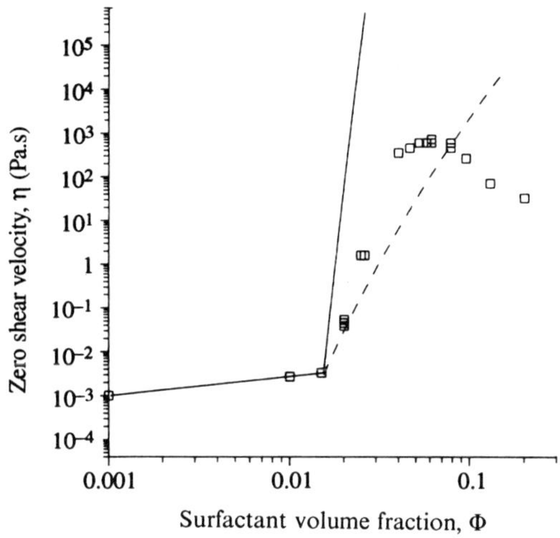

Figure 4.10 Variation of the zero-shear viscosity of 12-2-12,2Br$^-$ with the surfactant volume fraction in aqueous solution at 20°C. The lines going through the experimental results in the semidilute regime have been calculated from the variation of the micelle length assuming unbreakable rigid cylinders (full line) and breakable flexible cylinders (broken line). *Reproduced from Reference 75, with permission of the American Chemical Society.*

4.2.6 Phase behavior

There have been relatively few studies of phase diagrams of dimeric surfactants. However, the *m-s-m*,2Br$^-$ surfactants have been investigated systematically [63]. It was first shown that the pure dimeric surfactants do not give rise to thermotropic liquid crystals, a rather unexpected behavior, as the corresponding conventional monomeric surfactants, $C_mH_{2m+1}(C_{s/2}H_{s+1})$-$N^+(CH_3)_2Br^-$, show liquid crystalline thermotropic behavior. This was tentatively attributed to the geometric constraint introduced by the spacer on the arrangement of the charged groups [63].

For these surfactants the concentration range of the lyotropic mesophases was found to decrease as the spacer carbon number *s* increased and completely disappeared at $s = 10$ and 12. These surfactants only give micellar solutions, even at concentration as high as 90%. This may be of considerable interest in the use of surfactants. Lyotropic mesophases occurred again for $s = 16$. Notice that the corresponding monomeric surfactants show a similar behavior. The observed mesophases had the texture of the conventional lamellar and cylindrical phases. A detailed investigation of the 12-8-12,2Br$^-$/water mixtures led to the phase diagram shown in Figure 4.11 [63]. The alkyl chains are inside the cylinders in the cylindrical phase. The octanediyl spacer lies nearly flat and extended within the interface in both the cylindrical and lamellar phases.

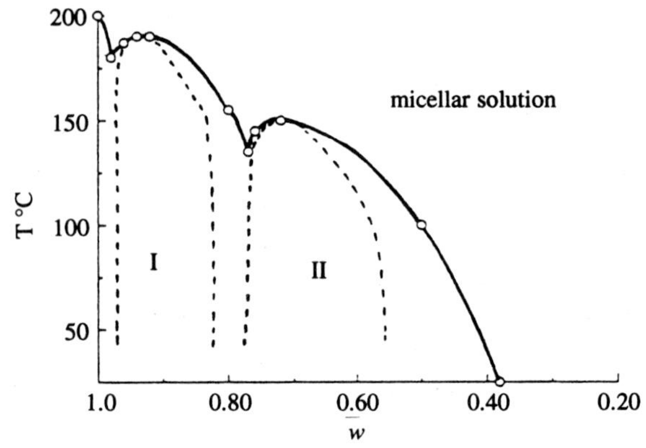

Figure 4.11 Phase diagram of the 12-8-12,$2Br^-$/water mixtures showing the ranges of the micellar solution, lamellar phase (I) and cylindrical phase (II). w is the surfactant weight percent. *Reproduced from Reference 63, with permission of the American Chemical Society.*

The dimeric surfactant $(CH_2)_6$-1,6-bis[$Br^-N^+(CH_3)_2(CH_2)_5O\Phi\Phi CN$] is similar to the preceding ones. Its mixtures with water showed no cylindrical phase, but two lamellar phases were detected [84].

Thus for these two ionic dimeric surfactants, the dimerization results in simpler phase diagrams than for their monomeric counterparts. However, the dimeric sugar surfactant investigated by Eastoe *et al.* [69] showed the phase sequence hydrated crystals → lamellar ($L\alpha$) → cubic (V_1) → hexagonal (H_1) → micellar (L_1) with decreasing surfactant concentration. This sequence is similar to that for conventional surfactants.

4.2.7 *Relation between the properties of dimeric surfactants and their applications*

(a) Solubilization. The solubility of the dye *trans*-azobenzene by micellar solution of *m-s-m*,$2Br^-$ surfactants with $s = 2$ to 12 and $m = 8$ to 16 has been investigated [57]. For the $m = 10$ surfactant series, the solubilizing capacity was found to be maximum for $s = 6$ to 8. A rather speculative interpretation of these results was put forward in terms of the micelle structure.

(b) Biological effects. The inhibitory action of dimeric quaternary ammonium surfactants, particularly of the 10-*s*-10,$2Br^-$ surfactant series, on photosynthesis [85] and bacterial activity [86,87] has been investigated. The photosynthesis by spinach chloroplasts was most inhibited when $s = 11$–13 [85]. The antimicrobial activity was tested for *Escherichia coli* and *Candida albicans* and found to be maximum for about the same values of *s* [86].

Other series of bisquaternary ammonium surfactant showed that the antibacterial activity is a maximum for $m = 10$–13 [87].

(c) Wetting, foaming and lime-soap dispersing ability. A large number of dimeric surfactants of varied chemical structure of the head groups and spacers were synthesized and tested for their ability to wet various materials, to foam and to disperse lime-soap [88–91]. The dimeric surfactants often performed better than conventional surfactants.

(d) Use of dimeric surfactants in electrokinetic chromatography. The micellar electrokinetic chromatography technique was developed for the separation of electrically neutral compounds by electrophoresis with an ionic micellar solution. Tanaka *et al.* [91] have shown that the use of a dimeric sodium sulfonate surfactant achieved much better separations than the often used sodium dodecyl sulfate.

4.2.8 Conclusions

The results reviewed clearly show the potential of dimeric surfactants, and probably also of the higher homologues, such as trimeric and pentameric surfactants, from the point of view of both surfactant applications and basic research. We agree with the statement of Rosen [53] that dimeric surfactants represent something 'new' in the field of surfactants and their properties make them candidates 'for being the next generation of surfactants'.

References

1. Fuoss, R.M. and Edelson, D. *J. Am. Chem. Soc.*, 1951, **73**, 269.
2. Yiv, S. and Zana, R. *J. Colloid Interface Sci.*, 1980, **7**, 449.
3. Rosen, M.J., Baum, M. and Kasher, F. *J. Am. Oil Chem. Soc.*, 1976, **53**, 742.
4. Danielsson, I. *Acta Academia Aboensis, Mathematica et Physica XX.15*, p. 86, Åbo Akademie, Åbo (Finland), 1956.
5. Ishikawa, M., Matsumura, K. *et al. J. Colloid Interface Sci.*, 1992, **151**, 70.
6. Attwood, D., Natarajan, R. *J. Pharm. Pharmacol.*, 1979, **32**, 460.
7. Menger, F.M. and Wrenn, S., *J. Phys. Chem.*, 1974, **78**, 1387.
8. Agharkar, S. and Lindenbaum, S. *J. Phys. Chem.*, 1975, **79**, 2068.
9. Meguro, K., Ikeda, K. *et al. J. Colloid Interface Sci.*, 1987, **118**, 372.
10. Adair, D.A., Reinsborough, V.C. *et al. Can. J. Chem.*, 1976, **54**, 1162.
11. Yiv, S., Kale, K.M. *et al. J. Phys. Chem.*, 1976, **80**, 2651.
12. Graber, E., Lang, J. and Zana, R. *Kolloid Z. Z. Polym.*, 1970, **238**, 470.
13. Skold, R. and Tunius, M. *J. Colloid Interface Sci.*, 1992, **152**, 183 and *J. Colloid Interface Sci.*, 1993, **155**, 43.
14. Mukerjee, P. *Adv. Colloid Interface Sci.*, 1967, **1**, 241.
15. Zana, R. *Langmuir*, 1996, **12**, 1208.
16. Zana, R. *J. Colloid Interface Sci.*, 1980, **78**, 330.
17. Ueno, M., Hikota, T. *et al. J. Am. Oil Chem. Soc.*, 1972, **49**, 250.
18. Zana, R., Yiv, S. and Kale, K.M. *J. Colloid Interface Sci.*, 1980, **77**, 456.
19. Yasuda, M., Ikeda, K. *et al. Langmuir*, 1990, **6**, 949.

20. Ikeda, K., Nakasima, T. *et al. Bull. Chem. Soc. Jap.*, 1989, **62**, 578.
21. Fuoss, R.M. and Chu, V.F. *J. Am. Chem. Soc.*, 1951, **73**, 949.
22. Eisenberg, H. and Fuoss, R.M. *J. Am. Chem. Soc.*, 953, **75**, 2914.
23. Rice, S.A., *J. Am. Chem. Soc.*, 1958, **80**, 3207.
24. Wong, T., Ikeda, K. *et al. J. Phys. Chem.*, 1989, **93**, 4861.
25. Fromherz, P. *J. Phys. Chem.*, 1989, **93**, 8383.
26. Ikeda, K., Ishikawa, M., *et al. Bull. Chem. Soc. Jap.*, 1989, **62**, 1032.
27. McKenzie, D., Bunton, C. *et al. J. Phys. Chem.*, 1987, **91**, 5709.
28. Lianos, P. and Zana, R. *J. Colloid Interface Sci.*, 1981, **84**, 100.
29. Ikeda, K., Esumi, K. *et al. J. Colloid Interface Sci.*, 1989, **130**, 290.
30. Zana, R., Muto, Y. *et al. J. Colloid Interface Sci.*, 1988, **123**, 502.
31. Attwood, D. and Patel, H.K. *J. Colloid Interface Sci.*, 1989, **129**, 222.
32. Ishikawa, M., Matsumura, K. *et al. J. Colloid Interface Sci.*, 1991, **141**, 10.
33. Gallot, B. and Skoulios, A. *Koll. Z. Z. Polym.*, 1968, **222**, 51.
34. Gallot, B. *Mol. Cryst. Liq. Cryst.*, 1971, **13**, 326.
35. Gutman, H., Luz, Z. *et al. Liq. Cryst.*, 1987, **2**, 739.
36. Ikeda, K., Khan, A. *et al. J. Colloid Interface Sci.*, 1988, **133**, 192.
37. Nagarajan, T. *Chem. Eng. Comm.*, 1987, **55**, 251.
38. Fuhrhop, J.H. and Fritsch, D. *Acc. Chem. Res.*, 1986, **19**, 130.
39. Fuhrhop, J.H., David, H.H. *et al. J. Am. Chem. Soc.*, 1986, **108**, 1785.
40. Fuhrhop, J.H., Liman, U. and Koesling, V. *J. Am. Chem. Soc.*, 1988, **110**, 6840.
41. Fuhrhop, J.H., Hungerbuhler, H. and Siegel, U. *Langmuir*, 1990, **6**, 1295.
42. Lo Nostro, P. and Gabrielli, G. *Colloids Surf.*, 1990, **44**, 119.
43. Fuhrhop, J.H., Spiroski, D. and Boettcher, C. *J. Am. Chem. Soc.*, 1993, **116**, 1601.
44. Fuhrhop, J.H., Fritsch, D. *et al. J. Am. Chem. Soc.*, 1984, **106**, 1988.
45. Zana, R. and Talmon, Y. *Nature*, 1993, **362**, 228.
46. Menger, F.M. and Littau, C.A. *J. Am. Chem. Soc.*, 1993, **115**, 10083; *J. Am. Chem. Soc.*, 1991, **113**, 1451.
47. Frindi, M., Michels, B. *et al.*, *Langmuir*, 1994, **10**, 1140.
48. Rosen, M.J., Gao, T. *et al. Colloids Surf., A. Physicochem. Eng. Asp.*, 1994, **88**, 1, and references therein.
49. Devinsky, F., Lacko, I. *et al. J. Colloid Interface Sci.*, 1986, **114**, 314.
50. Pinazzo, A., Diz, M. *et al. J. Am. Oil Chem. Soc.*, 1993, **70**, 37.
51. Zana, R., Benrraou, M. and Rueff, R. *Langmuir*, 1991, **7**, 1072.
52. Zhu, Y.P., Masuyama, A., *et al. J. Colloid Interface Sci.*, 1993, **158**, 40.
53. Rosen, M.J. *Chem. Technol.*, 1993, p. 30.
54. Deinega, Y., Ulberg, Z. *et al. Kolloidn. Zh.*, 1974, **36**, 649; Ulberg, Z. and Podolskaya, V., *Kolloidn. Zh.*, 1978, **40**, 292.
55. Parreira, H.C., Lukenbach, E.R. and Lindemann, M.K. *J. Am. Chem. Oil Soc.*, 1979, **56**, 1015.
56. Devinsky, F., Masarova, L. and Lacko, I. *J. Colloid Interface Sci.*, 1985, **105**, 235.
57. Devinsky, F.L., Lacko, I. and Imam, T. *Acta Fac. Pharm.*, 1990, **44**, 103 and *J. Colloid Interface Sci.*, 1991, **143**, 336.
58. Devinsky, F. and Lacko, I. *Tenside Surf. Det.*, 1990, **27**, 344.
59. Abid, S.K., Hamid, S.M. and Sherrington, D.C. *J. Colloid Interface Sci.*, 1987, **120**, 245.
60. Alami, E., Beinert, G. *et al. Langmuir*, 1993, **9**, 1465.
61. Zhu, Y.P., Masuyama, A. *et al. J. Jpn. Oil Chem. Soc.*, 1993, **42**, 86.
62. Gibbs law relates the surface area per surfactant molecule, a, to the rate of change $d\gamma/dlnC$, of the surface tension, γ, with the surfactant concentration, C, according to: $a = (2.303nRT/N_A)(d\gamma/dlnC)^{-1}$, where N_A is Avogadro's number and n a constant equal to 1 for a non-ionic surfactant or an ionic surfactant in the presence of a large excess of swamping electrolyte, whereas $n = 2$ and 3 for 1-1 and 2-1 ionic surfactants, respectively. Obviously, the value $n = 3$ should be used for dimeric surfactants. This has been done in many studies [46,55,59,60] but other authors have used $n = 2$, [49,56–58] on the assumption that, as for bolaform surfactants, there is binding of one counterion to each dimeric surfactant ion (see section on bolaforms).
63. Alami, E., Levy, H. *et al. Langmuir*, 1993, **9**, 940.
64. Diamant, H. and Andelman, D. *Langmuir*, 1994, **10**, 2910; 1995, **11**, 3605.

65. Rozycka-Roszak, B., Witek, S. and Przestalski, S. *J. Colloid Interface Sci.*, 1989, **131**, 189.
66. Danino, D., Kaplun, A. *et al.* In: Herb, C.A. and Prud'homme, R.K. (eds) *Structure and flow in surfactant solutions.* ACS Symp. Ser. No. 578, American Chemical Society, Washington, D.C., 1994.
67. Danino, D., Talmon, Y. and Zana, R. *Langmuir*, 1995, **11**, 1448.
68. Zana, R., Lévy H. *et al. Langmuir*, 1995, **11**, 3694.
69. Eastoe, J., Rogueda, P. *et al. Langmuir*, 1994, **10**, 4429.
70. Moss, R.A., Fujita, T. and Okumura, Y. *Langmuir*, 1991, **7**, 2415.
71. Moss, R.A. and Li, J.-M. *J. Am. Chem. Soc.*, 1992, **114**, 9227.
72. Kim, J.-M. and Thompson, D.H. *Langmuir*, 1992, **8**, 637.
73. Thompson, D.H., Wong, K.F. *et al. J. Am. Chem. Soc.*, 1992, **114**, 9035.
74. Zana, R., Papoutsi, D. and Szönyi, S. unpublished results.
75. Rosen, M., Zhu, Z. and Gao, T. *J. Colloid Interface Sci.*, 1993, **157**, 254.
76. Gao, T. and Rosen, M.J. *J. Am. Oil Chem. Soc.*, 1994, **71**, 771.
77. Schosseler, F., Anthony, A. *et al. Langmuir*, 1995, **11**, 3347.
78. Danino, D.A., Talmon, Y. and Zana, R. 1995 (manuscript submitted).
79. Kern, F., Lequeux, F. *et al. Langmuir*, 1994, **10**, 1714.
80. Karaborni, S., Esselink, K. *et al. Science*, 1994, **266**, 254.
81. Lequeux, F. *Europhys. Lett.*, 1992, **19**, 675.
82. Cates, M.E. *J. Phys. (Paris)*, 1988, **49**, 1593.
83. Granek, R. and Cates, M.E. *J. Chem. Phys.*, 1992, **96**, 4758.
84. Fuller, S., Hopwood, J. *et al. Liq. Cryst.*, 1992, **12**, 521.
85. Kralova, K. and Sersen, F. *Tenside Surf. Det.*, 1994, **31**, 192.
86. Imam, T., Devinsky, F. *et al. Pharmazie*, 1983, **38**, 308.
87. Devinsky, F., Lacko, I. *et al. Tenside Det.*, 1985, **22**, 10.
88. Zhu, Y.P., Masuyama, A. and Okahara, M. *J. Am. Oil Chem. Soc.*, 1990, **67**, 459; *J. Am. Oil Chem. Soc.*, 1991, **68**, 268.
89. Zhu, Y.P., Masuyama, A. *et al. J. Am. Oil Chem. Soc.*, 1991, **68**, 539.
90. Zhu, Y.P., Masuyama, A. *et al. J. Am. Oil Chem. Soc.*, 1992, **69**, 626.
91. Tanaka, M., Ishida, T. *et al. J. Chromatogr.*, 1993, **648**, 469.

5 Fluorinated and semi-fluorinated surfactants

P.D.I. FLETCHER

5.1 Introduction

Surfactants which contain one or more fluorinated or partially fluorinated hydrophobic groups (fluorinated surfactants or 'fluorosurfactants') can show dramatically different properties to those of hydrocarbon surfactants. The special properties of fluorosurfactants, which result from the unique properties of fluorocarbon chains, makes them irreplaceable in a wide range of applications and justifies Kissa's claim in his 1994 text on the subject that fluorinated surfactants are truly 'super surfactants' [1]. Rather than exhaustively review the huge literature on fluorosurfactants, the main aim of this chapter is to give an account of the special surfactant properties of fluorinated surfactants and to relate these properties to the differences in physical properties between fluorocarbon and hydrocarbon chains.

The chapter is organized as follows. The main structural types of common fluorosurfactants and some aspects of their synthesis are described briefly in sections 5.2 and 5.3. In section 5.4, the physical properties of fluorinated and partially fluorinated chains are discussed and compared with those of the corresponding hydrocarbons. As will be seen, the different properties of the two types of chain provide a good basis for the rationalization and interpretation of the special properties of fluorosurfactants. Adsorption and aggregation behavior of fluorosurfactants in water is discussed in section 5.5. Adsorption and aggregation in apolar oil solutions of fluorosurfactants is covered in section 5.6. Section 5.7 gives a description of their phase behavior in mixtures of oil and water. The wetting of solid surfaces by fluorosurfactant solutions is dealt with in section 5.8. Finally, section 5.9 gives a brief summary of the main applications of fluorosurfactants.

5.2 Structures of fluorinated surfactants

Like conventional surfactants, fluorosurfactants generally contain a hydrophilic group and a hydrophobic group within the same molecule. A selection of the wide range of fluorosurfactant molecular structures are described here. A more exhaustive list is given in chapter 1 of Reference 1.

The hydrophobic group of fluorinated or semifluorinated surfactants can be linear or branched and may contain an aromatic group or additional elements as O, N, Cl, S and Si. Some examples are listed below:

$C_nF_{(2n+1)}$-
$C_nF_{(2n+1)}C_mH_{(2m+1)}$-
$C_nF_{(2n+1)}OCF_2CF_2$-
$C_nF_{(2n+1)}OC_6H_4$-
$C_nF_{(2n+1)}CONH(CH_2)_3N<$
$C_8F_{17}CH_2CH_2Si(CH_3)_2$-

As with conventional hydrocarbon surfactants, fluorinated surfactants are classified according to the charge type of the hydrophilic group.

5.2.1 *Anionic*

Anionics are the most important class of fluorosurfactants. Because of their negative charge, they are sensitive to pH and electrolyte concentration and may precipitate in water containing di- or tri-valent metal ions. They include:

carboxylates $R_fCOO^-M^+$
sulfonates $R_fSO_3^-M^+$
sulfates $R_fSO_4^-M^+$
phosphates $R_fOP(O)O_2^{2-}M_2^+$

where R_f signifies a fluorinated or semi-fluorinated hydrophobe and M^+ is an inorganic or organic counterion.

5.2.2 *Cationic*

Like anionics, these are sensitive to the pH and electrolyte content of the solution. Since most solid surfaces are negatively charged, cationic fluorosurfactants commonly show strong adsorption to solid surfaces. The fluorinated hydrophobe may be attached directly or indirectly to a quaternary ammonium group, a protonated amino group or a heterocyclic base.

5.2.3 *Zwitterionic*

Over a pH range around their isolectric point these species have zero overall charge but are highly polar. Depending on the dissociation constants of the ionizing groups in the molecule, they may function as either anionic or cationics at extremes of pH. Common types include:

carboxybetaines $R_fCH_2CH(OOCH_3)CH_2N^+(CH_3)_2CH_2COO^-$,
sulfobetaines $R_fXYN^+(CH_3)_2(CH_2)_nSO_3^-$ and
sulfatobetaines $R_fXYN^+(CH_3)_2(CH_2)_nOSO_2O^-$

where $n = 1, 2$, or 3 and X and Y can be a variety of linker groups including ether, amide and thio-ether groups.

5.2.4 Non-ionic

Non-ionic hydrophilic groups include polyoxyethylene or polyoxypropylene segments and polyhydric alcohols. Some examples include:

oxyethylated alcohols $R_fCH_2O(CH_2CH_2O)_mH$
oxyethylated thiols $R_fCH_2CH_2S(CH_2CH_2O)_mH$
fluorinated polyhydric alcohols $R_fCH_2CH_2O[CH_2CH(CH_2OH)O]_mH$.

A range of polymeric fluorinated or semi-fluorinated surfactants has been investigated. Linear fluorocarbon chains are less flexible than hydrocarbon chains and thus have high melting points and show low solubility. Increased chain flexibility can be achieved by incorporating O atoms within the backbone. Fluorinated polyethers, such as oligomers of hexafluoropropene oxide (HFPO), are suitable hydrophobes for polymeric surfactants [2–5].

Surfactant properties in aqueous systems arise because of the antipathy of the hydrophobic group for water combined with the affinity of the hydrophilic group for water. Molecules containing a fluorocarbon chain attached to a hydrocarbon chain (*but with no hydrophilic group*) have a similar amphiphilic character in that the flurocarbon chain has an affinity for fluorocarbon solvents but an antipathy for hydrocarbons. The hydrocarbon chain segment has an affinity for hydrocarbons and an antipathy for fluorocarbons. Thus, semifluorinated alkanes (SFAs) of general structure $F(CF_2)_n(CH_2)_mH$ can show surfactant properties in organic apolar solvent systems in the absence of water. A range of SFAs has been synthesized in order to examine their properties as surfactants and thermotropic liquid crystals [6].

5.3 Synthesis

Direct fluorination with elemental fluorine is impractical for the commercial synthesis of fluorosurfactants owing to the very high reactivity of fluorine. The commercially important pathways to fluorosurfactants are electrochemical fluorination, telomerization and oligomerization of tetrafluorethylene.

In *electrochemical fluorination* [7–10], the organic substance to be fluorinated is dissolved or dispersed in liquid hydrogen fluoride. A direct electric current at a voltage below 8 V (necessary to prevent evolution of fluorine) is then passed through the reaction mixture. Hydrogen is evolved at the cathode and the organic substance is fluorinated. The anode is generally nickel and the cathode is usually iron or nickel. The mechanism is believed to involve the generation of F radical atoms at the anode which then react with the organic substance. All hydrogen atoms in the molecule are replaced by fluorine but some functional groups such as carboxylic halides and sulfonyl halides are retained. Fluorocarbons containing these functional groups form useful intermediates for a range of fluorosurfactants.

Telomerization [11–16] is defined as a process of reacting a molecule called a telogen with two or more ethylenically unsaturated molecules called taxogens.

YZ	+ nA	→ Y-(A)n-Z
Telogen	Taxogen	Telomer

Telogens are generally short-chain fluoroalkyliodides, such as trifluoromethyl iodide or pentafluoroethyl iodide, which can dissociate to yield free radicals capable of adding toxogen molecules. Commercial telomerization of tetrafluoroethylene with pentafluoroethyl iodide was developed by the Du Pont Company. The telogen, pentafluoroethyl iodide, is prepared by reacting iodine pentafluoride and iodine with tetrafluoroethylene.

$$5CF_2{=}CF_2 + IF_5 + 2I_2 \rightarrow 5CF_3CF_2I \quad \text{(catalysed by } SbF_5\text{)}$$

Telomerization, which can be regarded as polymerization to a small degree, of tetrafluoroethylene with pentafluoroethyl iodide produces a mixture of even-carbon-numbered telomers differing in their overall carbon chain length.

$$C_2F_5I + nC_2F_4 \rightarrow C_2F_5(C_2F_4)_nI$$

Numerous variations of this telomerization process have been patented [1]. Since perfluoroalkyl iodides do not react with nucleophiles, such as OH^- or NH_3, and hence cannot be readily converted to useful intermediates, they are generally reacted with ethylene.

$$C_2F_5(C_2F_4)_nI + CH_2CH_2 \rightarrow C_2F_5(C_2F_4)_nCH_2CH_2I$$

The resultant perfluoroalkyl ethyl iodide products can be readily converted to the corresponding alcohols, thiols and sulphonyl chlorides used as intermediates for fluorinated surfactants.

Oligomerization of tetrafluoroethylene [17–19], developed by ICI, is based on the anionic polymerization of tetrafluorethylene in the presence of a fluoride catalyst such as caesium, potassium or tetralkylammonium fluoride. Unlike the free radical polymerization (used to produce PTFE), the anionic polymerization produces highly branched oligomers, mainly isomeric variants of the tetramer, pentamer and hexamer. The overall process can be written as:

$$nCF_2{=}CF_2 \rightarrow \text{isomers of } (C_2F_4)_n \text{ where } n \text{ is generally 4–6}$$

The product is an unsaturated perfluorocarbon which is susceptible to nucleophilic attack. Some examples of reactions to produce fluorosurfactants are given below. Reaction of the pentamer product (the most abundant product) with phenol, followed by sulfonation, yields an anionic fluorosurfactant.

$$C_{10}F_{20} + C_6H_5OH \rightarrow C_{10}F_{19}OC_6H_5 \rightarrow C_{10}F_{19}OC_6H_4SO_3H$$

The base-catalyzed reaction of the oligomers with ethoxylated alcohols gives non-ionic fluorosurfactants.

$$C_{10}F_{20} + HO(CH_2CH_2O)_nR \rightarrow C_{10}F_{19}O(CH_2CH_2O)_nR$$

If R is H, the pentamer fluorocarbon product reacts with both ends of the ethoxylated di-alcohol and the reaction yields a non-ionic fluorosurfactant with a fluorocarbon chain at both ends of the molecule. If R is methyl, the product is a single-tailed, methyl-capped non-ionic fluorosurfactant.

This brief discussion attempts only to give an overview with illustrative examples of the production of fluorosurfactant intermediates and products. For more details, the reader should consult chapter 2 of Reference 1, and Reference 20.

5.4 Properties of fluorocarbon chains

The carbon–fluorine bond is amongst the most stable known covalent bonds with the heat of formation of the C–F bond in CF_4 being 486 kJ/mol [21]. This high bond strength is thought to result from an excellent match between the 2s and 2p orbitals of fluorine and the corresponding orbitals of carbon. Fluorine atoms (atomic radius 0.072 nm when covalently bound) are close to the optimum size to shield carbon atoms without steric stress. These factors combine to make fluorocarbon chains very thermally and chemically stable relative to hydrocarbons. For example, perfluoroalkanecarboxylic acids can be heated to 400°C in borosilicate glass without decomposition [22]. The good chemical stability of fluorosurfactants is an important consideration when choosing surfactants to operate in harsh environments such as extremes of pH, high temperatures or in combination with strong oxidizing or reducing agents.

An additional chemical difference between hydrocarbon and fluorocarbon chain surfactants results from the high electronegativity of F. When bonded to acidic groups such as carboxylic acid group, the electronegative induction effect of the fluorocarbon group leads to an increased acid strength. For example, the dissociation constants of the series of carboxylic acids CH_3COOH, CF_3COOH, CF_3CH_2COOH and $CF_3CH_2CH_2COOH$ are 1.8×10^{-5}, 5.5×10^{-1}, 1×10^{-3} and $1.56 \times 10^{-5} M^{-1}$ respectively [23–26]. Hence, acid groups attached to fluorocarbon chains are much stronger than the corresponding hydrocarbon molecules. The example

Table 5.1 Volume properties of liquid linear alkanes (H chains) and perfluoroalkanes (F chains) at 25°C. Data from Reference 27

Property	C_6H_{14}	C_6F_{14}	$C_{12}H_{26}$	$C_{12}F_{26}$
Density ($g\,cm^{-3}$)	0.6548	1.6717	0.74516	1.92*
Molar volume (cm^3)	131	201	228.5	336*
Molecular volume (nm^3)	0.2176	0.334	0.3796	0.558*

*Signifies that the data is extrapolated to 25°C from measurements made above the melting point.

demonstrates that even two methylene groups are not sufficient to completely shield the electronegative effect of the CF_3 group. Hence fluorosurfactants are expected to show differences in the degree of dissociation of charged hydrophilic groups (and concomitant changes in the hydrophilicity) as compared with hydrocarbon surfactants.

The density of perfluoroalkanes are much higher than those of the corresponding alkanes as shown in Table 5.1. The covalent radii for C, H and F are 0.0771, 0.030 and 0.0640 nm, respectively [28]. Thus, the bond lengths for C–C, C–H and C–F are 0.1542, 0.1071 and 0.1411 nm, respectively. Using a tetrahedral angle of 109°, the all-*trans* lengths (l) of linear alkanes and perfluoroalkanes can be expressed as:

$$l = 0.1251n_c + 0.2344 \qquad \text{alkanes}$$

$$l = 0.1251n_c + 0.3578 \qquad \text{perfluoroalkanes}$$

where l is in nm and n_c is the number of carbon atoms. Estimated in this way, the all-*trans* lengths of dodecane and perfluorododecane are 1.74 and 1.86 nm respectively. Dividing the molecular volumes of these species by these lengths gives the approximate cross sectional areas occupied per chain in the liquid state as 0.22 and 0.30 nm^2 for hydrocarbon and fluorocarbon chains respectively. The cross sectional areas are slightly higher than the corresponding values for the solid state as determined by diffraction. The larger volume and area filled by a perfluorocarbon chain is an important consideration in the adsorption and aggregation behavior of fluorosurfactants.

The melting points of linear perfluoroalkanes are higher than the corresponding hydrogenated alkanes and the difference increases with increasing chain length for chains containing more than five carbon atoms [20]. In contrast, for chain lengths greater than four, the normal boiling points of the perfluoroalkanes lie below those of the corresponding alkanes (Figure 5.1). The enthalpies and entropies associated with these transitions are compared for representative linear hydrocarbons and pefluoroalkanes in Table 5.2. The total intermolecular forces present in the liquids (as judged from the molar enthalpies of vaporization) are rather similar for the alkanes

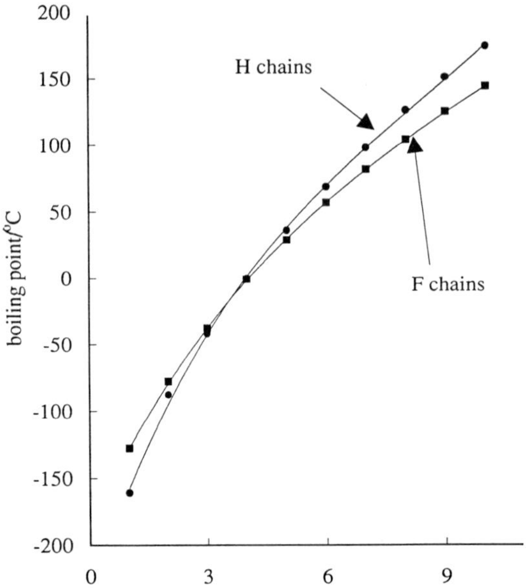

Figure 5.1 The normal boiling point of the linear alkanes (H chains) and the linear perfluoroalkanes (F chains). The data is from Reference 29 for the F chains and from Reference 30 for the H chains.

and perfluoroalkanes. The entropies of vaporization are again similar and are all close to the Trouton's rule value expected for non-structured liquids [32]. The enthalpies of fusion of the perfluoroalkanes are approximately half the values seen for the alkanes. The lower entropies associated with the fusion of the perfluoroalkanes suggest that a lower degree of disordering occurs in the transition from solid to liquid with the perfluorocarbon chains. This may be associated with the high rigidity of perfluorocarbon chains resulting in less conformational freedom in perfluoroalkane liquids as

Table 5.2 Thermodynamic parameters of melting and vaporization of linear hydrocarbons and perfluoroalkanes. Data from References 27, 29–31

Property	C_6H_{14}	C_6F_{14}	$C_{12}H_{26}$	$C_{12}F_{26}$
m.pt.(°C)	−95.3	−86	−9.6	75
b.pt.(°C)	69	57	216.3	184.5
ΔH_{fusion}(kJ mol^{-1})	13.0	6.60	36.8	23.7
ΔS_{fusion}(JK^{-1} mol^{-1})	73.1	35.3	140	68.1
ΔH_{vap}(kJ mol^{-1}) at b.pt.	28.8	28.2	43.5	45.7
ΔS_{vap}(JK^{-1} mol^{-1}) at b.pt.	84.2	85.4	89.0	99.9
ΔH_{vap}/molar volume (kJ cm^{-3}) at b.pt.	0.220	0.140	0.190	0.136

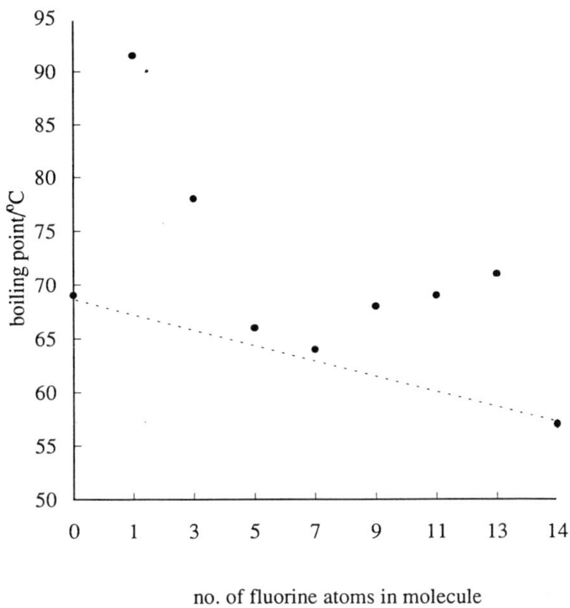

Figure 5.2 The normal boiling points of a series of linear partially fluorinated *n*-hexanes of structure $H(CH_2)_6H$, $F(CH_2)_6H$, $F(CF_2)(CH_2)_5H$, $F(CF_2)_2(CH_2)_4H$ ... $F(CF_2)_5(CH_2)H$, $F(CF_2)_6H$ and $F(CF_2)_6F$. The data is from Reference 27.

compared with alkane liquids. It is believed from X-ray data of liquid perfluoroalkanes that these molecules do not show free rotation about the C–C bonds in the liquid state [33]. The high rigidity of the perfluorocarbon chain is presumably a consequence of the larger covalent radius of F as compared with that of H.

Unlike fully hydrogenated or fluorinated chains, partially fluorinated chains have a significant degree of polarity which arises due to the presence of net C–C and C–F dipoles which are absent in perfluorocarbons (PFCs) and hydrocarbons (HCs). This point is illustrated in Figure 5.2 which shows the normal boiling points of a series of partially fluorinated hexanes. The boiling point is highest, indicating strongest intermolecular forces, for the 1-fluorohexane species containing only a single F atom on the terminal carbon.

An important determinant of the solvent properties of a liquid is the cohesive energy density which is the enthalpy required to vaporize a unit volume of the liquid. The comparison of values in the bottom row of Table 5.2 shows that perfluoroalkane liquids have significantly less cohesive energy *per unit volume* than the corresponding hydrocarbons (despite the molar enthalpies of vaporization being similar). The difference in cohesive energy densities between perfluorocarbon and hydrocarbon chains is, in part, responsible for their tendency to demix [34].

Perfluorocarbon chains are apolar and thus interact only by dispersion forces. To a reasonable approximation, the interaction energy V between like molecules is given by [35]:

$$V = -(\alpha^2 I/3)r^6 \tag{5.1}$$

where α is the polarizability volume of the molecule, I is the ionization energy and r is the distance of separation. The polarizability volume is proportional to $(n_{RI}^2 - 1)/(n_{RI}^2 + 2)$ where n_{RI} is the refractive index. Since perfluorocarbon liquids have the lowest known refractive index values, the low cohesive energy per unit volume of fluorocarbons is largely a result of their very low polarizabilities. Figure 5.3 shows a comparison of perfluoroalkane refractive indices with those for the corresponding hydrocarbons.

The low cohesion of fluorocarbons is also reflected in the values of the surface tensions. The values for linear PFCs and HCs are compared in Figure 5.4 where it can be seen that the fluorocarbon values are 6–8 $mN\,m^{-1}$ lower than the corresponding alkanes.

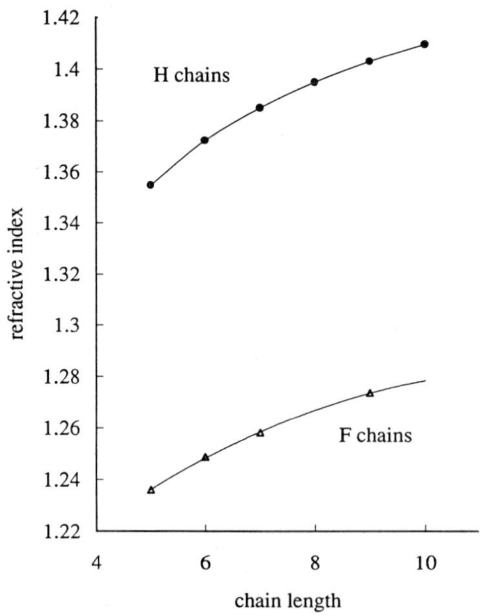

Figure 5.3 Comparison of refractive index for the linear alkanes (H chains) and perfluoroalkanes (F chains) at 25°C. The data refers to the sodium D line (589 nm) and is from Reference 36 for the F chains and from Reference 30 for the H chains.

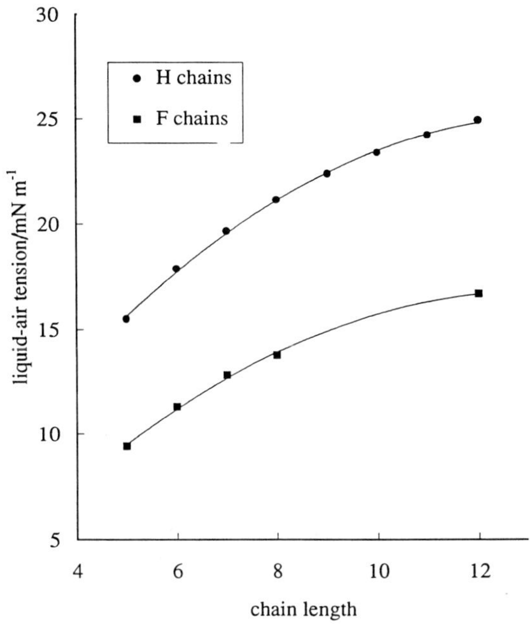

Figure 5.4 Liquid–air surface tensions for the linear alkanes (H chains) and perfluoroalkanes (F chains) at 25°C. The value for *n*-perfluorododecane is extrapolated from measurements above the melting point of this compound. The data is from Reference 29, 37–39 for the F chains and from Reference 30 for the H chains.

The interfacial tension of perfluorohexane with water at 25°C is 56.45 mN m^{-1} [40] somewhat higher than the value for the hexane–water interface (50.4 mN m^{-1}) [41]. The relative affinities for perfluorocarbons and hydrocarbons with water can be expressed in terms of the free energy required to separate unit area of oil–water contact. This free energy is the work of adhesion W_a and is equal to $\gamma_{oa} + \gamma_{wa} - \gamma_{ow}$ where γ_{oa}, γ_{wa} and γ_{ow} are the tensions of the oil–air, water–air and oil–water surfaces, respectively. For hexane and perfluorohexane, the works of adhesion at 25°C are 39.4 and 26.8 mN m^{-1} respectively, indicating that the perfluorocarbon chain has a weaker interaction with water than the hydrocarbon chain. Again, this is a consequence of the low polarizability of perfluorocarbon chains.

The different strengths of the interactions present in perfluorocarbons and hydrocarbons causes non-ideal behavior in their mixtures. For long chains and low temperatures, immiscibility between fluorinated and hydrogenated solvent pairs can result. Since the mixing of fluorocarbons and hydrocarbons is an endothermic process [34] the extent of this immiscibility decreases with increasing temperature until complete miscibility is achieved

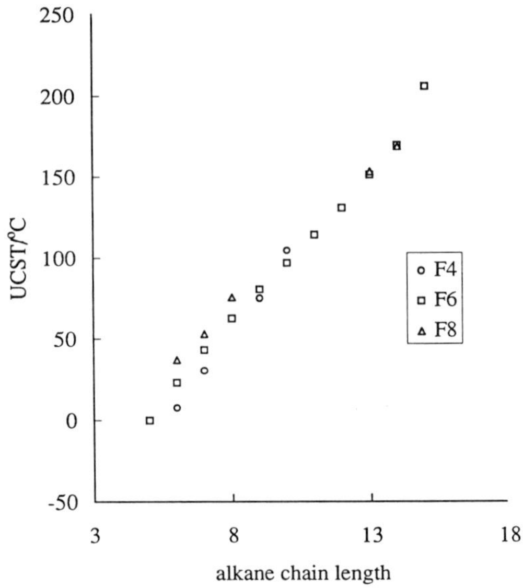

Figure 5.5 Upper critical solution temperatures for mixtures of linear alkanes with linear perfluoroalkanes versus the chain length of the alkane. Values for perfluorobutane (F4), perfluorohexane (F6) and perfluorooctane (F8) are shown. The data is from Reference 34.

at the upper critical solution temperature (UCST). The UCST provides a measure of the extent of the unfavourableness of the mixing. Figure 5.5 shows the variation of UCST with linear alkane chain length for mixtures with several linear perfluoroalkanes. It can be seen that the immiscibility increases strongly with alkane chain length but is much less sensitive to the chain length of the perfluoroalkane. Phase separations in systems containing fluorinated organic compounds have been extensively studied over the past few decades and the interested reader should consult the review by Lo Nostro [34]. As will be discussed, the unfavourableness of mixing perfluorocarbon and hydrocarbon chains can have important consequences for the behaviour of perflurocarbon and hydrocarbon surfactant mixtures and for the properties of semifluorinated alkanes in apolar solvents.

The main properties of fluorocarbon chains that are important in considering the behavior of fluorosurfactants are summarized below:

1. Fluorocarbon chains are chemically very stable.
2. Fluorocarbon chains have a higher molecular volume and a larger cross sectional area than the corresponding hydrocarbons.
3. Fluorocarbon liquids have a lower cohesive energy density than hydrocarbons as a result of their low polarizability.
4. Fluorocarbons show a lower affinity for water than is seen with hydrocarbons.

5. The mixing of hydrocarbon and fluorocarbon chains is unfavourable.
6. Partially fluorinated chains can show additional polar interactions not present in the fully fluorinated compounds.

5.5 Adsorption and aggregation of fluorinated surfactants in water

5.5.1 Fully fluorinated chain surfactants

Dilute solutions of fluorosurfactants in water generally show qualitatively similar behavior to that of hydrocarbon surfactants. At low concentrations, surfactant monomers are present in solution and co-exist at equilibrium with a monolayer adsorbed at the air–water surface. Increasing the solution concentration leads to increased adsorption and a concomitant reduction in surface tension. For a non-ionic surfactant in water, the excess surface concentration of the surfactant Γ is related to the variation of tension with surfactant concentration according to the Gibbs equation:

$$\Gamma = (-1/kT)(\mathrm{d}\gamma/\mathrm{dln}C) \tag{5.2}$$

where k is the Boltzmann constant, T is the absolute temperature and C is the surfactant concentration. The use of concentration instead of the surfactant activity is valid only at very low surfactant concentrations when non-ideality effects are negligible. The area occupied per surfactant A is simply equal to $1/\Gamma$. At concentrations in excess of the critical micelle concentration (CMC), virtually all additional surfactant forms non-adsorbing micelles and the surface tension remains approximately constant. The variation of surface tension with surfactant concentration is shown schematically in Figure 5.6.

For many practical applications, the effectiveness of a surfactant is primarily dependent on the minimum value of the surface tension (γ_{min}) that it is capable of achieving together with the surfactant concentration required to achieve this. Substitution of the hydrocarbon chain in a conventional surfactant for a fluorocarbon chain affects both these parameters. For a surfactant which forms micelles, the surfactant concentration required to achieve the lowest possible solution–vapour tension is the critical micelle concentration (CMC). (For the purposes of this discussion we neglect the small decrease in surface tension at concentrations higher than the CMC seen for some surfactants and thus we equate γ_{min} with the tension at the CMC γ_{CMC}.) Thus, an effective surfactant is one which has a low CMC, i.e. only a low concentration must be added to achieve the limiting tension. In these terms, fluorosurfactants are more effective than the corresponding hydrocarbon surfactants (of the same chain length) and the magnitude of this increase in effectiveness can be quantified as follows.

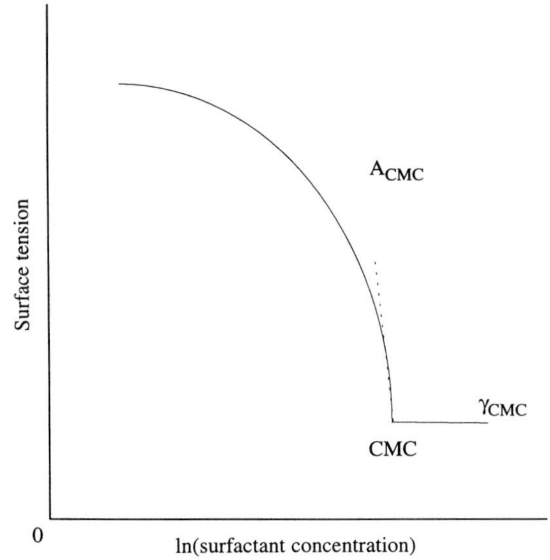

Figure 5.6 Schematic representation of the variation of surface tension with ln(surfactant concentration). The slope at the CMC (dashed line) yields the limiting area per surfactant in the monolayer at the CMC (A_{CMC}).

According to the phase separation model of micelle formation (valid in the limit of large aggregation number) the standard free energy of micellization for a non-ionic surfactant is given by

$$\Delta G^{\circ}_{\text{micellization}} \approx RT\ln(CMC) \tag{5.3}$$

where (for the CMC entered in units of M) $\Delta G^{\circ}_{\text{micellization}}$ refers to the free energy of transferring 1 mole of surfactant monomers from an ideal 1 M solution in water into micelles. It is generally believed that the total $\Delta G^{\circ}_{\text{micellization}}$ can be equated with the sum of the contributions from each group within the total molecule, i.e. that

$$\Delta G^{\circ}_{\text{micellization}} = \Delta G^{\circ}_{\text{mic}}(\text{headgroup}) + (n-1)\Delta G^{\circ}_{\text{mic}}(\text{—}CF_2\text{—}) + \Delta G^{\circ}_{\text{mic}}(\text{—}CF_3) \tag{5.4}$$

where n is the number of carbon atoms in the hydrophobic tail group. Hence, the contribution to this free energy per $-CH_2-$ or $-CF_2-$ group of the surfactant hydrophobe can be estimated from a plot of $\ln(CMC)$ versus chain length for a homologous series of surfactants containing the same hydrophilic headgroup. Figure 5.7 shows plots of this type for the homologous series $F-(CF_2)_n-CH_2-(OCH_2CH_2)_4-OH$ and $H-(CH_2)_n-(OCH_2CH_2)_4-OH$ for which n is varied. From the slopes of the plots, the contributions per $-CH_2-$ and $-CF_2-$ group are found to be -3.1

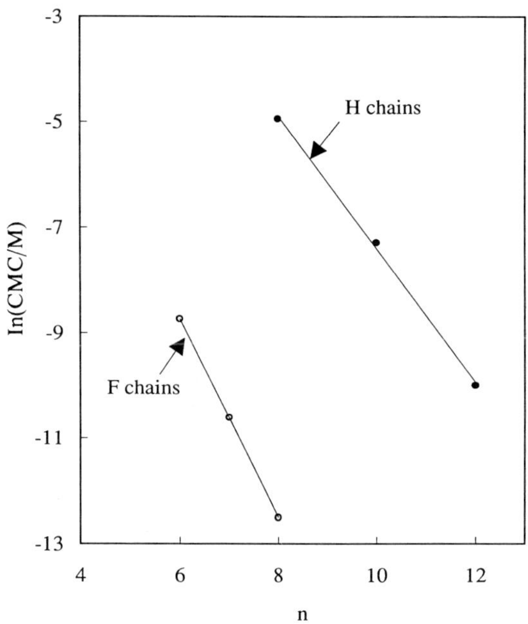

Figure 5.7 Variation of ln(CMC) with chain length for $F(CF_2)_nCH_2(OCH_2CH_2)_4OH$ (F chains) and $H(CH_2)_n(OCH_2CH_2)_4OH$ (H chains) at 25°C. The data is from Reference 42 for the F chains and from Reference 43 for the H chains.

and $-4.7\,kJ\,mol^{-1}$ respectively. From this comparison, it can be seen that a perfluorocarbon chain is more hydrophobic than a hydrocarbon chain and that one $-CF_2-$ group is approximately equivalent to 1.5 $-CH_2-$ units. Whereas the addition of two $-CH_2-$ units to a hydrocarbon surfactant reduces the CMC approximately 10-fold, two additional $-CF_2-$ units generally reduces the CMC by approximately 40-fold. A fluorosurfactant containing seven carbons behaves, in terms of the CMC, like a hydrocarbon surfactant containing 11 or 12 carbon atoms. Similar conclusions regarding the relative effectiveness of hydrocarbons and fluorocarbon chains have been reached for ionic surfactants [44].

Ravey *et al.* have developed this comparison in greater detail than is presented here [42, 45–48]. On the basis of a more detailed analysis of a wider range of surfactant structures than is discussed here, they conclude that a $-CF_2-$ unit is approximately equivalent to 1.7 $-CH_2-$ units (slightly higher than the factor 1.5 estimated above). They further show [42] that there is a good correlation between the magnitude of the free energy contributions of the different hydrophobic segments ($-CH_2-$, $-CF_2-$, $-CH_3$ and CF_3) and either the volume or surface area of the cavities created in water by their presence. This implies that the greater surfactant effectiveness of fluorocarbon chains arises mainly from their larger size.

Table 5.3 Minimum surface tensions and limiting areas per molecule for various ionic fluorinated surfactants in water at 25°C. Only systems for which the Krafft temperature is below 25°C are shown. The data is from References 44, 49

Surfactant	γ_{min} (mN.m^{-1})	A_{min} (nm^2)
n-$C_7F_{15}COOH$	15.2	0.415
n-$C_7F_{15}COONa$	24.6	0.42
n-$C_8F_{17}COOLi$	24.6	
n-$C_8F_{17}COONH_4$	14.8	
n-$C_{10}F_{21}COOLi$	20.5	
n-$C_8F_{17}SO_3Li$	29.8	0.552

We have seen so far that perfluorocarbon chains are capable of reducing the tension at lower concentrations than the corresponding hydrocarbon chains. In addition, the final minimum tension achieved at the CMC for perfluorocarbon chains is generally lower than for hydrocarbon chains. As seen in Table 5.3, the minimum tensions for various ionic fluorosurfactants are commonly in the range 15–25 mN m^{-1} as compared with hydrocarbon surfactants for which the values are generally 30–40 mN m^{-1}. Ravey *et al.* [42, 45–48] have shown that a range of non-ionic fluorosurfactants can similarly reduce the tension of water to less than 20 mN m^{-1}. If it is argued that the surface of an aqueous surfactant solution containing an adsorbed monolayer of a surfactant is similar in surface properties to that of a liquid made up of the surfactant hydrophobic tails (i.e. perfluorocarbons or alkanes), then the lower minimum tension of fluorosurfactant solutions is expected since the tensions of perfluoroalkanes are lower than those the alkanes (Figure 5.4) owing to the low cohesive energy density of perfluorocarbon chains. Alternatively, it can be argued that the lower tension for fluorosurfactant is a consequence of their stronger adsorption before the CMC is reached.

Compared to hydrocarbon surfactants of comparable chain length, fluorosurfactants are thus both more effective in achieving minimum tension *and* the minimum tension is lower. However, there is a drawback. The discussion of the CMC values presented above was prefaced by the statement that it applied only to surfactants capable of forming micelles. If the solubility of the monomeric surfactant is less than the CMC, increasing the surfactant concentration leads to the precipitation of a solid phase (generally hydrated to some degree). In this case, the lowest tension achievable is limited by the solubility and not by the CMC and may be much higher. At a particular temperature, known as the Krafft temperature, the monomer solubility is equal to the CMC. For surfactant solutions at a temperature below the Krafft temperature, the minimum tension is solubility-limited and high and no micelles are formed. Fluorosurfactants can reach minimum tensions of 15–25 mN m^{-1} *only* at temperatures above their Krafft temperatures.

Krafft temperatures of surfactants correlate with, but are not generally equal to, the melting points of the pure surfactants [44]. This is because the solid phase that separates at high concentrations below the Krafft temperature is generally a hydrated solid phase. As noted earlier, the melting points of perfluoroalkanes are generally higher than those of the corresponding alkanes, an effect which is thought to be a consequence of the higher rigidity of the perfluorocarbon chains. Thus, fluorosurfactants are expected to show high Krafft temperatures as shown in Figure 5.8. Krafft temperatures can be reduced either by choosing a more appropriate counterion (for ionic systems) [44, 49], by inserting flexible linkages such as Si or O into the fluorocarbon chain or by introducing branched chains.

There have been relatively few studies of the sizes of fluorosurfactant micelles. Light scattering studies are difficult because of the low difference in refractive index between fluorosurfactants and water which makes the scattered intensity very low. On the other hand, small angle neutron scattering (SANS) provides an excellent means to study fluorosurfactant micelles due to the strong contrast between fluorocarbons and H_2O.

Burkitt *et al.* [50] used SANS to investigate the sizes and shapes of micelles of ammonium salts of octanoic, decanoic and perfluorooctanoic acids. The aggregation number of the spherical micelles of ammonium

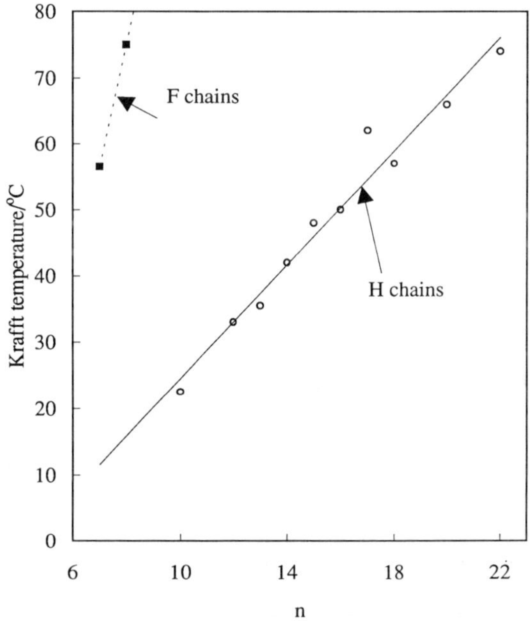

Figure 5.8 The effect of chain fluorination on the Krafft temperature. Values are for sodium alkylsulphonates with either hydrocarbon or perfluorocarbon chains. The data is from Reference 44.

octanoate is 7 at 0.36 M and increases to 12 at 0.48 M. At 0.12 M, ammonium perfluorooctanoate has an aggregation number of 43 and forms cylindrical micelles of diameter 2.0 nm and length 4.8 nm. The larger micellar size of the fluorosurfactant is consistent with the trend expected for a larger tail volume at constant area per headgroup.

Berr and Jones [51] investigated sodium perfluorooctanoate (SPFO) micelles using SANS and compared the results with data for the corresponding hydrocarbon surfactant. The SPFO micelles were found to be spherical and to have aggregation numbers which increased with increasing concentration (27 at 0.1 M increasing to 41 at 0.51 M) and which were more than twice those of sodium octanoate. The scattering curves were best accounted for using a model in which all of the perfluoralkyl tails reside in a water-free core. The smaller aggregation numbers for the sodium salt of the perfluorooctanoate ion allow the formation of spherical micelles, in contrast to the cylindrical micelles formed by the ammonium salt.

The lyotropic liquid crystal phase behaviour of binary non-ionic fluorosurfactant + water mixtures have been compared with those of the corresponding hydrocarbon surfactant systems by Ravey *et al.* [42, 45–48, 52]. The patterns of phase behavior are very similar except that there is a difference between the fluorosurfactant and hydrocarbon surfactant chain lengths required to achieve a particular phase behavior. For example, the diagram for $F(CF_2)_6CH_2(OCH_2CH_2)_4OH$ is broadly equivalent to that for $H(CH_2)_{12}(OCH_2CH_2)_3OH$ [42]. As with the CMC values, surfactant chain fluorination produces an effect similar to that of increasing the hydrocarbon surfactant chain length.

5.5.2 *Partially fluorinated chain surfactants*

Partial fluorination of the surfactant hydrophobic chain can strongly affect the ability of a surfactant to form micelles. This arises because of the necessity to mix fluorocarbon and hydrocarbon segments in the micelle core and partial fluorination may lead to a degree of polarity in the chain. Muller *et al.* [53–57] investigated a series of surfactants in which only the terminal methyl group was fluorinated and Table 5.4 compares (where the data is available) the CMC values of these materials with the corresponding fully hydrogenated and fully fluorinated compounds. It can be seen that the terminal fluorinated species show CMC values that are approximately twice as large as the hydrocarbon surfactants and considerably higher than the fully fluorinated species. F-NMR studies show that the terminal F_3C– experiences an environment approximately midway between that of bulk water or hydrocarbon suggesting that either water penetrates the micelle extensively or that the fluoro group spends a significant fraction of time at the micelle surface. The increased unfavourableness of micelle formation is a result of a combination of the unfavourable energy of mixing of hydrocar-

Table 5.4 CMC values (mM) of surfactants containing a hydrophobic tail in which the chain is either fully hydrogenated (fully H), fully fluorinated (fully F) or is hydrogenated except that the terminal CF_3– is fluorinated (terminal CF_3). Data from (a) References 53–57, (b) Reference 50, (c) Reference 43, (d) Reference 58 and (e) interpolated from data in Reference 49.

Surfactant	Fully H	Terminal CF_3	Fully F
C_9COONa	41(b)	167(a)	1.6(e)
$C_{11}COONa$		51(a)	
$C_{12}COONa$		24(a)	
$C_{12}SO_4Na$	8.2(c)	14.6(a)	
$C_8(OCH_2CH_2)_6OH$	9.9(c)	18(a)	
$C_{12}N(CH_3)_3Br$	13.7(d)	27.2(a)	
$C_{10}N(CH_3)_3Br$	64(d)	99	

bon and fluorocarbon segments or the increased polarity of the chain arising from the terminal CF_3 segment.

CMC values for $H(CF_2)_nCOOH$ and $F(CF_2)_nCOOH$ are compared in Reference 59. The 1-H substituted compounds for $n = 6$ and 8 show CMC values which are approximately three and five times higher than the fully fluorinated compounds.

A series of hybrid surfactants of general structure $F(CF_2)_n$.-$CHSO_4Na.(CH_2)_mH$ (abbreviated to $F_nSO_4H_m$) has been investigated by Guo *et al.* [60]. The corresponding series of hydrocarbon surfactants ($H_nSO_4H_m$) has been studied by Evans [61]. The perfluorocarbon chain surfactants show lower CMCs (Table 5.5) but the reduction is somewhat less than would be expected on the basis that a $-CF_2-$ group has the hydrophobicity of approximately 1.5–1.7 $-CH_2-$ groups.

5.5.3 *Mixtures of hydrocarbon and fluorocarbon surfactants*

Most product formulations containing surfactants actually contain a mixture of different surfactants, either as a result of a spread of molecular structures resulting from the manufacture ('impurities') or by design to optimize the performance of the product. The behavior of surfactant mixtures is therefore very actively researched and a number of specialist

Table 5.5 CMC values of two-tailed surfactants containing either one perfluorocarbon and one hydrocarbon chain (at 25°C) or two hydrocarbon chains (at 40°C). Data from Reference 60 for the mixed chain surfactant and from Reference 61 for the surfactants containing two hydrocarbon chains

n	m	CMC of $H_nSO_4H_m$ (mM)	CMC of $F_nSO_4H_m$ (mM)
7	6	9.7	0.78
8	6	6.65	0.26
7	2	49.5	4.5

texts are available [62, 63]. As discussed below, mixtures of hydrocarbon and fluorosurfactants show unusual behavior which is the result of the antipathy between hydrocarbon and fluorocarbon chains. The theoretical treatment of surfactant mixtures outlined below is drawn from References 62 and 63 where a fuller discussion can be found.

The simple theory of mixed surfactant solutions starts from the premise that the micelles can be treated as a separate phase (a 'pseudophase') in equilibrium with the surfactant monomer solution. This is known to be a good approximation for micelles with aggregation numbers larger than 50 or so. Consider a mixture of two pure nonionic surfactants, 1 and 2. The mole fraction of surfactant 1 is X_1 per total moles of the surfactant mixture, i.e. excluding the moles of solvent. If it is assumed that the activity coefficient for free surfactant monomers is unity (expected to be a reasonable assumption for systems with low CMC values), then it can be shown that the monomer concentration of surfactant 1 in the mixture (C_1^m) is given by:

$$C_1^m = f_1 x_1 CMC_1 \tag{5.5}$$

where CMC_1 is the CMC of pure surfactant 1 and f_1 and x_1 are the activity coefficient and mole fraction respectively of 1 *in the mixed micelles*. At the CMC of the mixture (denoted as CMC) C_1^m is also equal to $X_1 CMC_1$, where X_1 is the overall mole fraction of surfactant 1 (including both monomers and micelles if present). These equations, together with a simple mass balance equation, can be combined to give an expression for the CMC of the mixture in terms of the mole fractions, the CMC values of the pure surfactants and the activity coefficients of the surfactants in the mixed micelles.

$$1/CMC = (X_1/f_1 CMC_1) + (X_2/f_2 CMC_2). \tag{5.6}$$

For surfactants that mix ideally in the micelles, the activity coefficients are equal to unity. For non-ideal mixing an approximate expression for the activity coefficients can be obtained using regular solution theory in which it is assumed that the excess entropy of mixing (i.e. the mixing entropy for the real mixture less that calculated for the corresponding ideal mixture) is zero and the non-ideality results only from an enthalpic term. In this approximation, the activity coefficients are:

$$f_1 = \exp[\beta(1 - x_1)^2] \text{ and } f_2 = \exp[\beta(1 - x_2)^2] \tag{5.7}$$

The parameter β is related to molecular interactions in the mixed micelles by:

$$\beta = (W_{11} + W_{22} - 2W_{12})/kT \tag{5.8}$$

where W_{11} and W_{22} are the interaction energies between like molecules in the pure micelles and W_{12} refers to the interaction between unlike species in the mixed micelles. For mixtures of fluorocarbon and hydrocarbon chains, the interaction between unlike species is relatively weak and so β is expected to be positive, corresponding to mutual antipathy between the hydrocarbon and perfluorocarbon chains. If β is taken to be constant over the whole mole fraction range, then, at the mixed CMC

$$CMC = CMC_1 \exp[\beta(1 - x_1)^2]x_1 + CMC_2 \exp[\beta x_1^2](1 - x_1) \quad (5.9)$$

and:

$$\beta = \{\ln(X_1 CMC/x_1 CMC_1)\}/(1 - x_1)^2 \quad (5.10)$$

With experimental data of the mixed solution CMC at overall solution mole fraction X_1 and the CMC values of the two pure surfactants these two simultaneous equations can be solved numerically to obtain β and the mole fraction of component 1 in the micelles (x_1).

Figure 5.9 shows representative calculations of the variation of the mixed CMC with the overall mole fraction of surfactant 1 for values of β equal to 0 (corresponding to ideal mixing), 1, 2 and 3. The values of the CMC for the pure surfactants are both taken to be 5 mM in the calculations. Increasing values of β correspond to an increasing mutual antipathy between the two surfactants as expected for mixtures of hydrocarbon and

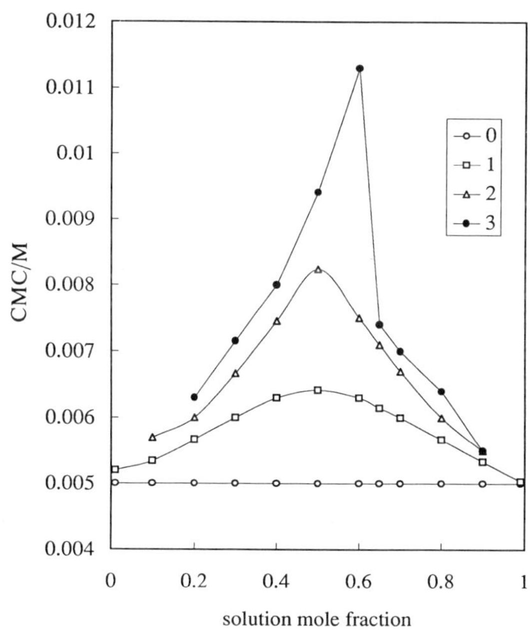

Figure 5.9 Mixed CMC values versus solution mole fraction calculated for a mixture of two non-ionic surfactants which each have a CMC in the pure state of 5 mM. The curves are calculated according to equation 5.6 for the β values shown on the figure.

fluorocarbon surfactants. With increasing β, the mixed CMC increases above that for the ideal mixtures (i.e. micellization becomes more unfavourable) due to the unfavourable mixing in the micelles. According to regular solution theory, the onset of immiscibility between a pair of liquids occurs for β values higher than 2. This is seen in the discontinuity in the CMC curve for $\beta = 3$ which corresponds to the formation of two separate populations of micelles, one rich in surfactant 1 and one rich in surfactant 2. The plots of x_1 versus solution mole fraction (Figure 5.10) become increasingly sigmoidal in shape with increasing β with the curve for $\beta = 2$ showing an inflection point. This marks the onset of the formation of separate populations of micelles, i.e. a miscibility gap. It is thus analogous to the upper critical solution temperature discussed earlier for liquid mixtures. For $\beta = 3$ the miscibility gap is well developed and the vertical portion of the curve corresponds to the formation of two separate populations of micelles with mole fractions of component 1 in the micelles of approximately 0.1 and 0.97 respectively.

As discussed in References 62 and 63, this treatment can be extended to the case of micelle formation by ionic micelles. Additionally, measurements of the surface tension for different compositions can be related to the compositions of the mixed micelles and the adsorbed monolayer formed at

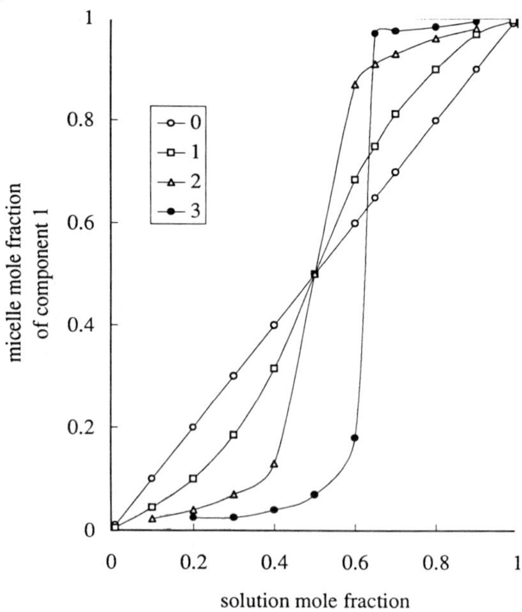

Figure 5.10 Calculated micelle compositions versus solution mole fraction calculated for the same systems as Figure 5.9.

the air–water interface. However, although it can provide a reasonable description of CMC values in mixed surfactants, it should be noted that the regular solution treatment is rather approximate since systems showing a non-zero enthalpy of mixing are not likely to show an ideal entropy of mixing. The parameters derived using it are best viewed as semi-empirical.

Detailed reviews of mixed surfactant behavior can be found in References 62 and 63. They also contain reviews devoted to mixtures of hydrocarbon and fluorocarbon surfactants. For the present limited discussion, we focus on the values of β estimated for hydrocarbon and perfluorocarbon surfactant mixtures and whether phase separation of separate populations of micelles occurs.

Experimental values of β have been estimated from the variation of the mixed CMC with composition for a range of different types of surfactant mixtures. For mixtures of hydrocarbon surfactants of the same charge and similar chain length β is generally close to zero (indicating ideal behaviour). For mixtures of ionics and non-ionics β is negative as the presence of the non-ionic spaces the charged headgroups making micellization more favourable. Mixtures of cationics with anionics show very strongly negative values of β due to the strong electrostatic attraction between the two surfactants in the mixture. For hydrocarbon/perfluorocarbon surfactant mixtures of the same charge, β values are generally positive corresponding to a repulsive interaction in the micelles. This has the consequence that hydrocarbon/perfluorocarbon surfactant mixtures can show CMC values which are higher than either pure surfactant alone as seen in the theoretical plots of Figure 5.9. For sodium perfluorooctanoate with sodium decyl sulphate, Shinoda and Nomura [64] estimated β to be 1.8, slightly less than the value of $+2$ required for micelle demixing. Related anionic surfactant mixtures of increased chain length generally show higher positive β values indicating that micelle demixing may occur in such systems although the experimental evidence available for demixing is contradictory in some cases (see, for example, chapter 5 in Reference 63). Ravey *et al.* [65] estimated β to be in the range 2.5–3.0 for mixtures of two non-ionics, $C_{12}E_6$ with $F(CF_2)_6$-C_2H_4-SC_2H_4-$(OC_2H_4)_2$-OH. Hydrocarbon/perfluorocarbon surfactant mixtures of different charges generally show negative values of β. For example, mixtures of lithium perfluorooctylsulfate with $C_{12}E_6$ have a β value of -4.8 [66]. This shows that the attractive interaction resulting from the spacing of the electrical charges is more than sufficient to overcome the mutual antipathy of the hydrocarbon and perfluorocarbon chains.

Hydrocarbon/perfluorocarbon surfactant mixtures of sufficient chain length may show β values which are sufficiently positive such that micelle demixing may occur. Funasaki *et al.* [67, 68] have used surface tension measurements to determine the composition range over which micelle demixing occurs for several systems and for different temperatures. As for hydrocarbon/perfluorocarbon bulk liquid mixtures, the immiscibility com-

Table 5.6 Values of T_c and X_c for micelle demixing in mixtures of a fluorosurfactant (FS) and a hydrocarbon furfactant (HS). The data is from References 67, 68

FS	HS	T_c (°C)	X_c of fluorosurfactant
NF	SDS	20.6	0.16
NF	StrS	36.8	0.288
NF	STS	52.2	0.535
NFE7M	DE7	40.1	0.45
NFE18M	DE7	33.8	0.70
NFE18M	DE9	25.9	0.73
NFE18M	DE25	<16.1	–
NFE18M	HE25	<25	–

NF = p-$[(CF_3)_2CF]_2C{=}C(CF_3)OC_6H_4SO_3Na$
SDS = $H(CH_2)_{12}SO_4Na$
STrS = $H(CH_2)_{13}SO_4Na$
STS = $H(CH_2)_{14}SO_4Na$
NFEmM = $[(CF_3)_2CF]_2C{=}C(CF_3)(OCH_2CH_2)_mOCH_3$
DEm = $H(CH_2)_{12}(OCH_2CH_2)_mOH$
HEm = $H(CH_2)_{16}(OCH_2CH_2)_mOH$

position range narrows with increasing temperature and thus the micelle systems show binodal curves terminating at a critical temperature and composition (T_c and X_c). T_c is analogous to the UCST discussed earlier for mixtures of alkanes and perfluoroalkanes (Figure 5.5). For the illustrative values of T_c and X_c for some mixed systems shown in Table 5.6, it can be seen that longer surfactant chain lengths promote micelle demixing (i.e. T_c increases) in a manner similar to that found for alkane/perfluoroalkane mixtures.

5.6 Adsorption and aggregation of fluorosurfactants in apolar oils

5.6.1 Fluorosurfactants containing a hydrophilic group

Hydrocarbon surfactants do not generally adsorb at the liquid–vapour surface of an apolar oil. In contrast, since fluorocarbon liquids have lower surface free energies than the corresponding hydrocarbons (Figure 5.4), fluorosurfactants are capable of adsorbing at the oil–vapour surface. Zisman *et al.* [69–71] investigated the adsorption of a range of oil soluble fluorocarbon derivatives at the surfaces of several organic liquids. The adsorption characteristics of a homologous series of dialkyl compounds are given in Table 5.7. It can be seen that these compounds adsorb more strongly from solution in mineral oil than from hexadecane, both in terms of the initial slope of the tension versus concentration plot and in terms of

Table 5.7 Adsorption parameters for dialkyl surfactants of general structure bis-(R_f) 3-methylglutarate at the liquid vapour surface of hexadecane and mineral oil at 20°C (data from Reference 71)

R_f	$-(d\gamma/dc)_{c=0}$ ($mN.m^{-1}.M^{-1}$)	A_{min} (nm^2)
Hexadecane as solvent		
$F(CF_2)_3CH_2$–	11	0.87
$F(CF_2)_5CH_2$–	33	0.83
$F(CF_2)_7CH_2$–	90	0.87
Mineral oil as solvent		
$F(CF_2)_3CH_2$–	19	0.39
$F(CF_2)_5CH_2$–	98	0.35
$F(CF_2)_7CH_2$–	225	0.37

the minimum area per surfactant achieved. For mineral oil as solvent, the limiting area per molecule is too small to accommodate both fluorocarbon chains of these diakyl species and thus it has been suggested that they adopt a configuration in the adsorbed monolayer in which one chain lies in the solvent and the second chain is directed towards air. For all the systems included in Table 5.7, the surface tension levels off at a constant value when the solubility limit is reached, i.e. the solubility limit is reached before micelle formation. The limiting tension reached is 23–25 $mN\,m^{-1}$ in all cases.

More recently, adsorption at the oil–air surface has been described for several novel surfactants. Abe *et al.* [72] showed that a surfactant of structure $C_6H_5CF(CF_3)OC_3F_7$ was capable of reducing the surface tension of toluene to approximately 10 $mN\,m^{-1}$ at 30°C. This unusually low value was achieved at a concentration in the mM range. The tension plot also showed a 'CMC-like' discontinuity. Related compounds containing no ether linkage showed only small surface activity. Sawada *et al.* [73] have synthesized a range of fluorinated silicon oligomers containing carboxy groups. These materials were found to be easily soluble in both water and a range of organic solvents. The oligomers were able to reduce the surface tension of both water and organic solvents effectively.

5.6.2 Fluorosurfactants without a hydrophilic group

For a molecule to operate as a surfactant in an apolar oil, there is no necessity for it to contain a hydrophilic group. In general, surfactant activity arises because of the amphiphilic nature of the molecule which contains both a 'solvophilic' and a 'solvophobic' block where the solvent does not have to be water. In hydrocarbon solvents the solvophobic block can be a fluorocarbon chain because of the antipathy of perfluorocarbon and hydrocarbon chains. The solvophilic block can be a hydrocarbon chain. Thus, semifluorinated alkanes (SFAs) of general structure $F\text{-}(CF_2)_n\text{-}(CH_2)_m\text{-}H$

(abbreviated to F_nH_m) have the basic di-block architecture required for surfactant properties in apolar oils. There is increasing interest in these novel 'primitive' surfactants (which contain no electrical charges, dipoles or hydrogen bonding) as the relative simplicity of the intermolecular interactions responsible for their surfactant properties are expected to be more amenable to rigorous computer modelling than are conventional surfactants in aqueous systems.

Because the surface tensions of fluorocarbons are lower than hydrocarbons, SFAs are expected to adsorb at the hydrocarbon–air surface but *not* at the fluorocarbon–air surface. This has been verified in our laboratory [74]. The adsorption of a range of linear SFAs at hydrocarbon–air surfaces has been investigated using surface tension methods [75–77]. Some typical surface tension plots are shown in Figure 5.11. The main findings for SFAs in linear alkane solvents can be summarized as follows:

1. The maximum tension reduction is typically around 5 $mN\,m^{-1}$. This is, of course, much less than the values of 40–55 $mN\,m^{-1}$ seen for hydrocarbon and fluorocarbon surfactants adsorbing at the air–water surface and is a consequence of the low tension difference between alkanes and perfluoroalkanes.

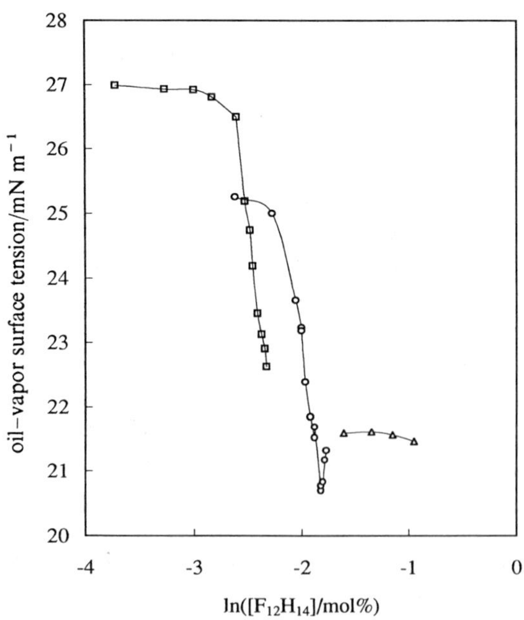

Figure 5.11 Variation of liquid–air tensions with concentration of the semi-fluorinated alkane $F_{12}H_{14}$ at 20°C. The data refer to *n*-pentadecane (squares), *n*-dodecane (circles) and *n*-octane (triangles) and were taken from Reference 77.

2. In favourable cases, adsorbed SFA monolayers at the alkane–air surface are close packed. For example, $F_{12}H_{14}$ at the dodecane–air interface at 20°C shows a limiting area per SFA molecule of 0.26 nm^2 [77], close to (or even slightly less than) the cross sectional area of an all-*trans* perfluorocarbon chain in a crystal phase (0.28 nm^2).
3. The surface activity effectiveness of SFAs (as judged from the concentrations in solution required to achieve maximum adsorption) increases with increasing perfluorocarbon chain length of the SFA, increasing chain length of the alkane solvent and with decreasing temperature. As discussed in section 5.4, each of these factors leads to an increased tendency of perfluorocarbon and hydrocarbon chains to demix in solution.
4. For the SFAs in hydrocarbon solvents, the minimum tension achieved is limited by the solubility of the SFA and not by micelle formation, i.e. the systems have high Krafft temperatures. At concentrations above their solubility limit, SFAs form gel structures which are thought to consist of solvent trapped within a matrix of fine microcrystallites of SFA [78]. In conventional surfactant systems, surfactant aggregation is driven by the attraction between solvophobic tailgroups but is opposed (and limited to the formation of finite aggregation number micelles) by the repulsive forces between solvophilic headgroups. It appears that linear SFA molecules lack sufficient repulsive forces between hydrocarbon chain 'headgroups' required to limit the aggregation to micelles and aggregation therefore proceeds to infinite aggregates, i.e. to solid phase formation. An additional factor responsible for the lack of micelle formation may be the greater tendency of perfluorocarbon chains to solidify, as compared with hydrocarbon chains, due to their higher rigidity.

The 'inverse' system of F_8H_{16} in perfluorooctane at 40°C has been reported to show micelle formation [79, 80] in which the aggregates contain a core of hydrocarbon chains and a shell of perfluorocarbon chains. The CMC is approximately 5 wt% and the micelles have been reported to have an aggregation number of either 4–6 [79] or approximately 90 [80]. For SFA micelles with a core of hydrocarbon chains, the SFA molecular geometry is likely to be more favourable for micelle formation than are putative SFA micelles in hydrocarbons in which the perfluorocarbon chains form the core, i.e. the thicker perfluorocarbon chain is on the exterior and the core is made up of flexible hydrocarbon chains.

5.7 Solubilization and microemulsion formation with fluorosurfactants in mixtures of oil and water

Aqueous micelles may solubilize oils and inverse micelles in oils may solubilize water. When sufficient oil or water solubilization occurs such that the dispersed material forms separate liquid cores within the aggregates, the resultant dispersions are known as *microemulsions*. Microemulsions may be defined as thermodynamically stable dispersions of oil-in-water (o/w) or water-in-oil (w/o) stablized by a surfactant. They contain aggregates in the size range 5–100 nm, are isotropic and generally scatter light only weakly.

For spherical, monodisperse microemulsion droplets, geometrical considerations give the following relationship between the droplet radius (r) and the composition.

$$r = (3V_{dc}/A_s)([\text{dispersed component}]/[\text{adsorbed surfactant}]) \quad (5.11)$$

where V_{dc} is the molecular volume of the dispersed component (water or oil) and A_s is the area occupied per surfactant molecule at the droplet surface. This simple model predicts that the droplet radius is proportional to the ratio of the molar concentration of the dispersed oil or water to that of the surfactant (=[dispersed component]/[adsorbed surfactant]), i.e. to the extent of solubilization of the dispersed component by the surfactant.

The free energy of formation of a liquid-in-liquid dispersion is made up of a positive term equal to the product of the oil–water tension and the increase in oil–water surface area and a negative term arising from the entropy of dispersion of the droplets. Thermodynamically stable microemulsions form when the oil–water tension is ultralow (less than approximately $0.1\,\text{mN}\,\text{m}^{-1}$) such that the favourable entropy term outweighs the unfavourable tension term. For highly curved oil–water monolayers with ultralow tensions such as those coating microemulsion droplets, the effects of monolayer curvature on the tension becomes significant. Microemulsion stability is thus dominated by the tendency of oil–water surfactant monolayers to curve. Qualitatively, a low tendency to curve together with an ultralow tension favours large microemulsion droplets and hence (as seen in equation 5.11) large solubilization of the dispersed component.

The tendency of a monolayer to curve is determined by the relative areas occupied by the surfactant headgroup (A_h) and tailgroup (A_t) *in situ* in the monolayer. When $A_h > A_t$, the preferred curvature is such that the headgroups form the exterior surface of the aggregate and o/w microemulsions are favoured. W/o microemulsions are favoured when $A_t > A_h$. Solubilization is maximized when A_h is approximately equal to A_t. The areas A_h and A_t are determined by the molecular structure of the surfactant but are also modified by factors such as penetration of the monolayer by water or oil, temperature and the addition of electrolyte which alter the interactions between adjacent surfactant head and tail groups in the monolayer. For example, consider an ionic surfactant with oil and water which forms an o/w microemulsion phase in equilibrium with excess oil. Addition of electrolyte to the aqueous phase screens the electrostatic repulsion between adjacent headgroups in the monolayer causing A_h to decrease. If A_h decreases to less than A_t, the system undergoes *phase inversion* of the microemulsion to give a w/o microemulsion phase in equilibrium with excess oil. In this way, the preferred monolayer curvature can be 'tuned' to optimize solubilization for a particular application.

Figure 5.12 gives a schematic representation of the effect of temperature on the microemulsion phase behaviour of a non-ionic surfactant. In this case, increasing temperature is thought to cause a dehydration of the headgroups causing A_h to shrink and thus leads to a phase inversion from o/w to w/o microemulsions in two-phase mixtures containing either excess oil or water. The temperature at which phase inversion occurs is denoted the *phase inversion temperature* (PIT).

Two factors are essential for microemulsion formation and thus for a good efficiency of solubilization. Firstly, the interfacial tension between the two liquids to be emulsified must be reduced to a very low value. Secondly, the preferred curvature of the surfactant monolayer must be 'tuned' to be near zero. From the limited number of observations available in the literature, it is believed to be difficult to form microemulsions with water and hydrocarbon oils using fluorosurfactants and to microemulsify water and fluorinated oils with hydrocarbon surfactants [81]. These effects are thought to be due to the lack of affinity between hydrocarbon and perfluorocarbon chains making it impossible to achieve very low interfacial tensions in these

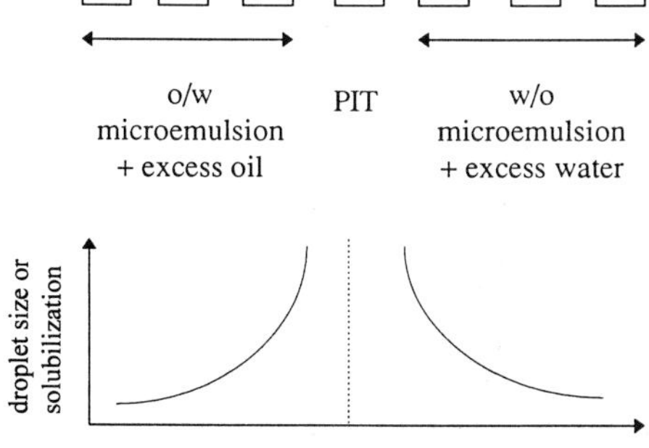

Figure 5.12 Schematic representation of the changes occurring with increasing temperature around microemulsion phase inversion for a mixture of oil + water + non-ionic surfactant. Below the PIT the o/w microemulsion droplets grow in size whereas, above the PIT, the w/o microemulsion drops decrease in size. Around the PIT, a third, surfactant-rich phase is formed in equilibrium with excess oil and water.

cases. Mukerjee and Handa [82] have compared the adsorption tendencies of a range of hydrocarbon and fluorosurfactants at the hexane–water and perfluorohexane–water surfaces. It can be seen (Table 5.8) that the affinity for the liquid–liquid interface increases when the surfactant and oil chain types are matched. It should be noted that this data refers to *dilute* adsorbed monolayers rather than the maximum adsorptions which are of more direct relevance to the formation of microemulsions.

Following interest in the possible use of fluorocarbon oil-in-water microemulsions as a potential blood substitute, there have been a number of investigations of microemulsion formation in mixtures containing fluorosurfactants with fluorocarbon oils [83–87]. The most systematic study has been made by Ravey and Stebe [42] who determined, *inter alia*, the PIT values for non-ionic fluorosurfactants with a range of fluorocarbon oils. The surfactant structures included $F(CF_2)_nCH_2(OCH_2CH_2)_mOH$ with $n = 6$–8 and $m = 3$–6. As for hydrocarbon surfactants, the PITs vary with the chain length of the oil in a systematic way consistent with the fact that short chain length oils penetrate the tail region of surfactant monolayers (thereby increasing A_t) more than long chain oils. They demonstrated that the experimental results for a range of fluorosurfactants and fluorocarbon oils could all be described in terms of an empirical equation based on the concept of hydrophile–lipophile balance (HLB) as developed by Griffin [88]. According to Griffin's formulation, the HLB of a surfactant is given by the wt% of hydrophilic groups (i.e. {(mass of hydrophilic groups)/(total mass of surfactant)} × 100) in the surfactant divided by 5. Ravey *et al.* found for fluorosurfactants with fluorocarbon oils:

$$\text{PIT (in °C)} \approx 31\text{HLB} + 16(6 + \text{ECN})^{1/2} - 270 \qquad (5.12)$$

where ECN (the 'equivalent carbon number') is equal to the number of carbon atoms in the oil for linear perfluoroalkenes but is assigned a different value for unsaturated, branched or cyclic species. The equivalent empirical relationship for non-ionic surfactants of the type $H(CH_2)_n(OCH_2CH_2)_mOH$ with hydrocarbon oils is

$$\text{PIT (in °C)} \approx 15.5\text{HLB} + 18(6 + \text{ECN})^{1/2} - 215 \qquad (5.13)$$

Table 5.8 Efficiency in tension lowering $-(d\gamma/dc)_{c\to 0}$ ($mN.m^{-1}.M^{-1}$) for hydrocarbon and fluorocarbon surfactant adsorption at the hexane–water and perfluorohexane–water interface at 25°C. Data from Reference 82.

Surfactant	Hexane–water	Perfluorohexane–water
$H(CH_2)_8SO_4Na$	5.00×10^3	1.39×10^3
$H(CH_2)_{10}SO_4Na$	19.8×10^3	4.41×10^3
$F(CF_2)_6COONa$	9.95×10^3	27.1×10^3
$F(CF_2)_8COONa$	82.5×10^3	281×10^3

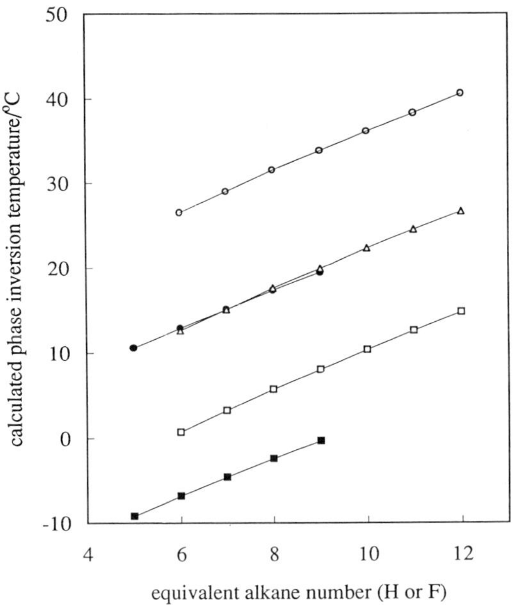

Figure 5.13 PIT values calculated from the empirical equations 5.12 and 5.13 plotted versus equivalent alkane number for the homologous series of surfactants $F(CF_2)_nCH_2$-$(OCH_2CH_2)_4OH$ (filled squares $n = 7$; filled circles $n = 6$) and $H(CH_2)_n(OCH_2CH_2)_4OH$ (open squares $n = 14$, open triangles $n = 12$; open circles $n = 10$).

Using these two empirical relationships, the PITs of hydrocarbon and fluorocarbon surfactant systems can be compared. Figure 5.13 shows a comparison of the PITs for $F(CF_2)_nCH_2(OCH_2CH_2)_6OH$ with *n*-perfluoroalkanes and $H(CH_2)_n(OCH_2CH_2)_6OH$ with *n*-alkanes. As for the CMC variation (Figure 5.7), it can be seen that a perfluorohexyl chain fluorosurfactant (with perfluoroalkanes) behaves like a hydrocarbon surfactant with a chain length of approximately 12. Ravey *et al.* further showed that the evolution of the ternary phase diagrams of the fluoro systems with temperature through the PIT bears a close similarity to that for the corresponding hydrocarbon system.

We have seen that the PIT marks a pivotal point in the phase progression and allows the rational optimization of solubilization properties for a particular application. Additionally, for conventional surfactants, there is a strong correlation between microemulsion phase behavior and the behaviour of *macro-emulsions* formed by emulsification of two-phase mixtures of a microemulsion with excess oil or water [89]. A system which forms an o/w microemulsion plus excess oil at equilibrium will generally form an o/w macro-emulsion on emulsification whereas a w/o microemulsion plus excess water will form a w/o emulsion. Thus, for oil–water volume ratios around unity, phase inversion of macro-emulsions occurs at approximately the same condition as phase inversion of the corresponding equilibrium microemulsions. Knowledge of the system PITs allows one to predict the emulsion type. It is further found that emulsions are generally easiest to prepare (because of the very low oil–water interfacial tension) but are maximally unstable around microemulsion phase inversion. Thus, it is expected that the correspondence between fluorocarbon and hydrocarbon surfactant systems found in the microemulsion phase behavior will also have relevance to macro-emulsion formation.

5.8 Wetting properties of fluorosurfactant solutions

5.8.1 Phenomenology and basic theory of wetting

As discussed in Reference 90, the wetting behaviour of a liquid on a solid surface is described in terms of the contact angle (θ) formed at the three phase contact line where the solid, test liquid and the third phase (which may be air or a second liquid) meet (Figure 5.14). When θ is zero the liquid spreads on the solid to form a liquid film and is then said to completely wet the solid. This is the situation for pure water on clean glass. When θ is 180° the liquid forms a near spherical drop leaving the surface unwetted. This situation is approached (but not realized completely) by, for example, mercury on a Teflon surface, for which θ is 150°. For finite contact angles between 0 and 180°, the liquid is said to partially wet the solid.

The equilibrium contact angle is related to the values of the three interfacial tensions in the system according to Young's equation:

$$\cos\theta = (\gamma_{S2} - \gamma_{S1})/\gamma_{12} \tag{5.14}$$

where γ_{S1}, γ_{S2} and γ_{12} are the interfacial tensions acting at the solid–test liquid, solid–surrounding phase and test liquid–surrounding liquid surfaces respectively (Figure 5.14).

For many practical purposes, it is helpful to have a means of characterizing the 'wettability' of solids. Zisman and co-workers [91] have found that the contact angles of liquids on low energy surfaces (defined as having a surface free energy of less than 100 mJ m^{-2}) such as hydrocarbons,

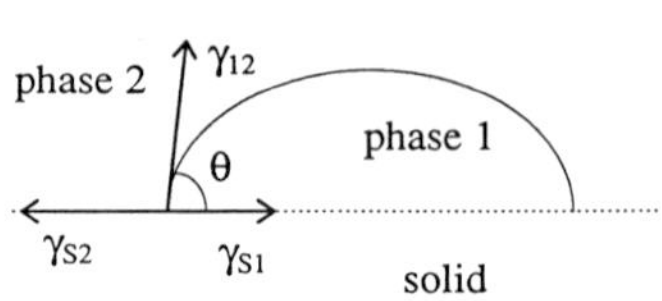

Figure 5.14 Tensions and contact angle geometry for a liquid drop on a solid.

fluorocarbons and some polymers depend primarily on the surface tension of the liquid γ_{12} (where phase 2 is air plus vapour). For a series of structurally related, pure test liquids, a plot of $\cos\theta$ versus the surface tension of the liquid is generally reasonably linear. The surface tension where the line cuts the $\cos\theta = 1$ axis is known as the critical wetting tension (CWT). It is the surface tension of a liquid that would just spread on the solid to give complete wetting. The critical wetting tensions of a range of low energy surfaces are tabulated in Table 5.9. Fluorocarbon surfaces have the lowest values of CWT due to their low cohesive energy. This makes fluorocarbons the most difficult surfaces to wet and gives rise to their use in many applications such as non-stick cookware and in waterproofing agents.

We now consider the wetting behavior of surfactant solutions on solids. In general, the addition of a surfactant may affect all three tensions that determine the contact angle. Combination of the equations of Young and Gibbs gives:

$$d(\gamma_{12}\cos\theta)/d\gamma_{12} = (\Gamma_{S2} - \Gamma_{S1})/\Gamma_{12} \tag{5.15}$$

where Γ is the surface excess concentration of adsorbed surfactant at the particular interface indicated in the subscript. For wetting of a low energy solid by a surfactant solution where phase 2 is air, it appears reasonable to assume that $\Gamma_{S2} = 0$. (We will see later that is commonly not the case for *high* energy solids.) With this assumption, the contact angle depends only on the liquid surface tension and the relative adsorption at the solid–solution and solution–air interfaces, Γ_{S1} and Γ_{12}. If it is further assumed that the ratio of adsorptions at the two interfaces is a constant for all surfactant concentrations, i.e. that Γ_{S1}/Γ_{12} is constant, then integration of equation 5.15 yields the following expression for the variation of $\cos\theta$ with γ_{12}. The constant of integration is found from the fact that the contact angle of the pure solvent (θ_s) is obtained when γ_{12} is equal to the tension of the pure solvent γ_s.

$$\cos\theta = (\gamma_s(\cos\theta_s + \Gamma_{S1}/\Gamma_{12})/\gamma_{12}) - \Gamma_{S1}/\Gamma_{12} \tag{5.16}$$

The variation of $\cos\theta$ with γ_{12} has been calculated according to equation

Table 5.9 Critical wetting tensions (CWT) of low energy surfaces at 20°C. Data from Reference 91

Surface constitution	CWT ($mN.m^{-1}$)
$-CF_3$	6
$-CF_2H$	15
$-CF_2-$	18
$-CFH-CH_2-$	28
$-CH_3$ (crystal)	22
$-CH_3$ (monolayer)	24
$-CH_2-$	31

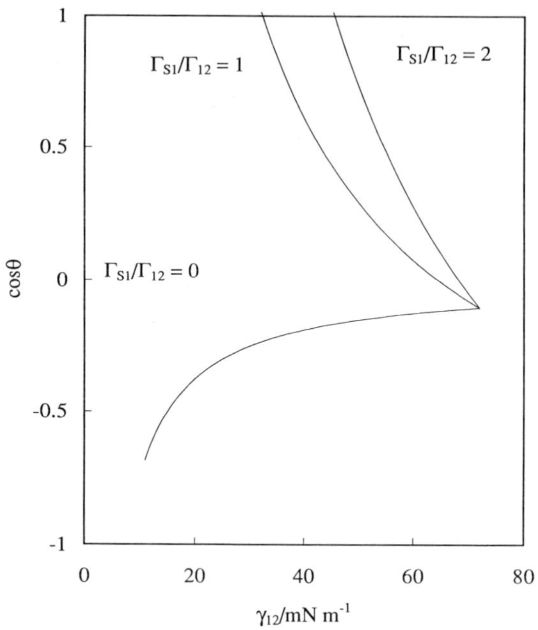

Figure 5.15 Variation of contact angle with γ_{12} calculated according to equation 5.16. The plots are calculated for $\gamma_s = 72\,\text{mN}\,\text{m}^{-1}$ and $\theta_s = 96°$ which corresponds approximately to water on a polythene surface.

5.16 for several values of the adsorption ratio Γ_{S1}/Γ_{12}. The plots (Figure 5.15) are calculated for $\gamma_s = 72\,\text{mN}\,\text{m}^{-1}$ and $\theta_s = 96°$, corresponding approximately to water on a polyethylene surface. For $\Gamma_{S1}/\Gamma_{12} = 1$, complete wetting occurs when γ_{12} is equal to the CWT. No adsorption at the S1 surface prevents wetting whereas increased relative adsorption at the solid surface causes complete wetting to occur at higher solution tensions corresponding to lower surfactant solutions. Thus, wetting by a surfactant solution is aided by surfactant adsorption at the solid–liquid interface.

With regard to experimental wetting studies, it should be noted that it is hard to ensure that the true equilibrium contact angle is measured experimentally. There are commonly significant differences between contact angles measured following either an advance or retraction of the three phase contact line. This contact angle hysteresis commonly arises because of surface roughness or heterogeneity.

5.8.2 Comparison of wetting properties of aqueous solutions of hydrocarbon and fluorocarbon surfactants on low energy solids

Because of their low minimum tension values, fluorosurfactants are expected to be better than hydrocarbon surfactants at promoting wetting of solids by aqueous solutions. Hydrocarbon surfactants are found to be unable to completely wet Teflon (with a CWT of 18 mN m^{-1}) whereas fluorosurfactant solutions can show complete wetting. A more quantitative comparison can be seen from the data shown in Table 5.10 which shows critical wetting tensions for three solid surfaces derived from contact angle measurements with either pure liquids or aqueous solutions of either the hydrocarbon surfactant AOT (sodium bis 2-ethylhexylsulphosuccinate) or the fluorosurfactant PFOA (perfluorooctanoic acid). For AOT on paraffin, the CWT for the surfactant solutions is similar to that estimated using pure liquids. This suggests that adsorption of this surfactant is similar at both the liquid–vapour and solid–liquid surfaces. For PFOA on paraffin, the surfactant solution CWT is considerably lower than the liquid value, indicating that the fluorosurfactant adsorption at the solid–liquid interface is significantly lower than for the liquid–air surface. The lower adsorption of the fluorosurfactant at the hydrocarbon surface may be a consequence of the antipathy between the fluorosurfactant chains and the hydrocarbon chains of the solid. For adsorption on Teflon, AOT shows a slightly higher CWT than found for pure liquids whereas PFOA shows a lower CWT suggesting that its adsorption on Teflon is again low. For the more polar surface of PMMA, it appears that the extents of adsorption at the solid–liquid surface of AOT and PFOA are both considerably less than for the air–water surface.

Rutherford [96] investigated contact angles of aqueous solutions of SDS (sodium dodecyl sulfate) and SPFO (sodium perfluorooctanoate) on glass surfaces which had been treated with silanizing agents producing either a hydrocarbon or fluorocarbon surface. For these systems, plots of $\gamma_{12}\cos\theta$ versus γ_{12} (equation 5.15) are linear and the slopes yield the values of $(\Gamma_{S1}/\Gamma_{12})$. SDS on the hydrocarbon surface gave a value of 0.98 whereas SPFO gave a value of 0.74. For the fluorocarbon surface, the value for SPFO is 0.87, slightly higher than for the hydrocarbon surface but still less than unity. Overall, the limited data available suggests adsorption of

Table 5.10 Critical wetting tensions (mN.m^{-1}) for various low energy solids with either pure liquids or aqueous solutions of either the hydrocarbon surfactant sodium bis-2-ethylhexylsulphosuccinate (AOT) or the fluorosurfactant perfluorooctanoic acid (PFOA) at 25°C. The data for the surfactant solutions is from Reference 92. For the pure liquids, the reference is given in the parentheses after the value

Liquid system	Solid		
	Paraffin	Teflon	PMMA
Pure liquids	25.5[93]	18[94]	40[95]
AOT solutions	23	22	26
PFOA solutions	15	13	11

PMMA is polymethyl methacrylate

fluorosurfactants from water to the hydrocarbon solid–water surface is less favourable than adsorption at the air–water surface. Fluorosurfactant adsorption at the fluorocarbon solid–water surface is more favourable than for the hydrocarbon solid but appears less favourable than for the air–water surface.

5.8.3 Modification of the wetting properties of high energy solids by adsorption

We now consider the alteration in the wetting properties of a high energy solid brought about by the adsorption of a monolayer film of surfactant. For clean, high energy surfaces (such as clean metals), Zisman and co-workers [97–100] have shown that monolayer adsorption can be achieved in two ways. Firstly, the solid can be contacted with the molten (non-wetting) adsorbate. Following contact over the surface, removal of the excess liquid leaves an adsorbed monolayer bound to the surface. In the second method (denoted by Zisman as the 'retraction method'), an aqueous solution of the adsorbate is contacted with the solid. When adsorption occurs such that the low energy, hydrophobic tails of the adsorbate are uppermost on the solid, the contact angle of the liquid may increase sufficiently for the liquid to de-wet from the surface leaving a dry surface with an adsorbed film. This type of 'autophobic' effect may also be observed for an aqueous solution of a cationic surfactant with clean glass where the opposite electrostatic charges of the surfactant and the solid cause a strong adsorption.

Zisman and co-workers [98–100] have extensively characterized the wetting properties of such monolayer-covered solids for which the molecular structure of the monolayers has been systematically varied. Much of the data compiled in Table 5.9 was determined using this type of adsorbed monolayer with suitable molecular structural variation. These data clearly show that adsorption of a fluorinated monolayer onto a high energy (i.e. initially easily wetted) solid is very effective in reducing the critical wetting tension thereby making the solid difficult to wet.

In summary, fluorosurfactants can be very effective *both* at promoting wetting of low energy surfaces *and* in rendering high energy surfaces difficult to wet. As will be seen in the next section, both these properties find many applications.

5.9 Applications of fluorosurfactants

Fluorosurfactants are expensive relative to conventional surfactants and so tend to be used in applications in which their performance cannot be matched by other surfactants. Hence, applications tend to be centred around the unique properties of fluorosurfactants. Firstly, they have high chemical stability which enables them to operate in harsh environments where conventional surfactants would degrade rapidly. Secondly, they have unmatched wetting properties, either in promoting wetting of low energy surfaces or (as a deposited film) in the inhibition of wetting of high energy solids by both water and oils. Finally, fluorocarbon liquids are capable of dissolving a high concentration of oxygen [101] and hence have potential use as biological oxygen carriers in artificial blood formulations consisting of fluorocarbon oil-in-water emulsions or microemulsions. Fluorosurfactants are required to stabilize such formulations. Since fluorosurfactant applications are extremely diverse, we shall give only illustrative examples of some important areas. A more comprehensive coverage can be found in chapter 8 of Reference 1 and references therein.

Applications exploiting the high chemical stability of fluorosurfactants include the following. Electroplating, etching and the flotation of minerals all involve operation at extremes of pH where conventional surfactants would degrade. Fluorosurfactants are added to electroplating baths to regulate foam production (from gas bubbles released at the electrodes) and have been shown to have beneficial effects in improving electroplated coatings. The speed of glass etching is increased by the addition of a fluorosurfactant which facilitates wetting of the glass surface by the HF etch solution. Wetting is accelerated and flotation collecting power improved by fluorosurfactants in ore flotation.

Applications based on wetting are very numerous. Fluorosurfactants are added to reduce the surface tension and to improve wetting in products such as inks, paints, polishes, and waxes, adhesives, coatings and in herbicide and insecticide formulations. In fire fighting foams, fluorosurfactants are added to ensure that the foam spreads over the surface of a burning oil in order to act as a barrier to inhibit vapour escape and re-ignition. In applications such as metal finishing, mould release agents, anti-misting and repellancy treatments, the fluorosurfactant is deposited at the solid surface leaving it both hydrophobic and oleophobic. Unlike silicone surfactants which can confer water repellancy properties onto a textile, a fluorosurfactant treatment can make the textile resistant to wetting and penetration by *both* oil and water. It is noteworthy that, in many wetting applications, the desired wetting performance can commonly be optimized by use of a combination of a hydrocarbon surfactant together with a small concentration of a (more expensive) fluorosurfactant.

The ability of fluorosurfactants to stabilize emulsions of fluorocarbons gives rise to their applications in emulsion polymerization of fluoropolymers and in preparing dispersion of fluorocarbons in water for use as bio-medical oxygen carriers ('artificial blood') where the oxygen solubility is a critical factor (see chapter 10 of Reference 1). The use of fluorosurfactants in bio-medical applications is hampered by their toxicity. Attempts have been made to overcome this by using biologically benign hydrocarbon surfactants in combination with semi-fluorinated alkanes (SFAs) to stablilize fluorocarbon-in-water emulsions. The stability enhancement caused by SFAs has been postulated to occur by the SFA acting as a 'molecular dowel' between the fluorocarbon liquid and the hydrocarbon surfactant monolayer.

In conclusion, it can be seen that the unique characteristics of fluorocarbon chains give fluorosurfactants their special performance properties. Their wide range of applications ensures that research and development of these fascinating materials will continue for the foreseeable future.

References

1. Kissa, E. *Fluorinated surfactants: synthesis, properties, applications.* Surfactant Science Series, Volume 50, Marcel Dekker, New York, 1994.
2. Elleutero, H.S., Milan Jr, A.S. and Moore Jr, E.P. (Du Pont), *U.S. Patent 3,250,808,* 1966.
3. Moore, E.P. (Du Pont), *U.S. Patent 3,322,826,* 1967.
4. Arbogast, F.L. (Du Pont), *U.S. Patent 3,412,148,* 1968.
5. Eleutero, H.S. *J. Macromol. Sci.*, 1972, **A6(6)**, 1972.
6. Rabolt, J.F. Russell, T.P. and Twieg, R.J. *Macromolecules*, 1984, **17**, 2786.
7. Simons, J.H. *et al.*, *J. Electrochem. Soc.*, 1949, **95**, 47.
8. Simons, J.H. (ed), *Fluorine chemistry*, Vol. 1, p. 414, Academic Press, New York, 1950.
9. Simons, J.H. and Brice, T.J. In: Simons, J.H. (ed.), *Fluorine chemistry*. Vol. II, p. 340, Academic Press, New York, 1954.
10. Simons, J.H. (3M), *U.S. Patent 2,519,983,* 1950.
11. Hanford, W.E. and Joyce Jr, R.M. (Du Pont), *U.S. Patent 2,440,800,* 1942.
12. Hanford, W.E. and Harmon, J. (Du Pont), *U.S. Patent 2,440,801,* 1942.
13. Hanford, W.E. and Roland, J.R. (Du Pont), *U.S. Patent 2,457,229,* 1946.
14. Joyce, R.M., Hanford, W.E. and Harmon J. *J. Am. Chem. Soc.*, 1948, **70**, 2529.
15. Starks, C.M. *Free radical telomerization.* Academic Press, Orlando, Florida, 1974.
16. Gordon III, B. and Loftus, J.E. In: *Encyclopedia of polymer science and engineering.* Vol. 16, p. 533. J. Wiley and Sons, New York, 1989.
17. Deem, W.R. (ICI) *Brit. Patent 1302350*, 1973.
18. Hutchinson, J. *Fette Seifen Anstrichm.*, 1974, **76**, 158.
19. Jones, R. *Fette Seifen Anstrichm.*, 1975, **77**, 455.
20. Hudlicky, M. *Chemistry of organic fluorine compounds*, 2nd edn (revised), Ellis Horwood, New York, 1992.
21. Sidgewick, N.V. *The chemical elements and their compounds*, Vol. II, p. 1099, Oxford University Press, Oxford, 1951.
22. Kauck, E.A. and Diesslin, A.R. *Ind. Eng. Chem.*, 1951, **43**, 2332.
23. Hudlicky, M. *Chemistry of Organic Fluorine Compounds.* Macmillan New York, 1962.
24. Henne, A.L. and Fox, C.J. *J. Am. Chem. Soc.*, 1951, **73**, 2323.
25. Henne, A.L. and Fox, C.J. *J. Am. Chem. Soc.*, 1953, **75**, 5750.
26. MacBee, E.T., Pierce, O.R. and Smith, D.D. *J. Am. Chem. Soc.*, 1954, **76**, 3722.
27. Mahler, W. Du Pont, Wilmington, USA, personal communication.
28. Pauling, L. *The nature of the chemical bond.* Cornell University Press, Ithaca, 1945.
29. Smart, B.E. In: Banks, R.E., Smart, B.E. and Tatlow, J.C. (eds), *Organofluorine chemistry. Principles and commercial applications*, Chap. 3, Plenum Press, New York, 1994.
30. *Selected Values of Properties of Hydrocarbons and Related Compounds*, API Research Project 44, Thermodynamics Research Centre, Texas A&M University, College Station, TX 1978.
31. Jin, Y., Boller, A. *et al. Thermochim. Acta,* 1994, **234**, 103.
32. See for example, Atkins, P.W. *Physical chemistry*, 5th edn, p. 134. Oxford University Press, Oxford, 1994.
33. Bastiansen, O. and Hadler, E. *Acta Chim. Scand.* 1952, **6**, 214.

34. Lo Nostro, P. *Adv. Colloid Interface Sci.*, 1995, **56**, 245.
35. Page 770 in Ref. 32.
36. Reed III, T.M. In: Simons, J.H. (ed.), *Fluorine chemistry*, Chap. 2. Vol. V, Academic Press, New York, 1964.
37. Green, S.W. *Chem. and Ind.*, Jan. 1969, 63.
38. Oliver, G.D., Blumkin, S. and Cunningham, C.W. *J. Am. Chem. Soc.*, 1951, **73**, 5722.
39. McLure, I.A., Soares, V.A.M. and Edmonds, B. *J. Chem. Soc. Faraday Trans. 1*, 1982, **78**, 2251.
40. Mukerjee, P. and Handa T. *J. Phys. Chem.*, 1981, **85**, 2298.
41. Aveyard, R. and Haydon, D.A. *Trans Faraday Soc.*, 1965, **61**, 2255.
42. Ravey, J.C. and Stebe, M.J. *Colloids and Surfaces A*, 1994, **84**, 11.
43. van Os, N.M., Haak, J.R. and Rupert, L.A.M. *Physico-chemical properties of selected anionic, cationic and nonionic surfactants*. Elsevier, Amsterdam, 1993.
44. Shinoda, K., Hato, M., and Hayashi, T. *J. Phys. Chem.*, 1972, **76**, 909.
45. Ravey, J.-C., Gherbi, A. and Stebe, M.J. *Prog. Colloid Polymer Sci.*, 1988, **76**, 234.
46. Matos, L., Ravey, J.-C. and Serratrice, G. *J. Colloid and Interface Sci.*, 1989, **128**, 341.
47. Selve, C. Ravey, J.-C. *et al. Tetrahedron*, 1991, **47**, 411.
48. Achilefu, S. Selve, C. *et al. Langmuir*, 1994, **10**, 2131.
49. Kunieda, H. and Shinoda, K. *J. Phys. Chem.*, 1976, **80**, 2468.
50. Burkitt, S.J., Ottewill, R.H. *et al. Colloid and Polymer Sci.*, 1987, **265**, 619.
51. Berr, S.S. and Jones, R.R.M. *J. Phys. Chem.*, 1989, **93**, 2555.
52. Ravey, J.-C. and Stebe, M.-J. *Progr. Colloid & Polymer Sci.*, 1987, **73**, 127.
53. Muller, N. and Birkhahn, R.H. *J. Phys. Chem.*, 1967, **71**, 957.
54. Muller, N. and Birkhahn, R.H. *J. Phys. Chem.*, 1968, **72**, 583.
55. Muller, N. and Johnson, T. W. *J. Phys. Chem.*, 1969, **73**, 2042.
56. Muller, N. and Platko, F.E. *J. Phys. Chem.*, 1971, **75**, 547.
57. Muller, N., Pellerin, J.H. and Chen, W.W. *J. Phys. Chem.*, 1972, **76**, 3012.
58. Aveyard R., Cooper, P. and Fletcher, P.D.I. *J. Chem. Soc. Faraday Trans.*, 1990, **86**, 3623.
59. Lin, I.J. *J. Phys. Chem.*, 1972, **76**, 2019.
60. Guo, W. Li, Z. *et al. J. Phys. Chem.*, 1992, **96**, 6738.
61. Evans, H.C. *J. Chem. Soc.*, 1956, 579.
62. Holland, P.M. and Rubingh, D.N. (eds) *Mixed surfactant systems*. ACS Symposium Series 501, ACS, Washington, 1992.
63. Ogino, K. and Abe, M. (eds), *Mixed surfactant systems*. Surfactant Science Series, vol. 46. Marcel Dekker, New York, 1993.
64. Shinoda K. and Nomura, T. *J. Phys. Chem.*, 1980, **84**, 365.
65. Ravey, J.-C., Gherbi, A. and Stebe, M.J. *Progr. Colloid Polym. Sci.*, 1991, **84**, 95.
66. Muto, Y., Esumi, K. *et al. J. Colloid Interface Sci.*, 1987, **120**, 162.
67. Funasaki, N. and Hada, S. *J. Phys. Chem.*, 1983, **87**, 342.
68. Funasaki, N., Hada, S. and Neya, S. *Bull. Chem. Soc. Japan*, 1983, **56**, 3839.
69. Jarvis, N.L. and Zisman, W.A. *J. Phys. Chem.*, 1959, **63**, 727.
70. Ellison, A.H. and Zisman, W.A. *J. Phys. Chem.*, 1959, **63**, 1121.
71. Jarvis, N.L. and Zisman, W.A. *J. Phys. Chem.*, 1960, **64**, 150.
72. Abe, M., Morikawa, K. and Ogino, K. *Langmuir*, 1992, **8**, 763.
73. Sawada, H., Itoh, N. *et al. Langmuir*, 1994, **10**, 994.
74. Binks, B.P., Fletcher, P.D.I. and Thompson, R.L. University of Hull, unpublished data.
75. Gaines Jr., G.L. *Langmuir*, 1991, **7**, 3054.
76. Napoli, M., Fraccaro, C. *et al. J. Fluorine Chem.*, 1991, **51**, 103.
77. Binks, B.P., Fletcher, P.D.I. *et al. Langmuir*, 1995, **11**, 977.
78. Hopken, J. and Moller, M. *Macromolecules*, 1992, **25**, 2482.
79. Turberg, M.P. and Brady, J.E. *J. Am. Chem. Soc.*, 1988, **110**, 7797.
80. Lo Nostro, P. and Chen, S.-H. *J. Phys. Chem.*, 1993, **97**, 6535.
81. Morita, M., Matsumoto, M. *et al. Colloids and Surfaces*, 1992, **67**, 81.
82. Mukerjee, P. and Handa, T. *J. Phys. Chem.*, 1981, **85**, 2298.
83. Robert, A. and Tondre, C. *J. Colloid Interface Sci.*, 1984, **98**, 515.
84. Stebe, M.-J., Serratrice, G., and Delpuech, J.-J. *J. Phys. Chem.*, 1985, **89**, 2837.
85. Serratrice, G. Matos, L. *et al. J. Chim. Phys.*, 1990, **87**, 1969.
86. Bongiovanni, R., Borgarello, E. *et al. Colloids and Surfaces*, 1990, **48**, 277.

87. Schubert, K.-V. and Kaler, E.W. *Colloids and Surfaces A*, 1994, **84**, 97.
88. Griffin, W.C. *J. Cosmet. Chem.*, 1949, **1**, 311.
89. Shinoda, K. and Friberg, S. *Emulsions and solubilization.* Wiley, New York, 1986.
90. Blake, T.D. In: Tadros, Th.F. (ed), *Surfactants*, Chap. 8. Academic Press, London, 1984.
91. Zisman, W.A. In: *Contact angle, wettability, and adhesion*, Advances in Chemistry Series, 43, Chap. 1. ACS, Washington, 1964.
92. Pyter, R.A., Zografi, G. and Mukerjee, P. *J. Colloid Interface Sci.*, 1982, **89**, 144.
93. Fowkes, F.M. *J. Phys. Chem.*, 1953, **57**, 98.
94. Burnett, M.K. and Zisman, W.A. *J. Phys. Chem.*, 1959, **63**, 1241.
95. Murphy, W.J., Roberts, M.W. and Ross, J.R.H. *J. Chem. Soc. Faraday Trans. 1*, 1972, **68**, 1190.
96. Rutherford, C.E. Ph.D. Thesis, University of Hull, 1995.
97. Hare, E.F., Shifrin, E.G. and Zisman, W.A. *J. Phys. Chem.*, 1954, **58**, 236.
98. Shafrin, E.G. and Zisman, W.A. *J. Phys. Chem.*, 1960, **64**, 519.
99. Shafrin, E.G. and Zisman, W.A. *J. Phys. Chem.* 1962, **66**, 740.
100. Bernett, M.K. and Zisman, W.A. *J. Phys. Chem.*, 1967, **71**, 2075.
101. Riess, J.G. and Le Blanc, M. In: Lowe, K.C. (ed.), *Blood substitutes.* VCH, Cambridge, 1988.

6 Siloxane surfactants

R.M. HILL

6.1 Introduction

Siloxane surfactants consist of a methylated siloxane hydrophobe coupled to one or more polar groups. This class of surfactants finds a variety of uses in areas where other types of surfactants are relatively ineffective. Substantial advances in our understanding of this class of surfactants have taken place in recent years including their aqueous phase behavior, their unusual, rapid wetting of hydrophobic substrates, and their use as agricultural adjuvants. This survey will attempt to describe the structure, preparation, and surfactancy properties of this fascinating class of surfactants incorporating these recent advances.

Siloxane surfactants are different from conventional hydrocarbon surfactants in three significant respects [1–3]:

1. They are surface active in non-aqueous media.
2. They are able to lower surface tension to 20 dynes/cm compared with about 30 for typical hydrocarbon surfactants, and
3. They are fluid to high molecular weights. This holds for polyoxyethylene groups shorter than about EO_{17} [4]. One implication of this is that this group of surfactants do not generally show either a Krafft point, or a 'gel' point for aqueous lamellar phase dispersions [5, 6].

Another characteristic of siloxane surfactants which contributes to their usefulness is the variety of synthetic routes available and the consequent diversity of possible structures [7].

The surface active character of the methylated siloxane portion of the molecule is essentially determined by the methyl groups—the siloxane backbone serves mostly as a flexible framework on which to attach the methyl groups [1, 8–11]. The surface energy of a methyl-saturated surface is about 20 dynes/cm [8]. This is also the lowest surface tension achievable using siloxane surfactants. In contrast, the surface energy of a methylene-saturated surface is much higher, about 32 dynes/cm [8]. Most hydrocarbon surfactants pack loosely at the air/liquid interface. The surface energy of such a surface is dominated by the methylene groups and therefore hydrocarbon surfactants typically achieve surface tensions of about 30 dynes/cm or higher. Thus, one fundamental difference between siloxane surfactants and hydrocarbon surfactants lies in the different surface energy of $-CH_3$ versus $-CH_2-$.

Siloxane surfactants are similar to hydrocarbon surfactants in many common features of surfactancy [1, 12–15]:

1. They show a break in the surface tension versus log activity (concentration) which is usually indicative of the onset of self-aggregation such as micelle formation.
2. Surface tension lowering and critical aggregation concentrations vary with molecular structure in the same way; larger hydrophobic groups lead to lower surface tension and smaller critical aggregation concentrations.
3. They show similar aggregation behavior in aqueous solution, forming the same types of aggregates and following the same trends with molecular structure.
4. Siloxane surfactants incorporating polyoxyethylene and mixed polyoxyethylene/polyoxypropylene groups also show inverse temperature solubility and cloud points.

Above their cloud points, some siloxane polyether copolymers act as defoamers [16]. However, the existence of a cloud point for a given siloxane surfactant should be viewed cautiously because many siloxane surfactants form cloudy dispersions in the dilute concentration range that are unrelated to the existence of a cloud point as it is usually understood [12].

Siloxane surfactants are used in a wide variety of applications including their use as foam stabilizers in plastic foams such as polyurethane foams, as wetting agents, emulsifiers, lubricants, release agents, and as antistatic agents. The largest uses are in the manufacture of polyurethane foam, textile treatments, personal care and cosmetic formulations, and as paint additives. They were introduced to the marketplace in the 1950s; their first (and still the largest) use was in the manufacture of polyurethane foam.

6.2 Molecular structures and nomenclature

Siloxane surfactants are amphiphilic materials containing a methylated siloxane hydrophobe coupled to one or more polar groups [17–19]. Polydimethylsiloxane (PDMS) is itself highly surface active [20] in organic solvents. Most siloxane surfactants are intermediate molecular weight (MW) copolymers with MWs of a few thousand. A variety of copolymers and terpolymers based on the PDMS backbone could be included under the umbrella of siloxane surfactants [4, 7]. Gruning and Koerner [7] suggest a fairly broad definition of siloxane surfactants not limited to molecules with well-defined and well-separated lyophilic and lyophobic parts. While I will try to direct the reader to some useful references describing these materials, this review will primarily focus on the structure, synthesis, and surfactant properties of materials which consist of a methylated siloxane portion

Table 6.1 Types and examples of polar groups in siloxane surfactants

Type of polar group	Examples
non-ionic	polyoxyethylene
	polyoxyethylene/polyoxypropylene
anionic	saccharide sulfate
cationic	quaternary ammonium salts
or zwitterionic	betaines

coupled to one or more polar or hydrophilic groups. Most siloxane surfactants are appropriately called copolymers, but one important group, the trisiloxane based surfactants, is not. The hydrophobic (or lyophobic) portion of these substances is the methylated siloxane moiety. The hydrophilic (or lyophilic) portion can be any of the groups indicated in Table 6.1.

There are three common types of siloxane surfactants: rake-type copolymers (also called comb- or graft-copolymers), ABA copolymers (in which the 'B' represents the siloxane portion), and trisiloxane surfactants. These are illustrated in Figures 6.1, 6.2 and 6.3. Structures incorporating branching (T-units in the siloxane backbone, see the notation below) and AB-type copolymers have also been made. Hydrophilic, higher alkyl, and fluorocarbon groups as well as combinations of these groups have been attached to the siloxane backbone to make materials that are surface active towards a variety of surfaces and interfaces.

Functional groups may be attached through either an Si-C or an Si-O-C linkage [21]. The most common hydrophilic group found in siloxane surfactants is the non-ionic polyoxyalkylene (PAO) group, either polyoxyethylene (PEO) or a random copolymer of polyoxyethylene and polyoxproylene (PPO). Note that PPO is hydrophobic and contributes to the hydrophobicity of the surfactant rather than to its hydrophilicity.

Siloxane polyoxyalkylene copolymers are known to industry by several names including siloxane polyethers (SPEs), polyalkyleneoxide silicone copolymers, silicone poly(oxyalkylene) copolymers, silicone glycol copolymers (or surfactants) [22]. The International Cosmetic Ingredient Nomenclature is dimethicone copolyol [23]. A technologically important variation

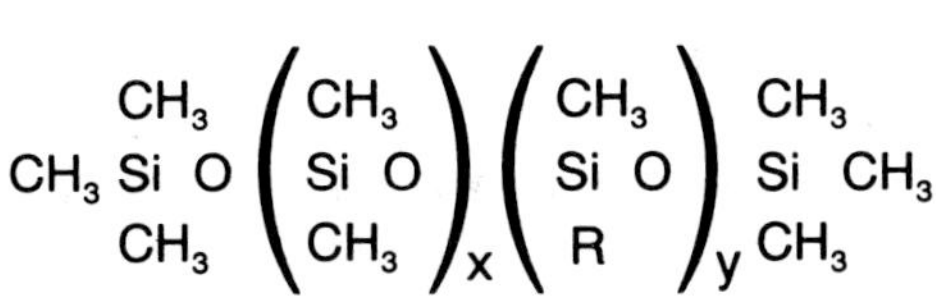

Figure 6.1 Rake-type siloxane surfactant. R stands for a polar organic group such as $-(OCH_2CH_2)_nOH$, an n-unit long polyoxyethylene chain, usually linked to the siloxane backbone through a $-(CH_2)_3-$ group.

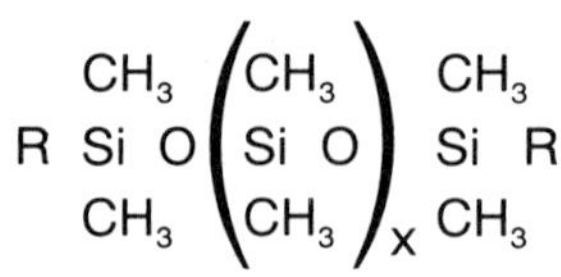

Figure 6.2 ABA type siloxane surfactant. R is the same as for the rake-type structures in Figure 6.1.

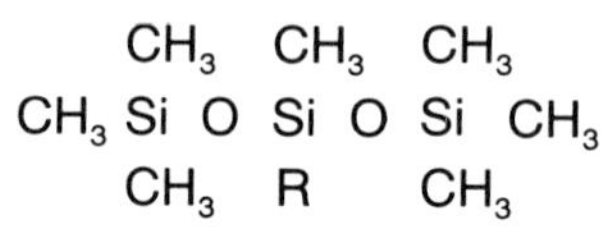

Figure 6.3 Trisiloxane surfactant. R is the same as for the rake-type structures in Figure 6.1.

of the comb-type structure is liquid crystal side chain polymers based on the polysiloxane backbone [24]. These form thermotropic liquid crystals which are of interest for liquid crystal display applications [25–27]. A few such polymers have been made which are water soluble and for which the lyotropic LC phase behavior has been reported [28, 29].

6.3 Synthesis of siloxane surfactants

Siloxane surfactants are prepared by attaching a polar organic group to a methylated siloxane species. There are three different ways to do this:

1. Directly by the reaction of SiOH and ROH (transesterification or esterification).
2. Directly by the reaction of SiH and $H_2C{=}CHCH_2R$ (hydrosilylation) where R is a polar organic group.
3. Indirectly by attaching a reactive group such as an epoxy to the siloxane backbone using reaction 2 and then using that reactive group to attach a polar group.

Non-ionic siloxane surfactants are usually prepared using the first and second routes, and ionic siloxane surfactants by the third route.

6.3.1 Preparation of the siloxane backbone

The first step in the synthesis of a siloxane surfactant is to prepare the methylated siloxane species with the appropriate SiOH or SiH functionality. There are two ways to do this, either cohydrolysis of the appropriate chlorosilanes or the equilibration reaction.

For example, to prepare a rake-type siloxane surfactant using route 2 above, the SiH functional siloxane backbone can be prepared by cohydrolysis of the appropriate chlorosilanes [21, 30]:

$$2(CH_3)_3SiCl + x(CH_3)_2SiCl_2 + y(CH_3)HSiCl_2 + H_2O \Leftrightarrow MD_xD'_yM + HCl \qquad (6.1)$$

This backbone could also be prepared by the equilibration of the appropriate proportions of:

$$MM + xD + yD' \Leftrightarrow MD_xD'_yM \qquad (6.2)$$

using either acid or base catalysis [21, 30]. The M, D, T, Q notation, commonly used among organosilicone chemists, stands for the following functional groups [31]:

M	$Me_3SiO_{1/2}-$
D	$-Me_2SiO-$
T	$-MeSiO_{3/2}-$
Q	$-SiO_2-$
M′	$Me_2(R)SiO_{1/2}-$
D′	$-Me(R)SiO-$
T′	$-(R)SiO_{3/2}-$
Me	$-CH_3$
R	H, or (after hydrosilylation) some non-methyl organic group such as $-CH_2CH_2CH_2(OCH_2CH_2)_nOH$

ABA structures, or mixed ABA and rake-type structures can be prepared by substituting M′M′ for MM in the above reaction. Branched siloxane backbones can be prepared by including T groups in this equilibration.

These reactions are well-known building blocks in the silicone chemistry set and are discussed in detail by Noll [21] and more recently by Clarson and Semlyn [30]. Aspects of these reactions especially relevant to preparing siloxane surfactants are discussed by Gruning and Koerner [7]. For a rake-type copolymer the equilibration reaction or co-hydrolysis yields three types of polydispersity: (a) a distribution of chain lengths, x + y; (b) a distribution in the number of D′ units per chain, y; and (c) random location of the D′ units along the chain. Note that (b) implies that some PDMS chains will be present with no D′ units in them. For an ABA type copolymer only (a) polydispersity is present. In contrast to much polymer chemistry, the molecular weight distribution produced by the equilibration reaction is determined by thermodynamics rather than by kinetics. This means that the resulting distribution is highly reproducible [7].

6.3.2 *Transesterification*

The first non-ionic siloxane surfactants were prepared by the reaction of alkoxymethylsiloxane polymers with hydroxy terminated polyoxyalkylenes [34].

$$\equiv SiOR_1 + R_2OH \Rightarrow \equiv SiOR_2 + R_1OH \qquad (6.3)$$

R_1 is usually $-CH_3$ or $-CH_2CH_3$, R_2 is a polyalkyleneoxide. This reaction yields products in which the hydrophilic groups are linked to the siloxane hydrophobe through an Si–O–C linkage. These materials proved to be useful in polyurethane foam manufacture, and other non-aqueous applications, but in water this linkage hydrolyzes rapidly to generate silanol and alcohol [34]. An esterification reaction can also be carried out on SiCl functional silanes yielding HCl as a by-product.

6.3.3 *Hydrosilylation*

More hydrolytically stable siloxane–polyoxyalkylene surfactants can be prepared by the hydrosilylation of methyl siloxanes containing some Si-H groups with vinyl functional polyoxyalkylenes [7,21,30,34]:

$$\equiv SiH + CH_2{=}CHCH_2(OCH_2CH_2)_nOR \Rightarrow \equiv Si(CH_2)_3(OCH_2CH_2)_nOR \tag{6.4}$$

This reaction is usually catalyzed using chloroplatinic acid (Spier's catalyst), but other forms of Pt and other metals have also been used [21]. Many examples of non-ionic siloxane polyether copolymer surfactants are available commercially, and are widely used in industry.

6.3.4 *Two-step synthesis using reactive intermediate*

Reactive, organofunctional siloxanes can be prepared by hydrosilylation of an appropriate Si-H functional siloxane with a suitable, vinyl-terminated functional organic compound. Common functional groups include halogens, amines, and epoxies [7,21,30,35]. These functional groups can then be converted into an appropriate lyophilic (usually hydrophilic) group by well-established chemistries. This is the preferred method of preparing ionic siloxane surfactants although non-ionic surfactants have also been prepared by this route [36]. For example, quaternization of halide functional siloxanes leads to cationic siloxane surfactants:

$$\equiv Si(CH_2)_3X + NR_3 \Rightarrow \equiv Si(CH_2)_3N^+R_3X^- \tag{6.5}$$

where X = halide [37 and references therein]. Anionic siloxane surfactants can be prepared by the sulfonation of epoxy-functional silicones as shown below [38]:

$$\begin{aligned} &\equiv Si(CH_2)_3OCH_2CH(O)CH_2 + NaHSO_3 \\ &\quad \Rightarrow \equiv Si(CH_2)_3OCH(OH)CH_2SO_3^- Na^+ \end{aligned} \tag{6.6}$$

Zwitterionic siloxane surfactants can be prepared by the ring opening of cyclic sultones by aminofunctional silicones as shown below [35]:

$$\equiv Si(CH_2)_3N(CH_3)_2 + O(CH_2)_3SO_2 \Rightarrow$$
$$\equiv Si(CH_2)_3N^+(CH_3)_2(CH_2CH_2CH_2SO_3^-) \qquad (6.7)$$

Other examples of reactions suitable for the preparation of ionic siloxane surfactants are given by Snow *et al.* [34,37,39,40], Maki *et al.* [41], and by Gruning and Koerner [7]. Many examples of such ionic surfactants have been prepared and investigated, and some are available commecially [16].

6.3.5 Organophilic siloxanes and terpolymers

Organophilic or fluorocarbon functionality can be incorporated into a siloxane surfactant using the same basic chemistry. Siloxane surfactants incorporating both hydrophilic and organophilic functionality have been prepared [7, 42]. The use of siloxane polymers with long chain alkyl groups ($>C_{10}$) attached to the siloxane backbone, called silicone waxes, in personal care applications is discussed by Schaefer [16]. Such waxes are somewhat soluble in organic media, and may be surface active in (for example) mineral oil. They are claimed to improve spreadability, lustre and pigment dispersion.

This brief review of the synthetic approaches to siloxane surfactants is by no means comprehensive, but it should serve to illustrate the primary chemistries involved. Gruning and Koerner [7] catalog a number of possible modifying groups.

6.4 Hydrolytic stability

The hydrolytic stability of siloxane surfactants, especially the trisiloxane surfactants, is mentioned by many papers in this field and is viewed as a serious problem by some seeking to make use of siloxane surfactants [43, 44 and references therein]. The Si–O–Si linkage is susceptible to hydrolysis in the presence of water [1,7,31,43]:

$$\equiv Si{-}O{-}Si \equiv \Leftrightarrow \equiv Si{-}OH + HO{-}Si \equiv \qquad (6.8)$$

This equilibrium is catalyzed by acid or base, and is slow near neutral pH [7, 43]; it is essentially a re-equilibration of the siloxane backbone (see equation 6.2 above). It is known in the silicone chemistry literature that residual acidity or basicity of glassware surfaces (from cleaning solutions, for example) can influence this reaction requiring that careful work be done in plasticware, or glassware rigorously treated with a hydrophobizing agent such as octyltrichlorosilane.

Nevertheless, there is very little actual analytical data published regarding the rate of hydrolysis of siloxane surfactants in aqueous media. For instance, Gradzielski *et al.* [1] observed that the trisiloxane surfactants they worked with hydrolyzed completely within a few weeks whereas the polymeric siloxane surfactants were stable for at least a few months. However, no analytical results were presented. Sturmer *et al.* [14] noted that the aqueous phase behavior of their $M(D'E_7OH)M$ changed somewhat after a short period of time which they attributed to hydrolysis. This change took substantially longer to occur in D_2O than in H_2O. Hill [13] and He [15], studying the same class of surfactants did not observe such short time changes.

Hill *et al.* [12] noted a substantial change in the observed phase behavior of one of four polymeric siloxane surfactants accompanied by only about 5% (determined by ^{29}Si NMR) hydrolysis. Apparently, such changes in behavior are not reliable guides to the extent of hydrolysis. It is vital that further discussion of the hydrolysis of this class of surfactants be based on quantitative analytical determination of actual hydrolysis rather than merely being inferred from changes in surface tension of aqueous phase behavior.

The effect that hydrolysis has on the surfactancy of a siloxane polyether copolymer will depend on (1) how much trimethyl species is present, and (2) how randomly cutting (and reassembling) the chain affects the overall hydrophile–lipophile balance of the mixture. For trisiloxane surfactants a large proportion of trimethyl species is present. Hydrolysis forms trimethyl silanol which recombines to form hexamethyldisiloxane, MM [7]. This species either segregates into the micelle core, or phase separates, which drives the reaction forward. Polymeric surfactants with long linear segments might be expected to form cyclosiloxanes. The rates of these reactions are not known but appear to be quite slow near $pH = 7$. Knoche *et al.* [43] determined a half life for dilute $M(D'E_{7.5}OMe)M$ of >40 days.

The rate of hydrolysis will obviously depend on the amount of contact between the aqueous solution and the siloxane portion of the surfactant molecules. Since surfactants self-associate to form aggregates in which water is largely excluded from the aggregate core, it is reasonable to assume that only unassociated surfactant molecules are susceptible to hydrolysis [4,7,43]. Thus the rate of hydrolysis should depend on the concentration of unassociated surfactant molecules which is always less than or equal to the critical aggregation concentration (CAC or CMC). An induction period is sometimes observed during which the concentration of surface active species decreases to the CAC [43]. Only after this has occurred do surface tension and wetting properties noticeably deteriorate. Thus concentrated solutions may be stable for long periods of time. Polymeric siloxane surfactants are generally more hydrolytically stable because (1) their CAC is very small, and (2) cutting and recombining the siloxane chains only slowly generates species which are either volatile, insoluble, or much more or much less hydrophobic than the starting material. Under strongly acid or basic conditions or at sustained temperatures $>70°C$ hydrolysis leads to rapid loss of surfactancy of most siloxane surfactants [43].

6.5 Surface tension lowering and adsorption

6.5.1 Non-aqueous media

The surface activity of PDMS in non-aqueous media such as the polyols used in polyurethane foam manufacture is a combination of the low surface energy of the methyl-rich siloxanes species and insolubility determined by molecular weight [7–11, 34]. It is well known that the solubility of a polymer in a given solvent is a combination of solvent–solute interactions and the molecular weight of the polymer [45]. Lower MW PDMS is soluble in such organic solvents as toluene. As the MW increases, the solubility decreases. In the MW range in which the polymer is just marginally soluble, it shows notable surface activity, even foam stabilization [2, 8]. The lyophobic character of the methylated siloxane portion of siloxane surfactants in non-aqueous media is partly determined by this molecular weight effect. This means that a higher molecular weight siloxane surfactant will have a greater amphiphilic character even with the same proportions of (for example) PDMS and polyoxyethylene. In non-aqueous media this can mean the difference between significant surface activity and solubility [7, 16].

Schmidt [49] presents some results illustrating the ability of siloxane polyether copolymers to lower the surface tension of a variety of organic liquids including mineral oil and several polyols from values of about 25–30 dyne/cm to values near 21 dynes/cm. Kendrick and Owen [10, 11] show how siloxane surfactants lower the surface tension of polyols used in the manufacture of polyurethane foams. They concluded that the polysiloxane chain is looped away from the surface with the ends of the loops anchored to the surface by the polyether segments. In order to lower surface tension to the 20–21 dyne/cm value it was found to be necessary to have dimethylsiloxane segments at least 10 monomer units long between each pendant polyether group, in other words the D/D′ ratio of the siloxane backbone must be at least 10.

6.5.2 Aqueous media

Many authors have presented aqueous surface tension results for siloxane surfactants. Three of these investigated a wide range of molecular structures and will be summarized here. Vick [22] presented surface tension results for three homologous series of rake-type polymeric non-ionic siloxane surfactants. He found that the surface tension (of 1% aqueous solutions) increased with increasing size of the siloxane hydrophobe, and with increasing polyoxyethylene chain length but not with weight fraction of polyoxyethylene. Siloxane surfactants containing mixed EO/PO groups generally gave higher surface tensions than all-EO surfactants. The lower surface tension values were about 21 dynes/cm, which is also the surface tension of a medium to high molecular weight silicone polymer [2]. This is considerably below values achievable using organic surfactants, but higher than fluorocarbon surfactants. Values for solutions of polymeric siloxane surfactants are typically between 20 and 32 dynes/cm.

Kanner *et al.* [34] prepared a number of low molecular weight siloxane polyoxyethylene surfactants and characterized their surfactancy, and wetting properties. These surfactants included the following siloxane hydrophobes: MM′, M′M′, MD3M′, MD6M′, MD9M′, MDD′DM, D3D′, and M3T′. EO hydrophilic groups 1–17.5 units long were attached to these hydrophobes. Most of the structures prepared had –OMe endcaps on the polyoxyethylene groups. (Not all of the possible permutations of these structures were actually prepared.) Many of these surfactants gave aqueous surface tensions <25 dynes/cm which is lower than can be achieved using organic surfactants. The most surface active siloxane surfactants were those with the smallest siloxane groups [34] having only 2–5 silicon atoms. Branching of the siloxane hydrophobe and variation of the polyoxyethylene chain length had relatively minor effects on the surface activity. Replacing methyl groups with longer alkyl groups resulted in higher surface tensions. For those surfactants forming clear aqueous solutions, aggregation numbers, deduced from vapor pressure osmometry, were smaller than is typical for organic surfactants. These authors attributed turbidity to the sample being above its cloud temperature. This usually means a dispersion of isotropic liquid surfactant in aqueous solution. We now know that in many cases (including many of the structures prepared by Kanner *et al.* [34]) these turbid solutions are dispersions of lamellar phase liquid crystal or vesicles [12].

Gentle and Snow [46] investigated the adsorption of a homologous series of trisiloxane surfactants $M(D'E_nOH)M$ with $n = 4$–20. They found that the CAC, surface tension at the CAC, and area per molecule varied with molecular structure in a way that was consistent with their ‘umbrella’ model for the shape of the trisiloxane hydrophobe at the air/water interface. The log(CAC) and the surface tension at the CAC both increased linearly with EO chain length.

6.5.3 *Interfacial tension lowering*

Interfacial tensions of aqueous solutions of trisiloxane surfactants against n-decane are about 1 dyne/cm [5, 14], similar to values given by hydrocarbon surfactants. Gradzielski [1] found the polymeric siloxane surfactants they investigated tended to give higher values against alkanes, up to 10 dynes/cm. Kanellopoulos and Owen [47] investigated the adsorption of a group of siloxane surfactants at the silicone oil/water interface and found interfacial tensions down to about 4 dyne/cm. Maki *et al.* [36] found interfacial tensions between silicone oils and aqueous siloxane surfactant solutions in the range 3–6 dynes/cm. Bailey *et al.* [18] found interfacial tensions 'which approach zero' for solutions of AB copolymer siloxane surfactants against silicone oils. Svitova *et al.* [48] measured interfacial tensions between a series of normal alkanes and 0.5 w% solutions of $M(D'E_8OH)M$ and found that the interfacial tension varied linearly between about 0.025 dynes/cm for hexane to 0.4 dynes/cm for hexadecane. These small values indicate that trisiloxane surfactants are very effective at lowering interfacial tensions against low energy hydrocarbon substrates, which is consistent with their ability to cause spreading over such substrates (see next section). Polymeric siloxane surfactants are less effective.

6.6 Superwetting

The ability of siloxane surfactants to enhance wetting plays an important role in many different aqueous [22, 49] and non-aqueous [44] applications. One common way to characterize wetting which is useful in textile and in household and personal care applications is the Draves wetting test in which the time to wet (or sink) of a weighted skein of cotton is measured. Cotton is a relatively hydrophilic surface. Vick [41] found that the time to wet (Draves test) depended on the size of the siloxane hydrophobe and the length of the EO group—the most rapid wetting was observed for the surfactants with the shortest siloxane groups and the smaller EO groups. Several authors [22, 50, 52] have noted that surface tension of surfactant solutions does not correlate with wetting time or degree of spreading.

Dilute aqueous solutions of certain siloxane surfactants are able to rapidly wet quite hydrophobic surfaces such as polyethylene or Parafilm [13, 31–33, 48, 50–57]. Kanner *et al.* [34] noted that siloxane surfactants based on hydrophobes containing 2–5 silicon atoms, and having a limited but finite solubility in water, were the 'best' wetting agents on polyethylene. This unusual phenomenon has been termed 'superwetting' or 'superspreading' and appeared to be unique to a very small group of siloxane surfactants [13, 31–34, 50–54]. All three commercially available trisiloxane superwetter surfactants are variations of the structure shown in Figure 6.3 with n = 7–8 and $-OH$, $-OCH_3$, and $-OCH(O)CH_3$ end capping groups [13].

Gradzielski *et al.* [1] attributed the rapid wetting properties of siloxane surfactants to low adhesive forces between individual molecules in interfacial films. Ananthapadmanabhan and Goddard [50–52] attributed the phenomenon to the unusual shape of this molecule compared with conventional hydrocarbon surfactants. Zhu *et al.* [53, 54] found a maximum in spreading rate as a function of surfactant concentration and showed that the spreading rate was linearly related to dispersion turbidity and sensitive to humidity. The requirement of high relative humidity was related to the need for a pre-existing water film on the surface [53–56]. Lin *et al.* [56] confirmed the concentration maximum using a quartz crystal microbalance (QCM) to measure spreading rates and showed that there was also a maximum in spreading rate as a function of substrate surface energy.

Recently, these characteristic features, (1) a maximum in spreading rate as a function of surfactant concentration and substrate surface energy, and (2) rapid spreading on relatively hydrophobic substrates, have been shown to be general features of surfactant enhanced spreading [57]. It is not necessary to have a turbid dispersion or high humidity. Only the trisoloxane surfactants are able to wet the most hydrophobic surfaces such as polyethylene or Parafilm®. While this characteristic profile is general, recent measurements of dynamic surface and interfacial tensions for siloxane and hydrocarbon surfactants suggest that the rate of spreading is influenced by the rate of surfactant adsorption which in turn is related to the type of surfactant microstructures present in the solution or dispersion [48, 58]. Working on liquid alkane surfaces, Svitova *et al.* [48, 58] have shown that rapid spreading correlates with a positive dynamic spreading coefficient at short time scales. Adsorption isotherms for trisiloxane superwetter surfactants have not been determined, but Tiberg and Cazabat [59] showed that the molecular orientation and organization of these surfactants is quite different on hydrophilic and hydrophobic surfaces. It is possible that the decrease in spreading rate with increasing hydrophilicity is due to this difference. The cause of the maximum as a function of concentration remains uncertain at this time.

Whether a liquid (A) will spread over a solid substrate or over another liquid (B) is determined by the sign of the spreading coefficient:

$$S = \gamma_B - \gamma_A - \gamma_{AB} \tag{6.9}$$

(S is defined in terms of A spreading over B). Thus the condition for spreading requires not only a low surface tension for the surfactant solution, but also a low interfacial tension between A and B as well. This is why solutions of fluorocarbon surfactant solutions, which have lower surface tensions than do solutions of siloxane surfactants, do not spread over hydrocarbon hydrophobic surfaces; the interfacial tension between the fluorocarbon surfactant solution and the hydrocarbon substrate is not small enough. The spreading coefficient is a thermodynamic condition for spreading; the dynamics of spreading are strongly dependent on hydrodynamics [60] and surfactant transport to the interfaces [61]. The spreading coefficient must be positive under dynamic conditions for spreading to occur but there is probably not a general correlation between the magnitude of the spreading coefficient and the rate of spreading.

Three fundamental issues remain unresolved with respect to this phenomenon: (i) the origin of the maximum spreading rate versus substrate surface energy, (ii) the origin of this maxinum versus surfactant concentration, and (iii) the humidity dependence [57].

6.7 Aqueous aggregation behavior

Most early papers on siloxane surfactants dealt with their surface properties, that is, such things as surface tension, adhesion, lubrication, and skin feel. Only recently have the aggregation properties of this class of surfactants in aqueous solution received much attention. It was occasionally argued [7, 32] that because of the fluidity of the siloxanes species, this class of surfactants should not form 'crystalline monolayers' or liquid crystal phases [1].

Surfactants in aqueous solution self-associate to form micelles or vesicles at low concentration and liquid crystal phases at higher concentrations [5, 62]. The type of aggregate or liquid crystal phase formed by a particular surfactant, and the progression of liquid crystal phases formed with increasing surfactant concentration, temperature, and salt level can be rationalized using a simple model based on molecular packing [63, 64]. Several studies of the aqueous aggregation properties of ionic and polyoxyethylene-based non-ionic siloxane surfactants have recently been reported [1, 12–15, 37, 40, 65–68] and will be summarized here.

Hill *et al.* [12, 13], Snow [37], and Sturmer *et al.* [14] observed globular and worm-like micelles, vesicles, and liquid crystal (LC) phases in solutions of siloxane surfactants. Figure 6.4 shows an example of a dispersion of a siloxane surfactant which forms an unusual diversity of globular and

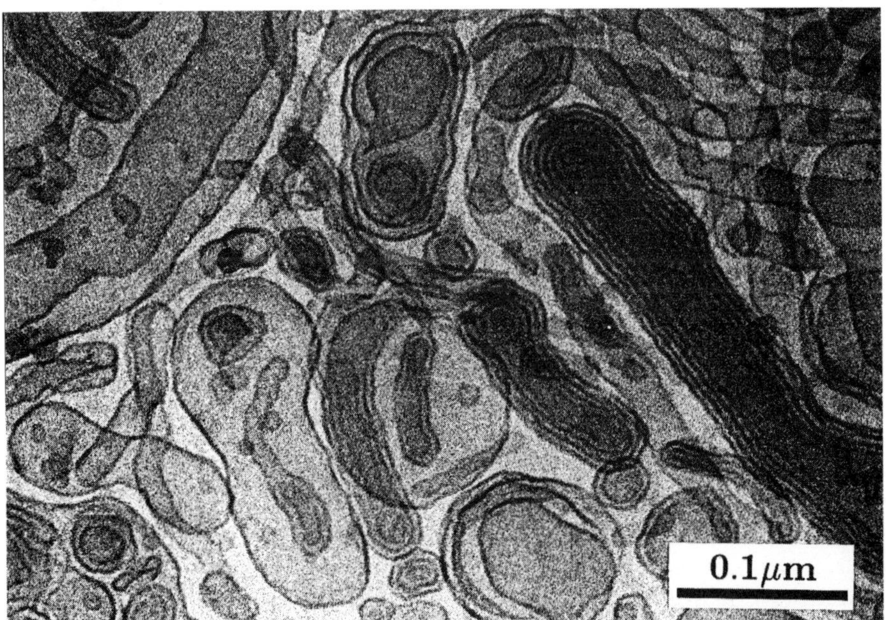

Figure 6.4 cryo-TEM micrograph of vesicles in a 10 wt% solution of the ABA siloxane polyoxyethylene copolymer surfactant $E_8D_{15}E_8$ (from Reference 12). Used with permission.

tube-like unilamellar and multilamellar vesicles [12, 66]. Gradzielski *et al.* [1] found globular micelles and hexagonal and lamellar LC phases in aqueous solutions of several anionic, cationic, and non-ionic siloxane surfactants. From micellar radii results they concluded that the siloxane chain must be coiled in the aggregates.

Lühmann and Finkelmann [28, 29] investigated the lyotropic LC phase behavior of several amphiphilic liquid crystal side chain polymers in which an amphiphilic molecule was attached to each monomer unit of a siloxane backbone. They found that these polymers formed the usual progression of cubic-hexagonal-lamellar phases. Yang and Wegner [69, 70] investigated several ABA-type siloxane surfactants and found a solubility boundary at high temperature, a broad mesophase region below this boundary, and the usual progression of cubic-hexagonal-lamellar phases. Thus, in spite of the random coiled configuration of the polymer chains, siloxane surfactants self-associate in water to form the usual micelles, vesicles, and ordered liquid crystal phases.

Like conventional hydrocarbon surfactants containing EO and PO groups, siloxane polyoxyalkylene surfactants become less soluble in water with increasing temperature [15, 22]; the temperature at which they become insoluble, and the solution becomes cloudy is called the cloud temperature, or cloud point [71, 72]. The general dependence of the cloud point on the weight fraction EO [22] and the endcap of the EO group [13] appears to be similar to hydrocarbon surfactants. The effects of electrolytes also follow the usual relationships [73]. Fink [74] found that the cloud temperature correlated better with EO chain length than with weight fraction EO. Above their cloud temperature, siloxane polyoxyalkylene surfactants function as antifoams [74]. Unlike hydrocarbon surfactants, these relationships cannot be simply related to molecular structure using the HLB system [22] although HLB values are sometimes quoted [23]. No simple empirical relationship analogous to the HLB system has been developed yet to predict the emulsifying or cloud point behavior of siloxane surfactants [22].

Fink found that the foam height generated by a series of siloxane surfactants in the Ross Miles test decreased rapidly with increasing size of the siloxane hydrophobe, and increased with increasing EO content. In all cases, the EO/PO containing siloxane surfactants gave a light, coarse foam, compared with the more dense, fine-celled foams generated by the all-EO surfactants.

6.8 Ionic siloxane surfactants

Gradzielski *et al.* [1] reported a binary phase diagram for an anionic trisiloxane surfactant. This surfactant formed a wide region of lamellar phase which was stable to temperature in the range they worked. Snow *et al.* [35, 37, 39, 40, 68] reported the synthesis and aqueous aggregation behavior of a series of trisiloxane cationic and zwitterionic surfactants. Snow found that these surfactants self-associate to form micelles and vesicles and form lamellar liquid crystal, and a hydrated bilayer structure, at high surfactant concentration. They also investigated the effect of several electrolytes on the phase behavior of these cationic trisiloxane surfactants, and found that salt promotes a transition from micelles to vesicles at low concentrations.

Maki *et al.* [41] reported the synthesis, surface tension, foaming, and bacteriostatic properties of several cationic organosilicon surfactants with methyl siloxane (Si–O–Si), methyl silmethylene, or carbosilane (Si–C–Si), and methyl silane (Si–Si) hydrophobic groups. Surface tensions were lowered to 22–23 dynes/cm, several of the synthesized materials were excellent foamers, and some exhibited bacteriostatic activity.

6.9 Non-ionic siloxane surfactants

The aqueous phase behavior of the homologous series of surfactants, $M(D'E_nOR)M$, has been studied for $n = 5$, 7.5/8, 12, 16, and 18 [1, 13–15, 65]. The $n = 5$ species was monodisperse [15], the others were polydisperse in the polyoxyethylene portion. For $n = 7.5/8$, R = H, CH_3, and $C(O)CH_3$ end capping groups were studied [13]. The hydrophobicity of the MD′M group has been shown to be comparable to that of a linear $C_{12}H_{25}$ group [1]. However, its shape is quite different—it is shorter and wider—its length is about 9.7 Å compared with 15 Å for $C_{12}H_{25}$ while its volume is larger, 530 Å^3 compared with 350 Å^3 for $C_{12}H_{25}$. The general patterns of phase behavior of the trisiloxane surfactants are similar to that of the C_iE_j series [5, 64], but because of the different size and shape of the MD′M hydrophobe, the phase behavior is shifted to longer EO groups. For instance, the phase behavior of $M(D'E_8OH)M$ is most similar to that of $C_{12}E_5$ [13]. Stürmer *et al.* [14] found evidence of a highly swollen lamellar phase at about 3–5% surfactant for the $M(D'E_8OH)M$/water system. The effect of polydispersity in the EO chain length is apparently to shift the phase behavior towards more hydrophilic, i.e. longer EO surfactants [15, 67]. The aqueous phase behavior of $MDM'E_8OH$, a 'linear' trisiloxane surfactant, was shown to be essentially identical to that of the 'branched' version, $M(D'E_8OH)M$ [13]. This was unexpected since the shape of these two molecules should be somewhat different. This illustrates the important role that the flexibility of the Si–O–Si chain plays in the behavior of this class of surfactants; the siloxane group tends to adopt a configuration which allows it to accommodate the preferred cross-sectional area of the polar group, in this case the coil of the hydrated polyoxyethylene chain.

The non-ionic trisiloxane surfactant M(D′E_{12}OH)M forms clear isotropic solutions in water at all compositions near room temperature [65]. Below 10°C a hexagonal (H_1) and a lamellar (L_α) liquid crystal phase are found. The minimum of the lower critical solution temperature (LCST) curve is at about 45°C. A small region of L_3, the 'sponge' phase is found inside the LCST curve [67]. He *et al.* [65] used a combination of electron microscopy, X-ray and neutron scattering, rheology, and NMR self-diffusion measurements to investigate the microstructure of the isotropic solutions formed by this system near room temperature. They conclude that this system forms small globular micelles at low concentrations which evolve to entangled worm-like micelles and then a random bilayer structure which persists to 100% surfactant.

Maki *et al.* [36, 75–79] reported the synthesis and surface activity of several non-ionic and cationic organosilicon surfactants with methyl siloxane, methyl silmethylene, and methyl silane hydrophobic groups. Non-ionic surfactants containing 1, 2, and 3 poly(dimethylsiloxane) chains as hydrophobic groups were synthesized to study the relation between the structures and the surface active properties [76, 78]. Surface tensions of the aqueous solutions were 20–21 dynes/cm. Interfacial areas/molecule were calculated and the results were discussed in terms of chain configurations. Similar cationic surfactants were also prepared and characterized [79]. Non-ionic surfactants were prepared from glycidyl ethers containing a series of dimethylpolysiloxanes [36]. Surface tensions in aqueous solution were 22–23 dynes/cm. They were also surface active in polyols. Some of the surfactants were effective in lowering the interfacial tension between water and a silicone oil, and in wetting low energy hydrophobic surfaces such as a polyethylene film. In contrast to the methyl siloxane derivatives, the methyl silmethylene surfactants only lowered aqueous surface tension to 25–26 dynes/cm, were less surface active in polyols, and spread less readily [77]. However, interfacial tensions between water and a silicone oil were somewhat lower than the siloxane derivatives.

Kanner *et al.* [34] reported solubility data for several polymeric and non-polymeric siloxane surfactants. More detailed aqueous phase behavior studies have been reported for a rather diverse group of rake-type and ABA polymeric siloxane surfactants [1, 12, 14, 66]. Dilute solutions of these surfactants contain worm-like micelles and vesicles [12, 66]. Both globular and tubular vesicles have been observed [12]. Vesicle surfaces often appear rough suggesting a high degree of bilayer flexibility. The ability of such branched and polydisperse molecular structures to form well-defined surfactant bilayers (and liquid crystal phases) is thought to be due to the unusual flexibility of the siloxane backbone [12]. At higher concentrations, liquid crystal phases are formed ranging from the normal hexagonal phase, H_1, to the lamellar phase, L_α, to the inverse hexagonal phase, H_2 [1, 12, 14]. As is the case for the trisiloxane surfactants, the trends in the aqueous phase behavior of polymeric siloxane surfactants can be rationalized in terms of a simple molecular packing calculation [12].

Stürmer *et al.* [14] reported the aqueous aggregation behavior of four relatively low MW rake type polymeric siloxane surfactants, including two non-ionics, a zwitterionic and a cationic example. Three of these surfactants formed bilayer structures. The phase boundaries of the cationic and the zwitterionic were insensitive to temperature, while the lamellar phase formed by the non-ionic example melted at about 50°C. The more hydrophilic non-ionic formed globular micelles and a hexagonal phase [1]. Sturmer *et al.* [14] point out that an associated polysiloxane surfactant molecule in water should adopt a conformation in which the siloxane backbone folds up to minimize hydrophobic contact with water. This is illustrated in Figure 6.5. The flexibility of the siloxane backbone would be expected to facilitate this folding. Addition of a second molecule increases the volume of the hydrophobic core, and the number of polar groups on the surface. Self-association of such molecules might be expected to proceed until the addition of another molecule does not add sufficient volume to generate the surface area required to accommodate its polar groups. The volume/area trade-off depends on the geometry of the aggregate and can be used to predict in a rough and ready way the expected phase behavior of polymeric siloxane surfactants [12].

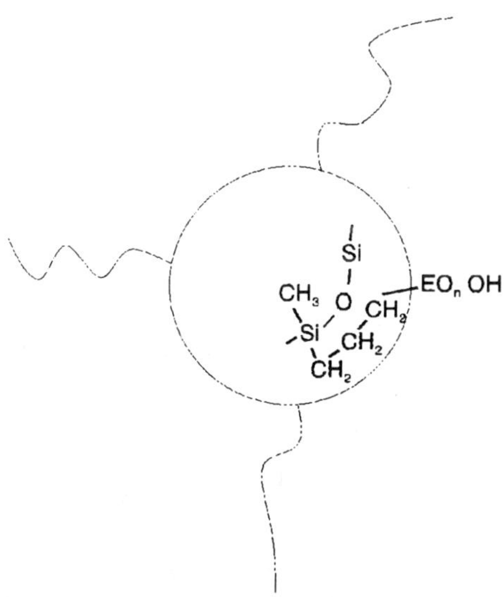

Figure 6.5 Schematic representation of a polysiloxane surfactant molecule in water. Adapted from Reference 14, used with permission.

6.10 EO/PO based siloxane surfactants

Siloxane surfactants incorporating copolymers (usually random) of polyoxyethylene and polyoxypropylene are widely used in polyurethane foam manufacture and in personal care applications, but to date no studies have been reported of their aggregation behavior. The phase map of Dow Corning® 190 is shown in Figure 6.6. Below the cloud temperature, this surfactant forms clear isotropic solutions with water at all concentrations. This behavior is apparently characteristic of siloxane surfactants containing EO/PO groups [80].

The ternary phase behavior of mixtures of a rake-type polymeric siloxane surfactant with decane and decanol was reported by Stürmer *et al.* [14]. They found a ringing gel and regions of hexagonal and lamellar phase. These authors did not discuss the relationship between the behavior of this system and that of ternary systems based on hydrocarbon surfactants.

6.11 Applications

Applications of siloxane surfactants exploit in one way or another the unusual features of this class of surfactants discussed above; foam stabilization in a non-aqueous medium, wetting/spreading, lubricity, and the ability to impart a silky feel to skin and hair.

6.11.1 Polyurethane foam manufacture

The largest commercial application of silicone surfactants is their use as additives for the production of polyurethane foam. Worldwide volume for siloxanes surfactants in polyurethane foam manufacture was estimated to be about 30 000 metric tons/year in 1994 [81].

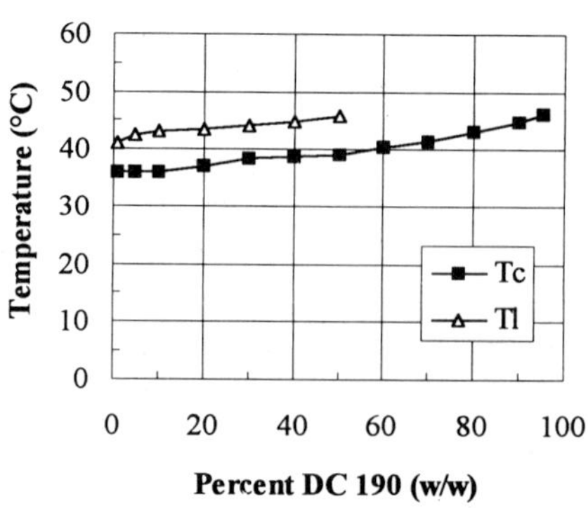

Figure 6.6 Phase map of Dow Corning® 190 in DI water. T_c marks the cloud temperature, T_l the temperature at which the cloudy dispersion begins to rapidly separate into two layers.

Polyurethane foams were first commercialized in the 1950s in Germany [82]. Fritz Hostettler of Union Carbide Corp. first patented the use of siloxane polyether copolymers in the manufacture of polyurethane foam [83–85]. Literature references describing this technology began to appear in 1960 [86, 87].

Polyurethane foam can range from a flexible porous elastomer to a rigid resin, and siloxane surfactants are available which have been optimized for each type [82]. There have been numerous reviews of the uses of siloxane polyether copolymers in polyurethane foam manufacture [2, 34, 82, 88–91]. A very diverse range of molecular structures are used matching the wide range of foams manufactured. The relevant molecular structure parameters are the:

molecular weight of the siloxane portion, MW_s
molecular weight of the polyether portion, MW_p
ratio of MW_s to MW_p
ratio of polyoxyethylene to polyoxyproylene in the polyether portion
degree of branching
end-capping of the siloxane or the polyether portion

The four most common structures are graft (rake or comb) copolymers, highly branched structures, ABA copolymers, and PDMS [82]. The synthesis of these polymers has been reviewed [7, 21, 30, 34]. There are three basic types of polyurethane foam: (a) flexible slabstock, (b) flexible molded foam, and (c) rigid foam [82]. The surfactants used for the manufacture of flexible slabstock foams have high molecular weights. The primary concern in these systems is foam stability. Polyethers capped with esters or ethers are used to prevent degradation of the surfactant. Polyethers with EO/PO ratios close to 1 are optimal. Surfactants in which the polyether is linked to the siloxane backbone through SiOC and SiC linkages are both used.

Flexible molded foam surfactants tend to be small molecules with molecular weights < 1500 Da. The function of the surfactant is primarily cell size control and stability of the foam near the mold surface. PDMS can be used. Cell opening is critical and special surfactants optimized for this function have been developed [82].

Rigid foam surfactants are usually intermediate in molecular weight between the flexible slabstock and flexible molded foam surfactants and are mostly based on molecular structures in which the polyether is linked to the siloxane backbone through an SiC linkage. Cell size control and emulsification are the primary functions of the surfactant. Emulsification is critical because the ingredients of rigid foam formulations are not compatible and do not mix well. Higher ratios of EO and PO are used, and polyethers are mostly uncapped.

Obviously, surfactants are used because of their surface activity, their ability to stabilize bubbles and foam. The equilibrium and dynamic surface tensions, and the surface viscoelasticity of siloxane polyether copolymers in polyols mimicking polyurethane foam systems have been much studied and significant correlations have been found, especially between dynamic surface properties and foam formation [82]. Approximately 10–20 D-units are required (on average) between each polyether group to achieve sufficient surface activity [82]. Higher branching leads to higher surface tensions. A higher percentage of siloxane causes more rapid migration to the surface and lower dynamic surface tensions.

A number of good treatments of polyurethane foam processing exist [92–95]. Polyurethane foams are made starting with a mixture of polyol, isocyanate, water, blowing agents, and catalysts. The two primary reactions involve the formation of urethane (the 'gelling' reaction) and the formation of urea (the 'blowing' reaction) which generates CO_2. During the course of these reactions the temperature in the foam rises rapidly, the viscosity increases dramatically, urea begins to phase separate, and bubbles nucleate and rapidly expand. Clearly this is an extremely complex and dynamic system, and fundamental questions remain about the relationship between the composition and molecular structure of siloxane surfactants, and their function in polyurethane foam manufacture.

6.11.2 *Textile and fiber industry*

Schmidt [4] briefly describes four applications of siloxane surfactants relevant to textile and fiber processing—silicone oil in water emulsifiers, improving the adhesion of adhesives to silicone treated fabrics, fiber lubricants, and release agents. During fiber production, siloxane surfactants enable the lubricant to spread quickly and completely even at very low pickup amounts [49]. These functions are related to the ability of the siloxane surfactant to lower the surface tension of the organic oil [49]. The types of siloxane surfactants useful for this application are tabulated and discussed by Schmidt [49]. A wide variety of uses exist, some of which do not seem obviously related to the surfactancy of these materials. Use of siloxane surfactants as wetting agents in aqueous textile treatments has been discussed only in terms of laboratory results, little published information is available about industrial applications. Organophilic siloxanes of various structures have been shown to restore adhesion of hot-melt adhesives to silicone coated rainwear. Siloxane surfactants improve spinability of certain thermoplastic fibers. Siloxane surfactants can, in certain instances, migrate to the surfaces of fibers during and after extrusion to modify their properties [49].

6.11.3 *Personal care and cosmetic applications*

Schaefer [16] reviewed the application of siloxane surfactants in personal care products. Oil soluble siloxane surfactants improve the spreading ability of organic oils and waxes and as such are used in make-up and skin care formulations. More hydrophilic siloxanes surfactants are used in aqueous applications such as shampoos and shower gels. Cationic and zwitterionic functionality improve substantivity and are used in hair care products. Substantivity of non-ionic siloxane surfactants is determined partly by their water solubility; low solubility, as evidenced by a low cloud point, leads to improved substantivity [16]. SPEs are claimed to reduce eye irritation of anionic surfactants [96]. They are known for their low physiological risk in cosmetic formulations [1]. In hair care preparations SPEs improve combing and gloss and impart a silky feel [16]. In lotions they impart smoothness and softness to the skin, and defoam minimizing the whitening on rub-out. SPEs were found to help prevent cracking in soap and syndet bars [16, 23]. Under non-neutral pH conditions, the trisiloxane surfactants show only limited hydrolytic stability which has hampered their wide application [16]. Cationics and zwitterionics have found some application in hair care products [7, 16].

Vick [22] discussed the structure/property relationships that are responsible for the behavior of siloxane surfactants in cosmetic applications. Vick analyzed the behavior of siloxane surfactants in terms of surface tension lowering, wetting, emulsification, and foam characteristics, all features which he asserts play a key role in their performance in personal care applications.

6.11.4 Agricultural adjuvancy

The use of organosilicone surfactants as adjuvants to enhance the efficacy of agrochemicals such as herbicides, fungicides, and foliage applied fertilizer has recently been reviewed extensively [44, 97, 98]. Such adjuvants allow agrochemicals to be used at lower levels, which is desirable for both economic and environmental reasons [44]. These reviews have focussed on equilibrium and dynamic surface tensions, contact angles on hydrophobic surfaces, and spread areas relative to water (after some fixed time period). They also correlated these surface properties with greenhouse and field performance, with studies of foliar uptake and the penetration of certain agrochemicals into plant leaf surfaces. Many siloxane surfactants have been studied, but the primary interest is in the trisiloxane surfactants discussed above, $M(D'E_nOR)M$ with $n = 7$–8 and various end caps.

Surfactant effects on foliar uptake and field performance of agrochemicals are species and compound specific. Knoche [44] tabulates the effect of siloxane surfactants on the foliar uptake of a variety of agrochemicals by different plant species, as well as field and greenhouse performance. It is believed that certain trisiloxane surfactants enhance transport of spray solutions into plants, either through stomatal pores or by way of cuticular penetration. Performance in field trials and greenhouse studies parallels these considerations. The best way to understand the seemingly confusing effects on different species is to recognize that efficacy of foliage applied agrochemicals generally requires penetration of the active ingredient into the plant. The efficacy of agrochemicals such as contact fungicides which do not need to penetrate the plant is directly improved by enhancing spreading. For other types of agrochemicals, spreading is only useful if it correlates with penetration. Highly efficient spreading of the aqueous vehicle over the leaf can lead to rapid evaporation and even roll-off thereby inhibiting penetration [44]. The wetting efficacy of surfactants as ag-adjuvants is often discussed in terms of spread areas relative to water, the ratio of the area of a spread droplet of the surfactant solution after some period of time (45 s, for example) to the area of a (non-spreading) droplet of water. Spread areas are much larger for certain trisiloxane surfactants than for conventional oganic surfactants, polymeric siloxane surfactants, or fluorocarbon surfactants [52]. Although dynamic and equilibrium surface tensions are frequently discussed in this context, there is little correlation between either of these quantities and ag-adjuvancy. The efficient wetting of hydrophobic leaf surfaces by aqueous pesticide formulations containing organosilicone surfactants is attributed to their ability to reduce surface tension to values well below what is possible with conventional organic surfactants [52]. The unusual shape of the trisiloxane surfactant is invoked as an additional factor to explain why fluorocarbon surfactants (which give even lower surface tensions) do not also spread. However, Hill *et al.* [13] showed that a 'linear' trisiloxane surfactant, which lacks this shape, spreads just as well, and Stoebe *et al.* [57] have shown recently that a linear alkyl ethoxylate (C12E3) exhibits a similar rate of spreading on slightly less hydrophobic surfaces. As was mentioned in the discussion above about the superwetting phenomenon, the contribution to the spreading coefficient from the interfacial tension between the liquid phase and the solid substrate must also be considered and this has not been done. Presumably the fluorocarbon surfactants do not sufficiently lower the interfacial tension between the aqueous formulation and the leaf surface.

A general synergy has been claimed for mixtures of siloxane surfactants with organic surfactants [99–101]. Hill [3] reported mixed CMC results for several siloxane surfactants with anionic, cationic, and non-ionic organic surfactants, and also reported detergency measurements for some siloxane surfactants in combination with organic surfactants. Most agrochemical formulations contain organic surfactants in addition to the siloxane surfactant, but it has been shown that some organic surfactants are antagonistic to both surface tension lowering and spreading [52, 102].

6.11.5 *Emulsion stabilization*

Siloxane surfactants have not to date found many applications as emulsifiers. This has been attributed to the fact that because the siloxane portion of siloxane surfactants is not able to form solid or crystalline phases, neither can siloxane surfactants, unlike many organic surfactants [7]. These liquid crystal phases play a key role in some types of emulsion stabilization. However, it has been shown recently that many siloxane surfactants do form liquid crystal phases [12, 13] suggesting that they may have more potential in this area than has been realized. Schmidt [4, 49] and Vick [22] note that calculating or measuring HLB values for siloxane surfactants yields quantities which do not correctly predict emulsifiability, at least when siloxane surfactants are used to emulsify hydrocarbon oils. Siloxane surfactants, including some non-ionic examples and one anionic example, have been used to emulsify silicone finishes for textile treatment purposes [49].

Stabilization of water-in-oil (w/o) emulsions using certain siloxane terpolymer surfactants has recently received considerable attention [7, 16] partly due to the emergence of transparent deodorant and antiperspirant products based on such emulsions [42]. Schaefer [16] attributed the efficacy of such a surfactant to the properties of a film of the terpolymer situated at the water/oil interface and its immobility due to the silicone functionality and its molecular weight. He stated that, "To some extent a three phase system consisting of water, siloxane and oil phase is formed." [16].

6.11.6 Paints and coatings

Fink [74] has reviewed the use of siloxane surfactants in paint and coating applications. A wide range of structures are used in this area, including structures which incorporate hydrophilic, organophilic, and fluorocarbon functionality, and even combinations of these. These additives are used as wetting agents (substrate adhesion), flow promoters (prevent formation of blisters and craters), lubricants (glide and scratch control), and as antifoams and de-aerators. Obviously, these uses each involve distinct and not necessarily related surface phenomena. Each use has a group of siloxane surfactants which have been developed for that purpose, and specifically tailored for the type of coating system involved, solvent or water based, acrylic, etc. There is little insight available at the present time concerning the mode of action in these varied systems, but Fink's review is a useful starting point for the technologist seeking help with specific problems.

6.12 Summary

Siloxane surfactants are a class of novel surfactants consisting of a methylated siloxane group with one or more hydrophilic groups attached. Most siloxane surfactants are intermediate MW copolymers. This class of surfactants is highly surface active in both aqueous and non-aqueous media, lowering surface tensions to nearly 20 dynes/cm. They are used in a wide range of applications in which their unique properties are required, including polyurethane foam and textile and fiber manufacture, in personal care products, in paints and coatings, and as wetting agents for pesticides. It remains difficult in most of these areas to correlate specific features of molecular architecture with performance. Recent work on the aqueous aggregation behavior of this class of surfactants shows that siloxane surfactants ranging from trisiloxanes to polymeric siloxanes form the usual micelles, vesicles, and normal and inverse liquid crystal phases which is surprising since they would be expected to adopt rather disordered coiled configurations in solution. This information is useful for understanding the physical properties of concentrated surfactant solutions, and may be particularly useful for understanding interactions between siloxane surfactants and conventional organic surfactants. Since the type of aggregate a surfactant forms is indicative of the intrinsic curvature it prefers at an interface, knowledge of aggregation behavior also helps understand emulsification and wetting. The unusual rapid wetting properties of dilute solutions of certain trisiloxane surfactants turns out to be a special case of surfactant enhanced wetting. Many possible siloxane surfactant structures remain to be characterized, and much work remains to be done, especially in specific application areas.

References

1. Gradzielski, M., Hoffman, H. *et al. Tenside Surf. Det.*, 1990, **27**, 366.
2. Owen, M.J., *Ind. Eng. Chem. Prod. Res. Dev.*, 1980, **19**, 97.
3. Hill, R.M. In: Holland, P.M. and Rubingh, D.N. (eds). *Mixed surfactant systems, ACS Symp. Ser. Vol. 501*, p. 278. Am. Chem. Soc., New York, 1992.
4. Schmidt, G.L.F. In: Karsa, D.R. (ed.). *Industrial applications of surfactants*, p. 24. Royal Soc. Chem., London, 1987.
5. Laughlin, R.G. *The aqueous phase behavior of surfactants*. Academic Press, New York, 1994.
6. Lasic, D.D. *Liposomes: from physics to applications*. Elsevier, Amsterdam, 1993.
7. Gruning, B. and Koerner, G. *Tenside Surf. Det.*, 1989, **26**, 312.
8. Owen, M.J. In: Mittal, K.L. and Bothorel, P. (eds) *Surfactants in solution Vol. 6*, p. 1557. Plenum Press, New York, 1986.
9. Owen, M.J. and Kendrick, T.C. *Macromolecules*, 1970, **3**, 458.
10. Kendrick, T.C., Kingston, B.M. *et al. J. Coll. Interface Sci.*, 1967, **24**, 135.
11. Kendrick, T.C. and Owen, M.J. *Chem. Phys. Appl. Prat. Surface (Spain) 5th*, 1968, **2**, 571.
12. Hill, R.M., He, M. *et al. Langmuir*, 1993, **9**, 2789.
13. Hill, R.M., He, M. *et al. Langmuir*, 1994, **10**, 1724.
14. Stürmer, A., Thunig, C. *et al. Tenside Surf. Det.*, 1994, **31**, 90.
15. He, M., Hill, R.M. *et al. J. Phys. Chem.*, 1993, **97**, 8820.
16. Schaefer, D. *Tenside Surf. Set.*, 1990, **27**, 154.
17. Schwarz, E.G. and Reid, W.G. *Ind. Eng. Chem.*, 1963, **56**, 26.
18. Bailey, D.L., Peterson, I.H. and Reid, W.G. *Chem. Phys. Appl. Surface Active Substances Proc. 4th Int. Congr.*, 1967, **1**, 173.
19. Kanner, B., Reid, W.G. and Petersen, I.H. *Ind. Eng. Chem. Prod. Res. Dev.*, 1967, **6**, 88.
20. Ross, S. and Nishioka, G. *Coll. Polym. Sci.*, 1977, **255**, 560.

21. Noll, W. *The chemistry and technology of silicones.* Academic Press, New York, 1968.
22. Vick, S.C. *Soap/Cosmetics/Chemical Specialties*, 1984, **May**, 36.
23. Gould, C. *Spec. Chem.* 1991, **August**, 354.
24. Hsieh, C.-J., Hsu, C.-S. and Hsiue, G.-H. *Polymer. Mater. Sci. Eng.*, 1988, **59**, 1014.
25. Finkelmann, H. and Rehage, G. *Adv. Polymer Sci.* 1984, **60/61**, 99.
26. Gray, G.W. In: McArdle, C.B. (ed.). *Side chain liquid crystal polymers*, p. 106. Blackie, Glasgow, UK, 1989.
27. Gray, G.W., Hill, J.S. and Lacey, D. *Mol. Cryst. Liq. Cryst.*, 1991, **197**, 43. See also Gray, G.W., Hawthorne, W.D. *et al. Polymer*, 1989, **30**, 964.
28. Lühmann, B. and Finkelmann, H. *Colloid Polym. Sci.*, 1987, **265**, 506.
29. Lühmann, B. and Finkelmann, H. *Makromol. Chem.*, 1985, **186**, 1059.
30. Clarson, S.J. and Semlyen, J.A. (eds). *Siloxane polymers.* PTR Prentice Hall, New York, 1993.
31. Bailey, D.L. US 3299112, Jan. 17, 1967, to Union Carbide Corp.
32. Bahr, B.C., Petroff, L.J. and Romenesko, D.J. US 5145977, Sept. 8, 1992, to Dow Corning Corp.
33. Ekeland, R.A., Petroff, L.J. and Romenesko, D.J. US 5145978, Sept. 8, 1992, to Dow Corning Corp.
34. Kanner, B., Reid, W.G. and Petersen, I.H. *Ind. Eng. Chem. Prod. Res. Dev.*, 1967, **6**, 88.
35. Snow, S.A. Fenton, W.N. and Owen, M.J. *Langmuir*, 1990, **6**, 385.
36. Maki, H., Murakami, Y. *et al. Kogyo Kagaku Zasshi (J.C.S. Jap. Ind. Chem. Section)*, 1968, **71**, 1675.
37. Snow, S.A. *Langmuir*, 1993, **9**, 424.
38. Morehouse, E.L. US 3660452, May 2, 1972, to Union Carbide Corp.
39. Snow, S.A., Fenton, W.N. and Owen, M.J. *Langmuir*, 1991, **7**, 868.
40. Lin, Z., He, M. *et al. J. Phys. Chem.*, 1993, **97**, 3571.
41. Maki, H., Horiguchi, Y. *et al. Yukagaku*, 1970, **19**, 1029.
42. a. Roidl, J. *Parfumerie und Kosmetik*, 1986, **67**, 232. b. Wolfes, W. *Parfumerie und Kosmetik* 1987, **68**, 195. c. Rentsch, S.F. US 5387417, Feb. 7, 1995, to Dow Corning Corp.
43. Knoche, M., Tamura, H. and Bukovac, M.J. *J. Agric. Food Chem.*, 1991, **39**, 202.
44. Knoche, M. *Weed Research*, 1994, **34**, 221.
45. Stevens, M.P. *Polymer chemistry, an introduction, 2nd ed.* Oxford University Press, Oxford, UK, 1990.
46. Gentle, T.E. and Snow, S.A. *Langmuir*, 1995, **11**, 2905.
47. Kanellopoulos, A.G. and Owen, M.J. *J. Coll. Interface Sci.*, 1971, **35**, 120.
48. Svitova, T., Hoffman, H. and Hill, R.M. submitted to *Langmuir.*
49. Schmidt, G. *Tenside Surf. Det.*, 1990, **27**, 324.
50. Ananthapadmanabhan, K.P., Goddard, E.D. and Chandar, P. *Colloids Surf.*, 1990, **44**, 281.
51. Goddard, E.D. and Padmanabhan, K.P.A. In: Foy, C.L. (ed.). *Adjuvants agrichemistry*, p. 373. CRC Press, New York, 1992.
52. Murphy, G.J., Policello, G.A. and Ruckle, R.E. *Brighton Crop Prot. Conf.–weeds*, 1991, **1**, 355.
53. Zhu, X., Ph.D. Thesis, University of Minnesota, 1992.
54. Zhu, X., Miller, W.G. *et al. Colloids Surf. A*, 1994, **90**, 63.
55. Lin, Z., Hill, R.M. *et al. Langmuir*, 1994, **10**, 4060.
56. Lin, Z., Hill, R.M. *et al. Langmuir*, 1996, **12**, 345.
57. Stoebe, T., Lin, Z. *et al. Langmuir*, 1996, **12**, 337.
58. Svitova, T., Smirnova, Y. and Yakubov, G. *Colloids Surf A*, 1995, **101**, 251.
59. a. Tiberg, F. and Cazabat, A.M. *Europhys. Lett.*, 1994, **25**, 205, b. Tiberg, F. and Cazabat, A.M. *Langmuir*, 1994, **10**, 2301.
60. de Gennes, P.G. *Rev. Mod. Phys.*, 1985, **57**, 827.
61. Dukhin, S.S., Kretzschmar, G. and Miller, R. *Dynamics of adsorption at liquid interfaces.* Elsevier, New York, 1995.
62. Tiddy, G.J.T. *Phys. Rep.*, 1980, **57**, 1.
63. Israelachvili, J.N., Mitchell, D.J. and Ninham, B.W. *J. Chem. Soc., Faraday Trans. 2*, 1976, **72**, 1525.
64. Mitchell, J.D., Tiddy, G.J.T. *et al. J. Chem. Soc., Faraday Trans. 1*, 1983, **79**, 975.

65. He, M. Hill, R.M. *et al.* In: Herb, C. and Prudhomme, R.K. (eds). *Structure and flow in surfactant solutions, vol. 578*, p. 192. Am. Chem. Soc., New York, 1994.
66. Lin, Z., Hill, R.M. *et al. Langmuir*, 1994, **10**, 1008.
67. Doumaux, H. Ph.D. Thesis, University of Minnesota, 1995.
68. He, M., Lin, Z. *et al. J. Phys. Chem.*, 1994, **98**, 6148.
69. Yang, J. and Wegner, G. *Macromolecules*, 1992, **25**, 1786.
70. Yang, J. and Wegner, G. *Macromolecules*, 1992, **25**, 1791.
71. Rosen, M.J. *Surfactants and interfacial phenomena, 2nd ed.*, p. 191–195. John Wiley & Sons, New York, 1989.
72. Nakagawa, T. In: Schick, M.J. (ed.) *Nonionic surfactants. Surfactant Sci. Ser. Vol. 1*, p. 572. Marcel Dekker, New York, 1966.
73. Hill, R.M. unpublished results, Dow Corning Corp.
74. Fink, H.F. *Tenside Surf. Det.*, 1991, **28**, 306.
75. Maki, H., Saeki, S. *et al. J. Am. Oil Chem. Soc.*, 1969, **46**, 635.
76. Maki, H., Komatsu, M. and Komori, S. *Kogyo Kagaku Zasshi (J.C.S. Jap. Ind. Chem. Section)*, 1967, **70**, 1771.
77. Maki, H., Murakami, Y. *et al. Kogyo Kagaku Zasshi (J.C.S. Jap. Ind. Chem. Section)*, 1968, **71**, 1679.
78. Maki, H., Horiguchi, Y. and Komori, S. *Kogyo Kagaku Zasshi (J.C.S. Jap. Ind. Chem. Section)*, 1970, **73**, 1142.
79. Maki, H., Suga, T. *et al. Yukagaku*, 1970, **19**, 245.
80. Hill, R.M. unpublished results, Dow Corning Corp.
81. Reed, D. *Urethanes Technology*, 1995, **Jan./Feb.**, 22.
82. Snow, S.A. and Stevens, R.G. manuscript in preparation.
83. Hostettler, F. *German Patent 1091324*, 1960.
84. Hostettler, F. US 3194770, 1965.
85. Hostettler, F. US 3194773, 1965.
86. Sandridge, R.L., Morecroft, A.S. *et al. J. Chem. Eng. Data*, 1960, **5**, 459.
87. Erner, W.E., Farkas, A. and Hill, P.W. *Modern Plastics*, 1960, **Feb.**, 107.
88. Boudreau, R.J. *Modern Plastics*, 1967, **January**, 133 and 234.
89. Schwarz, E.G. *Applied Polymer Symposia No. 14: Silicone Technology*, p. 71. J. Wiley and Sons, New York, 1970.
90. Kanner, B. and Prokai, B. *Adv. Ureth. Sci. Techn.*, 1973, **2**, 221.
91. Kanner, B., Goddard, E.D. and Kulkarni, R.D. *Proceedings 5th Intl. SPI Conf.*, 1980, 647.
92. Bruins, P.F. (ed.) *Polyurethane technology*, Interscience. New York, 1969.
93. Frisch, K.C. Saunders, J.H. (eds). *)Plastic foams parts I and II.* Marcel Dekker, New York, 1972.
94. Woods, G. *Flexible polyurethane foams.* Applied Science Publishers, New Jersey, 1982.
95. Oertel, G. (ed.). *Polyurethane handbook.* Carl Hanser Verlag, Munich, 1985.
96. Starch, M. and De Vries, C. *Parfumerie und Kosmetik*, 1986, **67**, 148.
97. Roggenbuck, F.C., Rowe, L. *et al. Weed Technol.*, 1990, **4**, 576.
98. Stevens, P.J.G. *Pestic. Sci.*, 1993, **38**, 103.
99. Petroff, L.J. US 4784799, 1988.
100. Bailey, D.L. US 3562786, 1971.
101. Policello, G.A. US 5104647, 1992.
102. Murphy, D.S., Policello, G.A. *et al.* In: Devisetty, B.N., Chasin, D.G. and Berger, P.D. (eds). *Pesticide Formulations and Applications Systems: 12th Volume, ASTM STP 1146*, p. 45. ASTM, Philadelphia, 1993.

7 Alkyl polyglucosides

D. BALZER

7.1 Introduction

The search for surfactants that are toxicologically safe and readily biodegradable has placed into the foreground feedstocks which in this arca hitherto lived more in the shade: the carbohydrates. Whereas fats and oils, renewable raw materials, have been in use from the very beginning of surfactant production, employing carbohydrates has been an exception, restricted to fields of specialty applications such as pharmaceuticals, foods and cosmetics. This occurred although sufficient quantities (worldwide about 3 million tons per year [1]) of low-cost and high-purity carbohydrates such as starch, sucrose, glucose or sorbitol are available as raw materials.

Carbohydrate-derived surfactants that are produced and used on an industrial scale are sorbitan esters (**1**), sucrose esters (**2**), and what is rather new, the fatty acyl glucamides (**3**) and alkyl polyglucoside (**4**) (Figure 7.1). All are non-ionics.

Whereas the esterification of the multifunctional carbohydrates with fatty acids forming sobitan esters [2] (worldwide production figure about 10 000 tons per year) and sucrose esters [3] (production not higher than 2000 tons/y) leads to complex products, more uniform products are achieved by the alternative routes of amidation and glycosylation. An advantageous property of amides and glucosides, in contrast to esters, is that they are less sensitive to hydrolysis.

Fatty acyl glucamides [4, 5] are obtained by reductive amination of glucose in the presence of ammonia or short-chain primary amines followed by acylation of the glucamine with fatty acid methyl esters. *N*-methyllauroyl glucamide has recently experienced a great increase in interest and is mainly used in dishwashing compositions. Its consumption figure is believed to be in the range of 5000 tons per year. The environmental compatibility of *N*-methyl-lauroyl glucamide is stated to be very good [6]. Doubts about toxicological safety have not yet been satisfactorily resolved because *N*-methylglucamine may form nitrosamine which has shown carcinogenic potential in rats [5, 7].

Alkyl polyglucosides, with a worldwide production capacity recently increased to about 60 000 tons per year, are by far the most important sugar-based surfactant. They are produced by proton-catalized acetalization of carbohydrates, preferably of glucose, or by *trans*-acetalization of butyl polyglucoside with fatty alcohol.

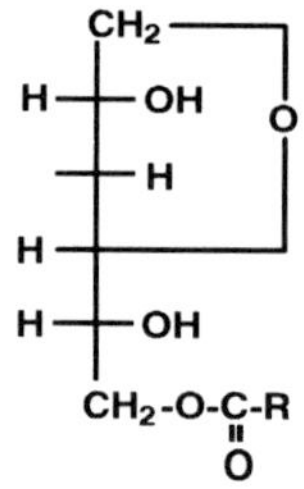

1

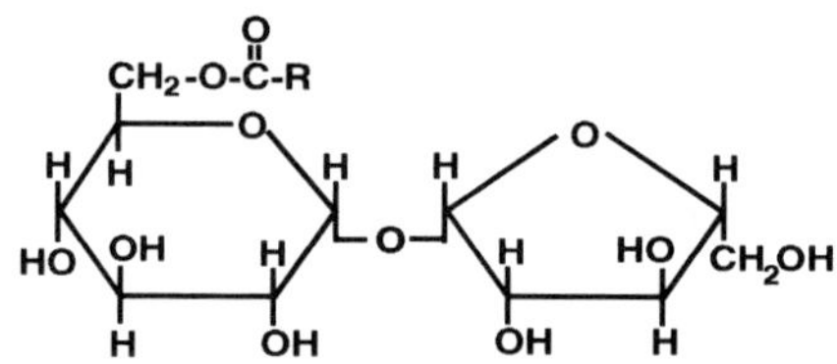

2

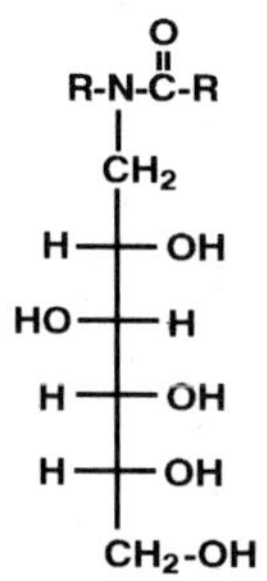

3

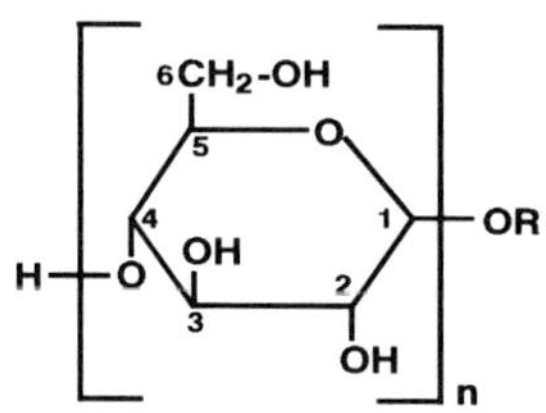

4

Figure 7.1 Sugar based surfactants: sorbitan esters (**1**), sucrose esters (**2**), fatty acyl glucamides (**3**) and alkyl polyglucosides (**4**).

Strictly speaking, they are not new surfactants but were described 100 years ago by Fischer [8]. However, it was not until 1934 that their potential as surface-active agents was appreciated in a patent granted to H. Th. Böhme AG of Chemnitz [9]. They then fell into obscurity for a long time, probably because they were not only difficult to manufacture, but also because of the other surfactants already in production. There were some developments of C_8C_{10}-alkyl polyglucosides in the mid-1970s by Rohm & Haas in the United States and by Seppic in France, who also introduced the products (Triton BG 10, Triton CG 110), but the response was not very great.

It was not until the first half of the 1980s that this old class of surfactants was unearthed again, against a background of increasing environmental concern, and this was also because their wide application was recognized. Companies such as A.E. Staley, Procter & Gamble, Henkel, Kao Corp., and Hüls launched R&D programs, the results of which are reflected in a large

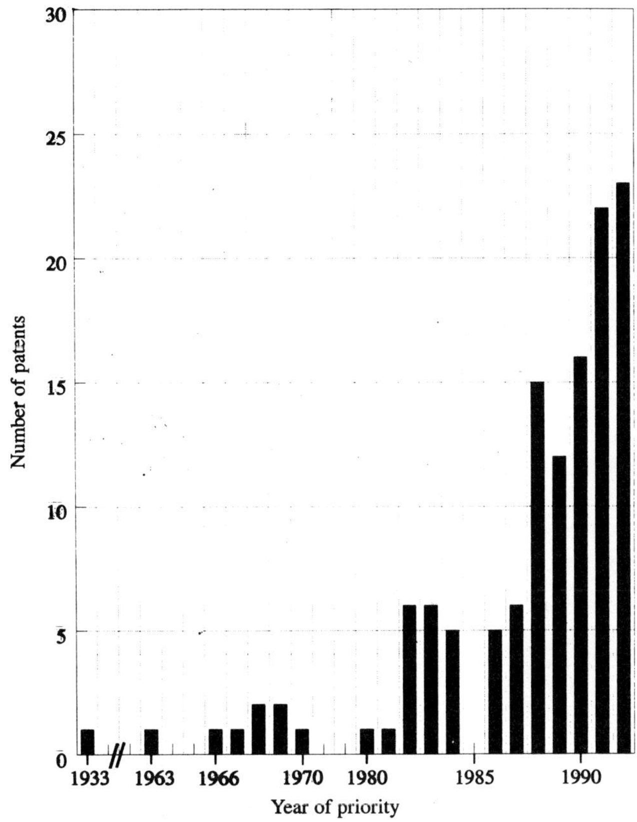

Figure 7.2 Production related patents versus year of priority. (Figure reproduced with permission from Hanser Verlag, Munich.)

number of patent applications. Whereas in the synthesis of alkyl polyglucosides just 10 patents were issued up to 1980, more than 120 have been applied for since 1981, and the rate of publication has probably not yet reached its peak (Figure 7.2). The trend in patents related to the application of alkyl polyglucosides is even more dramatic. Thirteen patents were issued before 1980 compared to more than 300 patents or applications between 1981 and 1992, figures which emphasize the great interest in this old–new class of surfactants.

Such a flood of patent applications has not occurred in the case of any other sugar derived surfactant and is a result of the special properties of alkyl polyglucosides. In addition to their low environmental impact these have a wide application, good compatibility and synergistic effects when combined with many other surfactants and ingredients of surfactant based formulations. An amazingly economical cost–performance ratio and high quality have already been attained, although the production process is very difficult to control. Hence, there seems to be a good chance that alkyl polyglucosides will play a major role alongside traditional surfactants as we move through the second half of the 1990s into the 21st century.

7.2 Definition

Alkyl polyglucosides are alkyl polyglycosides based not on any sugar but on glucose. Chemically, the alkyl polyglycosides are acetals, which for glucosides can be formally described in terms of structure **4** in Figure 7.1, where R denotes fatty alcohol chains of 8 to 18 carbon atoms and *n* the degree of glucosidation. Because the Fischer synthesis, the only route technically feasible, leads to a distribution of species with different degrees of glucosidation, *n* is a mean value. *n* is defined as the ratio of the total molar amount of glucose in the alkyl polyglucoside to the molar amount of fatty alcohol, taking into account the average molecular weight when fatty alcohol blends are employed. By this definition, *n* is 1 for pure alkyl monoglucoside and 2 in the case of pure alkyl maltoside. Most of the alkyl polyglucosides of importance to applications have a mean *n* value of between 1.1 and 1.8, and consist of the alkyl monoglucoside as the main component, and also of alkyl diglucoside (alkyl maltosides), alkyl triglucoside (alkyl maltotriosides), etc. up to alkyl octaglucoside [9]. For a mean *n* value $\leqslant 2$ the term alkyl oligoglucoside would be more correct than alkyl polyglucoside, but the latter gained acceptance in the literature quite a long time ago [11].

The distribution of molecular weights of the alkyl polyglucosides is analogous to the fatty alcohol ethoxylates which can be described unambiguously in terms of a distribution function. Such a simple approach is not possible for alkyl polyglucosides because three different types of isomerism result in a much more complex range of products than is formed by fatty alcohol ethoxylates. Stereoisomerism results in α-(**5**) and β-anomers (**6**), ring isomerism in pyranosides (**5**, **6**) and furanosides (**7**), and binding isomerism results in 1,2- and 1,3-interglycosidic linkages (Figure 7.3) as well as in the greatly preferred 1,6- and 1,4-linkages (**8**). Accordingly, even alkyl monoglucoside exists in four different isomer forms, viz: alkyl-α-D-glucopyranoside (**5**), the main component of the Fischer synthesis, alkyl-β-D-glucopyranoside (**6**), alkyl-α-D-glucofuranoside (**7**) and alkyl-β-D-glucofuranoside. There are more than 30 different alkyl diglucosides, several hundreds of alkyl triglucosides, several thousands of alkyl tetraglucosides, etc. Thus, alkyl polyglucosides are very complex mixtures, which cannot be described quantitatively, in the strict sense of the word, by any analytical method.

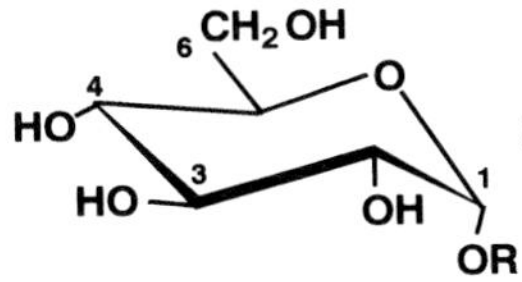

5

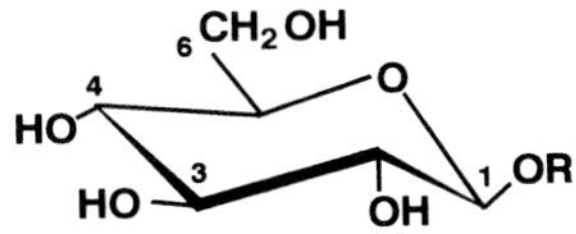

6

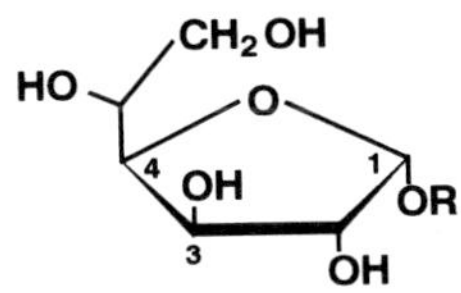

7

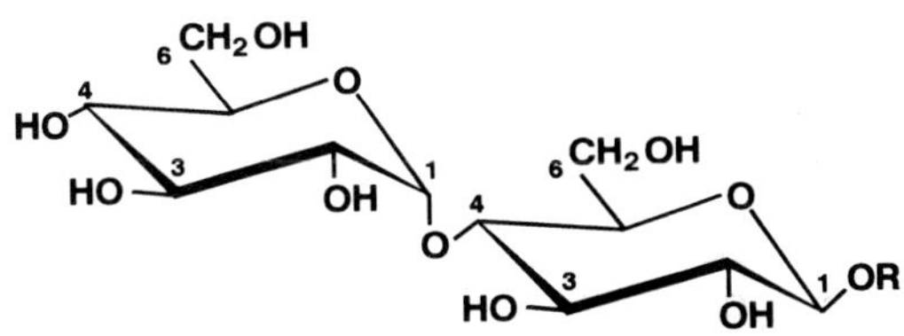

8

Figure 7.3 Typical isomers of alkyl polyglucosides: alkyl α-D-gluco-pyranoside (**5**), alkyl β-D-glucopyranoside (**6**), alkyl α-D-gluco-furanoside (**7**) and alkyl β-D-maltoside (**8**).

7.3 Synthesis

There are two basic routes to the synthesis of alkyl glycosides or alkyl polyglucosides. One, Koenigs–Knorr glycosilation is a rather complicated laboratory method of preparing pure stereoselective alkyl glycosides. The other one is the Fischer synthesis which can be employed to produce alkyl polyglucosides even on a technical scale of many thousand tons. For both routes various improvements have been established during the last few years.

7.3.1 Koenigs–Knorr synthesis

The Koenigs–Knorr glycosilation process [12] is a sophisticated four-step synthesis by which defined alkyl glycosides with high stereoselectivity (usually alkyl-β-anomers) and only small amounts of side products are obtained. It makes use of an effective protecting stategy for the many OH groups, special solvents and heavy metal salt promoters.

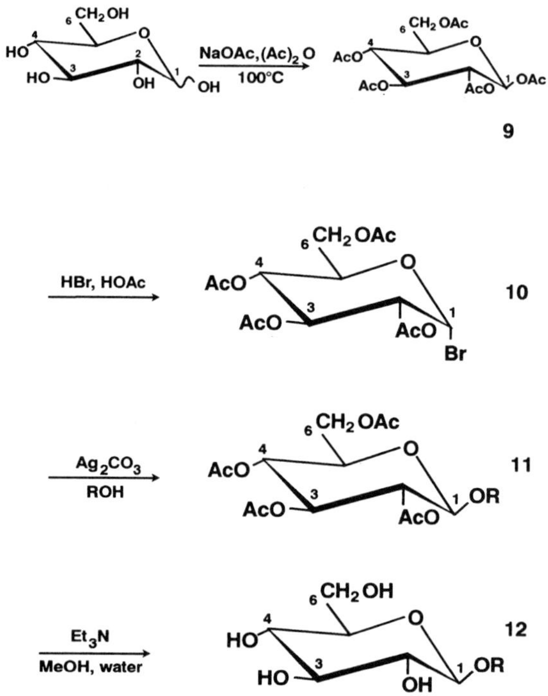

Figure 7.4 Koenigs–Knorr glycosilation applied to glucose.

In the first step (Figure 7.4) sugars such as D-glucose are peracylated with sodium acetate in acetic anhydride to form β-pentacetylglucose (**9**), which is converted with HBr in acetic acid to α-tetra-acetyl-1-bromoglucose (**10**). The condensation process with a long-chain alcohol is a rather sluggish reaction and needs a large excess of alcohol and also a silver or mercury salt as promotor for the nucleophilic attack. β-Tetra-acetyl-1-alkylglucose (**11**) has then to be de-esterified, which could be done by means of sodium methylate [12] or another base such as triethanol amine [14], to give β-1-alkylglucose or alkyl-β-D-glucopyranoside (**12**). The final product has to be purified, which could be done by liquid–solid chromatography on a silica column [13] or by using strongly basic ion exchange resins [14, 15]. The yield over all steps is about 50%, and the anomer ratio about 85% β- and 15% α-anomer for glucose and C_8- to C_{12}-fatty alcohols. For galactose, for example, the anomer ratio is 100% β-anomer [16].

A variation of the Koenigs–Knorr synthesis is the glycosilation of peracylated sugars with fatty alcohol in the presence of Lewis acid catalysts such as stannic tetrachloride and CH_2Cl_2 as solvent. The kinetically controlled nucleophilic substitution leads, at relatively low temperatures and short reaction times, to the peracylated β-alkyl glucoside, which at higher temperatures and longer reaction times passes to the thermodynamically more stable peracylated α-anomer [14, 15]. The products can be de-esterified with sodium methylate.

The glycosilations mentioned are rather time-consuming laboratory methods used to synthesize specific alkyl glucosides, a necessary tool for the analytical description of the alkyl polyglucoside. For production they are too complex and also environmentally critical, if one considers the heavy metal salt generated and the use of toxic solvents, to be employed under practical conditions.

7.3.2 *Fischer synthesis*

Compared to these reactions the Fischer synthesis, designed to produce alkyl polyglycosides [8], appears to be far less complex. But the reaction, a proton catalyzed acetalization, called glycosidation is rather non-selective. To permit its extension to a technical scale, the process parameters have to be controlled very carefully as is indicated by the large number of patent applications during the last few years (Figure 7.2). The Fischer reaction is thermodynamically controlled. Hence it preferentially leads to α-anomers [16].

There are two variants of the Fischer process [25]. One is the direct synthesis, in which a substantially dry (e.g. anhydrous or hydrated crystalline) monosaccharide such as D-glucose is reacted with a fatty alcohol. The other route, a two-stage synthesis, starts from low-cost starch or any glucose syrup, which is reacted in a first step with a lower alcohol such as *n*-butanol to form butyl glucoside, which in turn is *trans*-glycolized during the second step with the fatty alcohol (Figure 7.5). Both routes are equilibrium reactions. Hence alcohols have to be employed in great excess, and subsequently removed preferably by vacuum stripping at elevated temperatures. The alkyl polyglucosides produced in this way are slightly coloured. Hence, to meet the application requirements of surfactants, final refining steps are necessary. Both routes have been successfully implemented during the last few years on an industrial scale.

(a) Two-stage synthesis. For a long time it was generally believed that the Fischer glycosidation process is suitable only for the preparation of glucosides of lower alcohols ($\leqslant C_4$). The reasons for this are related to the fact that the solubility of glucose in alcohols declines sharply with increasing hydrocarbon chain length. Hence with higher alcohols the reaction mixture is separated into two phases. The resulting mass transfer limitations lead to a sharp slow-down of the reaction rate, and are highly favourable to side

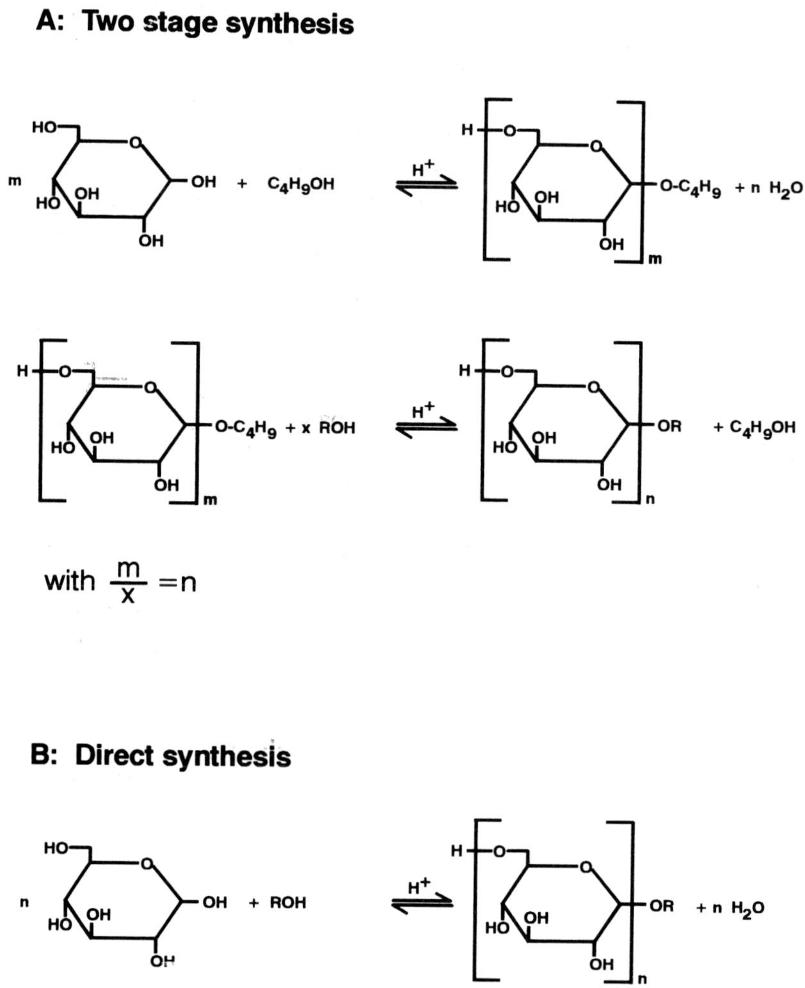

Figure 7.5 Fischer synthesis routes for alkyl polyglucosides.

reactions. Increasing the temperature to speed up the rate even makes the situation deteriorate because of the sensitivity of glucose to heat.

A successful approach designed to overcome the solubility difficulties in the preparation of higher alkyl polyglucosides was found by Boettner [17]. His process consists of dissolving glucose in methanol, adding *n*-butanol in the presence of an acidic catalyst while the methanol is removed by distillation. Then the higher alcohol is added while the butanol is removed. By employing a double alcohol interchange, Boettner was successful in preparing higher alkyl polyglucosides. His process is a first draft of the two-stage synthesis.

As already mentioned, the acid catalyzed Fischer reaction is not very selective. With the acetalization process forming the alkyl polyglucosides a proton catalyzed homopolymerization of the glucose also competes as the main side reaction. The formation of oligo- or polyglucose may predominate when separation into an aqueous saccharide solution phase and an alcohol–saccharide–water phase takes place, and, as frequently occurs, the acidic catalyst simultaneously stays in the saccharide solution phase [18]. The solubility problem was encountered with higher boiling solvents such as aliphatic glycols [19], aliphatic ether alcohols such as 2-ethoxyethanol [20], dimethylformamide or dimethyl sulphoxide [17], and *N*-methyl-2-pyrrolidone [21]. All these solvents increase the miscibility of the saccharide and the alcohol but they are difficult to remove quantitatively from the final product.

A much more effective method of circumventing the miscibility problem makes use of low-boiling alcohols such as *n*-butanol as solvent and as reactant. Although the solubility of saccharides such as glucose or dextrose syrup in butanol amounts to only a few percent, the situation changes drastically in the presence of butyl polyglucoside [22]. Phase studies of the pseudo-ternary system consisting of dextrose syrup (71.5% dextrose/28.5% water), *n*-butanol and butyl glucoside have proved that mixtures exhibit a single phase at conditions favorable to the Fischer reaction (see below). Hence, at any alcohol:glucose ratio the mixture is homogeneous containing at least 30% of butyl polyglucoside [23] (Figure 7.6). The boundary line between the single-phase and the two-phase region depends on the water content of the mixture. With growing water concentrations the boundary

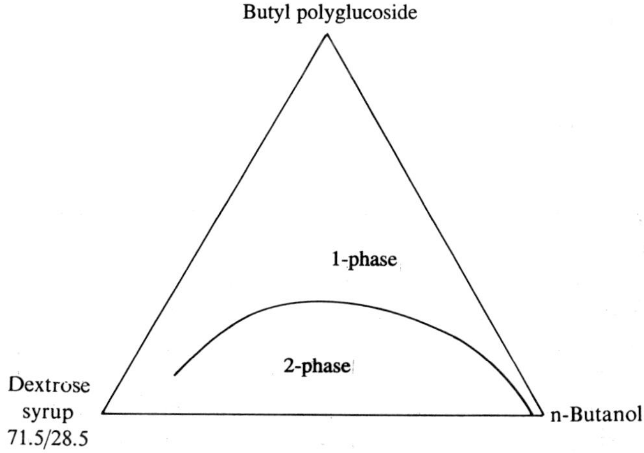

Figure 7.6 Pseudo-ternary phase diagram of dextrose syrup, *n*-butanol and butyl polyglucoside, 110°C (partly after Reference 23).

line is shifted to higher contents of butyl polyglucoside. Hence, in order to keep the reaction mixture homogeneous a certain content of butyl polyglucoside has to be maintained. This finding is the basis of the synthesis of butyl polyglucoside as the first stage of the *trans*-glucosidation route for alkyl polyglucoside. Since the reaction mixture in the second step, viz. butanol, butyl polyglucoside, fatty alcohol and alkyl polyglucoside (provided that its degree of glucosidation is not too high) is also a single-phase solution, the two-stage synthesis can be conducted under completely homogeneous conditions.

A further important point about the Fischer reaction is that the acetalization stage is an equilibrium reaction. Great excess of butanol and also a rapid removal of the water formed in the reaction as well as that introduced by the saccharide syrup favors the formation of the glucoside. A high rate of water elimination can be achieved by distillation. Because water and *n*-butanol form an azeotropic mixture, the alcohol in the azeotrope must be recycled into the reaction after separating from the water. A very suitable set-up for the synthesis of butyl polyglucoside is the combination of a stirred reactor and a reaction column [24]. The product leaving the column bottom for the second step of the synthesis, the transglucosidation, consists of about 35% of butyl polyglucoside dissolved in butanol [5].

Corresponding considerations are applicable for the transglucosidation stage (Figure 7.5) in which the rapid and quantitative elimination of the *n*-butanol reduces the back reaction, and shifts the equilibrium towards the alkyl polyglucoside.

A third critical point about the Fischer reaction is the thermal stress on the heat-sensitive carbohydrate reactants. For high reaction rates, which are desired, high temperatures have to be applied, but thermal stress, i.e. higher temperature and time of exposure, leads to oligomerization, ether formation, degradation and, especially, charring reactions. Whereas polymerization and ether formation are reactions competing with the acetalization in the presence of protons, the colour deepening is ubiquitous whenever the thermal stress is too high and may require bleaching, a rather subtle process. Hence, thermal stress has to be minimized in all steps of the process, in the course of which the distinct influence of temperature and residence time, respectively, have to be considered.

A last important point has to do with the catalyst and its concentration. Suitable catalysts are strong acids, such as sulfuric acid [26] or organic sulfonic acids [27]. Other catalysts mentioned in the literature are hydroxy polycarboxylic acids [28] and acidic cation exchange resins [17]. The same catalyst is usually used in both stages of the synthesis.

There is a considerable influence of the catalyst concentration upon the degree of glucosidation of the alkyl polyglucoside. With increasing acid concentration, *n* increases, leading to more viscous reaction mixtures [29]. In cases of relatively low fatty alcohol/butyl polyglucoside ratios the reaction mixture may become paste-like and difficult to handle. High contents of oligomers may also result in precipitation [5]. The points mentioned above indicate that the Fischer synthesis is complex and not simple to operate.

An outline of the two-stage synthesis operating continuously is as follows: This route consists of five steps (Figure 7.7):

- synthesis of butyl polyglucoside
- transglucosidation
- neutralization
- elimination of excess fatty alcohol and
- refining.

Synthesis of butyl polyglucoside. This can be performed in a stirred reactor/reaction column system at butanol:glucose molar ratios of 3:1 to 6:1 at temperatures of 90–120°C in the presence of a strong acid such as 0.1% H_2SO_4 [30]. The rapid removal of water at slightly reduced or atmospheric pressure reduces the reaction time and minimizes the formation of by-products.

Instead of using dextrose syrup, starch with a certain moisture content can also be employed [31]. In this case a butanolic degradation of the polysaccharide, requiring substantially more drastic conditions such as temperatures ⩾160°C, has to be performed simultaneously. Consequently, the butyl polyglucoside will be less pure and rather dark in colour compared to the situation in which a dextrose syrup is used.

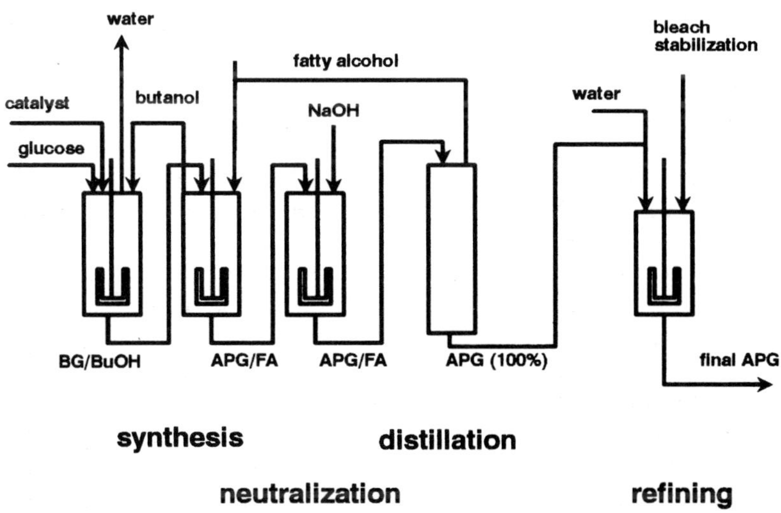

Figure 7.7 Processing of two-stage synthesis of alkyl polyglucosides.

Trans-glucosidation. The acidic butanol solution of butylpolyglucoside and fatty alcohol, in a molar ratio of 1 to 3–6, is fed into the transglucosidation step and reacted at a high rate at elevated temperatures (110–130°C) accompanied by vacuum distillation of the butanol, which is recycled into the first stage [32]. The degree of glucosidation *n* of the resulting alkyl polyglucoside, usually differing from the *m* of the butyl polyglucoside (Figure 7.5), can be tailored by the ratio of the reactants. For high figures of *x*, low values *n* are achieved. Another important factor is the residence time in the *trans*-glucosidation stage. As shown in Figure 7.8, with the strong decrease in the butyl glucoside concentration the amount of alkyl polyglucoside initially increases, but passes a maximum and decreases again in favor of the formation of polyglucose. Because water is absent, this formation is irreversible [5]. Hence, the residence time for the *trans*-glucosidation stage must be optimized in respect of the desired constituents.

Neutralization. The removal of excess fatty alcohol by distillation under catalytic conditions would shift the equilibrium to the left-hand side. Hence the catalyst has to be deactivated completely by neutralization with a base favorably performed in a continuous process [33]. Suitable bases are

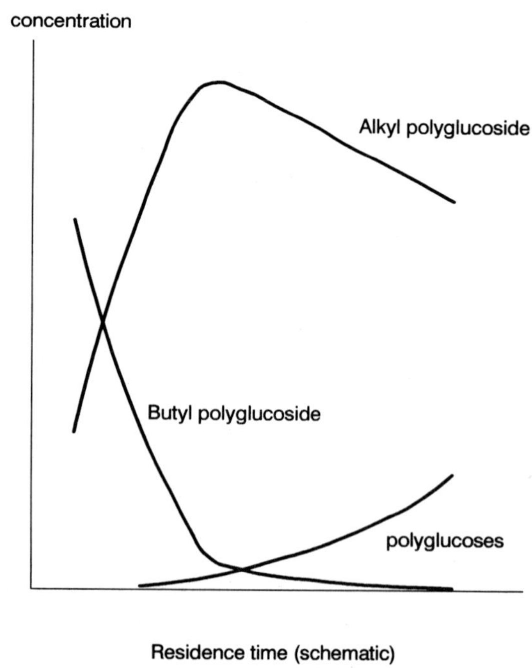

Figure 7.8 Concentration of starting material, product and by-product in the transglucosidation stage dependent on the residence time (schematic).

inorganic or organic alkali or alkaline earth compounds such as hydroxides, oxides, and alkylates.

Elimination of excess fatty alcohol. The subsequent multistage distillation process [34] involves initial removal of low-boiling products such as water and, depending on the neutralization step, possibly lower alcohols, at moderate temperatures and at slightly reduced pressure. The fatty alcohol is then eliminated under very low pressure at a comparatively moderate temperature in order not to affect the product adversely. Hence wipe-film or thin-film evaporators should be employed. The separated fatty alcohol is recycled into the *trans*-glucosidation step. The resulting solid alkyl polyglucoside is rather brownish colored, sticky, glass-like and difficult to handle. Hence, in order to minimize the thermal stress further the hot melt from the distillation stage is cooled by dissolving it rapidly in cold water, particularly since the following refining is also accomplished using an aqueous solution [5].

Refining. To meet the application requirements for the surfactants used in detergents and personal care products the final alkyl polyglucoside has to be light-colored and should not darken under long-term storage conditions (3 to 6 months at temperatures up to 40°C). Quite a lot of patents deal with these problems and demonstrate that their solutions are not simple. The reasons for the color are not clear, although humin structures are discussed [35]. The low concentration of colored materials probably combined with a similar complex isomerism, of the kind that is typical for the glucosides, makes clarification extremely difficult. Some of the procedures used to improve the color, mentioned in the patent literature, are based on process variants to protect the product against thermal stress [36] or adding small amounts of reducing agents such as hyposulfurous acid [37] or hypophosphoric acid [37, 38]. Others claim procedures for extracting the colored materials with polar solvents [39] or by adsorption on solid surfaces [32]. A great many methods of achieving a pale color are based on a final bleaching step, in which hydrogen peroxide particularly, but also ozone [40], sulfur dioxide, peroxy salts [41] and catalytic hydrogenation [35] are used as bleaching agents. The most efficient processes are probably those in which hydrogen peroxide is engaged as the bleaching agent [42]. Its reaction with aqueous alkyl polyglucoside of high concentration at elevated temperatures is accompanied by a strong decrease in pH [43] and entails the risk of hydrolysis. Therefore the bleaching should be performed under sufficiently alkaline conditions. A quite efficient variant uses the booster effect of low alkaline earth ion concentrations [44]. The refining process can be followed by stabilization to prevent microbial degradation. Because of their environmental safety, alkyl polyglucosides in aqueous solutions, of not excessively high concentrations ($\leqslant$ approx. 50%), may suffer from the disadvantage of biodegradation. For its prevention, mild antiseptic or fungicidal agents such as benzoic acid or sorbic acid in low concentrations can be employed. But even alkaline conditions, well known for their incompatibility with bacteria, fungi and yeasts, and their combination with water soluble alcohols create high stability to microbial attack [45, 46]. Generally, the alkyl polyglucosides are sold as aqueous mixtures containing approx. 30 to 80% by weight of active matter.

Direct synthesis. Most of the points discussed for the two-stage synthesis are also applicable to the direct route, but the situation seems to be much more difficult. Fischer glycosidation proceeds in a two-phase process here, and great efforts are necessary for a favourable space–time yield and the removal of the water from the reaction mixture in order to reduce the competitive polymerization forming polyglucose. As the reaction advances the mixture becomes more homogeneous [18].

Low-cost starch or glucose syrups are unsuitable raw material. Instead dry forms of solid glucose, e.g. dextrose monohydrate or dextrose anhydride, have to be used. Because the reaction is heterogeneous, the specific surface area of the solid has to be as large as possible. Hence finely divided glucose, often involving considerable difficulties in handling, has to be employed. The reaction mixture needs to be as anhydrous as possible. This means that water formed during the reaction has to be removed as quickly as it is formed in order to maintain the lowest possible water content [47–49]. In order to minimize the tendency towards biphase conditions favoring polymerization, claims are also made for hydrophobic catalysts which do not accumulate in the aqueous phase [50, 51]. It seems to be difficult to conduct the reaction continuously, batch-wise being preferred, while working-up and refining are usually continuous [5, 50]. The route of the direct synthesis also consists of five steps, viz:

- preparation of the glucose slurry
- glucosidation
- neutralization
- elimination of excess fatty alcohol and
- refining.

The first and sometimes the second steps differ from the course of the two-stage process.

Glucose slurry. To prepare the glucose slurry, finely divided anhydrous glucose, dextrose monohydrate for example, is suspended in fatty alcohol, and the quantity of water present is carefully reduced by heating under reduced pressure. In order to keep the high surface solid dispersed, high shear is applied [49].

Glucosidation. After adjusting the water level so that no undissolved water is present in the circulating slurry, the catalyst is added, and the reaction carried out at elevated temperature and reduced pressure to permit the water formed to be rapidly separated from the reaction mixture. The reaction rate depends on the chain length of the fatty alcohol, being higher with shorter-chain alcohols. The reaction temperatures have to be selected correspondingly. Typical values are between 90 and 140°C. The neutralization, elimination of excess fatty alcohols and refining are the same as in section 7.3.2(a).

7.3.3 *Composition of the alkyl polyglucosides*

Considering the different procedures and the very complex spectrum of the constituents, one should expect different products resulting from the two different routes. But as is shown for three commercial products manufactured by different companies for $C_{12}C_{16}$-alkyl polyglucosides of similar degrees of glucosidation the differences are not so dramatic (Table 7.1). The main components are the monoglucosides which contain a minor amount of alkyl furanosides. The next important groups are alkyl diglucosides and

Table 7.1 Composition (%) of $C_{12}C_{16}$-alkyl polyglucosides

Constituents	Synthesized via direct route		Via two stage process
	A	B	C
Water	48	45	48
Active matter	52	55	52
Surfactant composition			
G1	29(2)	28(1)	34(1)
G2	8	7	8
G3	4	4	3
G4	3	2	1
PG	8	12	5
FA	0.7	1.1	1.0
G	⩽0.1	0.1	0.1
BG	—	—	0.2
n	1.36	1.33	1.23
α/β	69/31	69/31	68/32

Analytical methods

G1, G2,... alkyl monoglucosides including furanosides as given in brackets, alkyl diglucosides,... by high temperature GC and thin layer chromatography (ThC); PG Polyglucose with small amounts of higher alkyl polyglucosides by precipitation with ethanol of aqueous/methanolic solutions, ThC; FA fatty alcohol by GC; G glucose by ThC or GC; BG butyl polyglucoside by GC; *n*, degree of glucosidation by high temperature GC; α/β-anomers by GC. (For details see Reference 52.)

triglucosides. Of higher oligomers only tetra- and in some cases pentaglucosides can be detected by high-temperature GC [52]. The content of polymerized glucose (PG) is quite considerable. As expected its content is higher for the products prepared via direct synthesis. Surprisingly, the content of butyl polyglucoside (BG) in product C is rather low and can be adjusted even lower. Hence the transglucosidation stage can be conducted quantitatively. The butanol content of about 10 ppm in this specimen is typical of natural fatty alcohols themselves. The anomeric ratio is the same for the three products.

7.4 Properties

7.4.1 Chemical properties

Alkyl polyglucosides are non-ionic compounds which, when synthesized, due to the Fischer glycosidation process contain minor portions of anionic constituents which are probably isomeric alkyl glucuronic acids [43, 60]. As acetals, the alkyl polyglucosides exhibit the typical property of hydrolysis to fatty alcohol and glucose at elevated temperatures and in an acidic environment. At pH $\geqslant 4$ and temperatures of 40°C shelf life tests have shown stability is >4 months [10], as is shown in Figure 7.9.

Similar to carbohydrates, alkyl polyglucosides contain primary and secondary OH groups which can be functionalized to form non-ionic, anionic or cationic compounds [14]. The primary OH groups are more reactive than the secondary ones and can easily be oxidized or sulfated.

7.4.2 Physical properties

(a) General properties. Solid state properties of defined alkyl glycosides are frequently discussed in the literature. As early as 1911 Fischer described the double melting point behavior of β-cetyl-D-glucopyranoside [8]. Later it was demonstrated that this is due to the formation of thermotropic liquid crystals [53], a property characteristic of sugar moieties linked to an *n*-alkyl chain containing more than six carbon atoms [54]. The two-stage melting process can be explained by low-temperature disengagement of the hydrocarbon chains from the three-dimensional crystal lattice (melting point) and higher temperature cleavage of the hydrogen bonds forming the strongly cohesive bonds between the carbohydrate moieties (clearing point). In between, often covering a temperature difference of 40 to 70°C is the range of thermotropic liquid crystals. This behavior is also shown by alkyl polyglucosides [55].

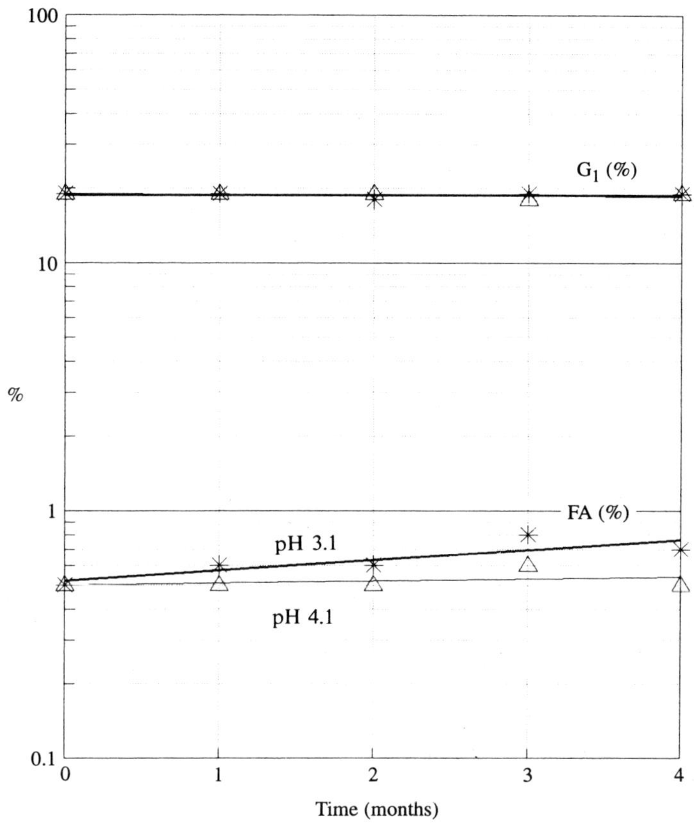

Figure 7.9 Shelf life at 40°C of $C_{10}C_{12}$-alkyl polyglucosides, conc. 30%, at pH 3.1 and 4.1, content of alkyl monoglucoside (G_1) and fatty alcohol (FA) [10].

Alkyl glucosides and alkyl polyglucosides with longer *n*-alkyl chain lengths show relatively low solubility in organic solvents and water. That causes them to concentrate in the interface and to display high interfacial activity. The best solvents are aromatics such as xylene or terpenes, and water. β-Anomers are more highly soluble in water than α-anomers [13, 56–58], and oligoglucosides have greater solubility then monoglucosides [11]; there is a strong reduction in solubility with longer hydrocarbon chains. Lower alcohols and glycols, which often increase the solubility in aqueous solutions, are less effective solvents. Other solvents such as formamide, and ketones such as cyclohexanone, are also less effective. In the presence of water and water-insoluble solvents middle phase microemulsions are frequently formed.

(b) Aqueous solution properties

Phase diagrams. The phase behavior of alkyl polyglucosides in water differs appreciably from those of the customary surfactants. Hexgonal liquid crystalline phases have not been observed although the micelles have a pronounced rod-shape structure. At higher concentrations lamellar phases appear, easy to identify by their characteristic texture [59]. The phase diagram of the alkyl polyglucosides is considerably influenced by the chemical structure, the effect of the hydrocarbon chain length being stronger than that of the degree of glucosidation.

For relatively short chain lengths like C_8 or C_{10} the diagrams are determined by a pronounced micellar phase (L_1) which, at higher concentrations, is followed after a two-phase region by a lamellar phase (L_α) as shown for $C_{10}C_{12}$-$APG_{1.3}$ (degree of glucosidation 1.3) in Figure 7.10. The Krafft point curve remains constant at about 1°C up to concentrations >60%.

This rather simple phase behavior changes drastically with longer chain alkyl groups as shown in Figure 7.11 for $C_{12}C_{14}$-$APG_{1.3}$. The Krafft point

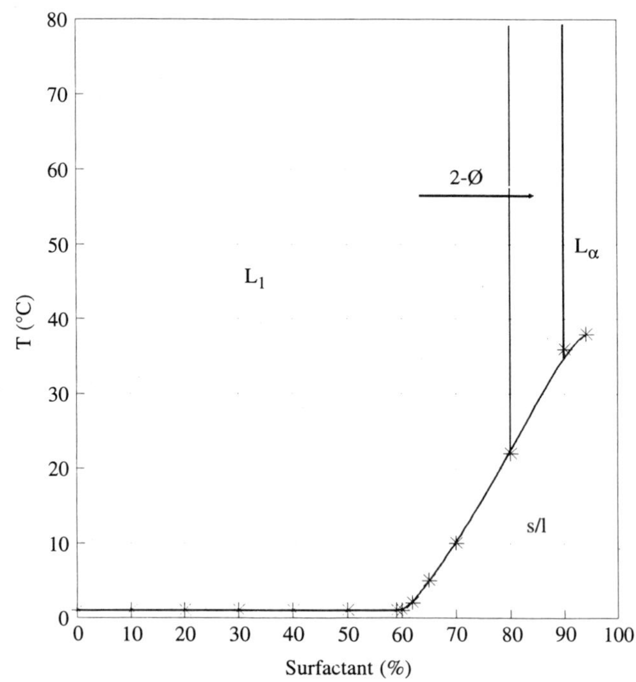

Figure 7.10 Phase diagram of $C_{10}C_{12}APG_{1.3}$, L_1 and L_α indicate the micellar and the lamellar liquid crystalline phase separated by a two-phase region, s/l solid–liquid region [60].

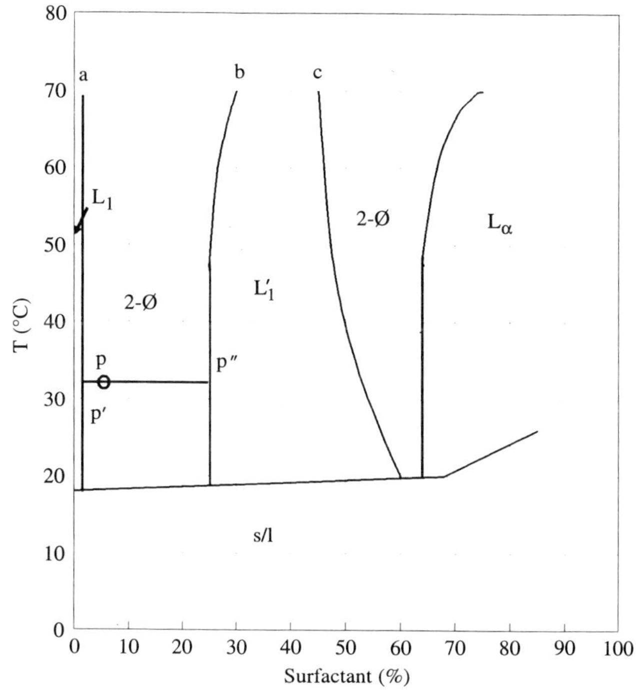

Figure 7.11 Phase diagram of $C_{12}C_{14}$-$APG_{1.3}$, L_1 and L'_1 indicate the micellar and L_α the lamellar liquid crystalline region, L'_1 and L_α are separated by a two phase region in which p is a certain point. At p the surfactant solution separates into an upper phase of concentration p′ and a lower phase of concentration p″ [60].

curve again shows relatively little dependence on concentration and remains near room temperature. The broad micellar range is separated by a pronounced miscibility gap into a low-concentration region L_1, and a high-concentration region, L'_1. In the range of coacervation between 1% and 25%, rapid phase separation is observed as indicated by the tie-line p′p″. The phase boundary line (a) corresponds to the equilibrium concentration in the upper phase, and (b) to that in the lower phase. In both L_1 and L'_1 rod-like micelles were observed, in L_1 by means of electrical birefringence and conductivity measurements in orthogonal directions under Couette flow, and in L'_1 by the latter technique [60]. The single-phase region is followed, after the usual miscibility gap, by the liquid crystalline region. This is not, as would be expected from the rod-like micellar structure, a hexagonal phase but a lamellar L_α phase, which means that near the phase boundary as in the case of $C_{10}C_{12}$-$APG_{1.3}$ a transition from rod-like to disk-like micelles should occur.

The slopes of the phase boundary lines shown in Figure 7.11 are quite independent of the temperature, but for line (c) this is only true if the level of the remaining fatty alcohol is low. The more this exceeds the normal value the more the slope of (c) decreases, thus increasing the two-phase region between L_1' and L_α, a fact that may be of importance for storing the product.

A notable point about the phase diagram is the broad miscibility gap in the concentration range between about 1% and 20%. No cloud point or lower consolute temperature, as is so characteristic of ethoxylates, is observed. Nevertheless, the coacervation behavior (cloud point) of the alkyl polyglucosides in Figure 7.11 is reminiscent of the clouding of the fatty alcohol ethoxylates, and the lower critical consolute point may be concealed by the heterogeneous region below the Krafft curve. This has been proved by the addition of increasing amounts of $C_{10}C_{12}$-$APG_{1.3}$ [60]. As shown in Figure 7.12 the initially wide miscibility gap becomes narrower and finally the coacervation region separates from the region below the Krafft point curve, and a lower critical consolute point appears. A similar effect can be visualized by increasing the degree of glucosidation [60]. Thus, alkyl

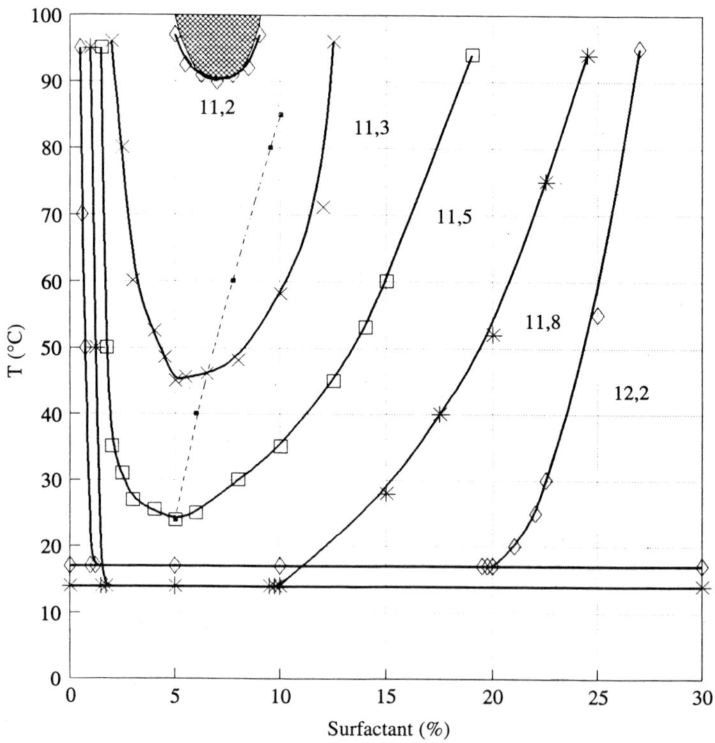

Figure 7.12 Phase diagram at lower concentrations of alkyl polyglucoside (degree of glucosidation 1.3) versus mean alkyl chain length (indicated by numbers). Critical point P_C of $C_{11.5}$-$APG_{1.3}$: c_c:5%, T_C:25°C [60].

polyglucosides show normal cloud points, but their dependence on changes in the chemical structure is, because of a much stronger interaction with the water, much more sensitive than for fatty alcohol ethoxylates. Changing the degree of glucosidation from 1.75 to only 1.85 for $C_{12}C_{14}$APG increases the cloud point from 27°C to 87°C. Bringing about a similar effect in the case of $C_{12}C_{14}$-ethoxylates means increasing the degree of ethoxylation by more than 4 EO units. This much higher effect can be used as a very sensitive analytical tool to characterize alkyl polyglucosides [61].

The cloud points of the alkyl polyglucosides are increased by the addition of alkali and anionic surfactants but decreased by salts. Even alkyl polyglucosides with fairly short alkyl chains such as C_8C_{10} have cloud points on addition of salts at lower surfactant concentrations (< approx. 1%) [60]. As a result of adding cationic surfactants, which form mixed micelles with alkyl polyglucosides, the cloud point initially decreases, passes through a minimum and then increases. This can be explained on the basis of charge effects [60]. The original small negative charge of the APG micelle (approx. 5%) is initially neutralized by the cationic surfactant and thereafter becomes positively charged. This increase in charge leads, as for ethoxylates, to a strong increase of the cloud point [62].

Rheology. As already mentioned, even at low concentrations alkyl polyglucosides form strongly anisometric rod-shaped micelles. By increasing the hydrocarbon chain length the axis ratio is greatly increased up to about 70 for $C_{12}C_{14}$-$APG_{1.3}$ [59]. These micellar structures are reflected in the viscosity data shown in Figure 7.13. Although the aqueous solutions of 50% active matter are almost Newtonian, even for $C_{12}C_{14}$-alkyl, the viscosity is greatly increased as a result of increasing the hydrocarbon chain lengths.

In blends with anionic surfactants such as $C_{12}C_{14}$-fatty alcohol ether sulfates with 2 EO (FAES) the behavior of alkyl polyglucosides depends on their chain lengths. For longer chain lengths they intensify the shear thinning behavior of FAES in the presence of salt, whereas C_8C_{10}-$APG_{1.3}$ eliminates the shear-thinning property of FAES (Figure 7.14).

Blends of relatively long alkyl polyglucosides with FAES or other anionic surfactants, mostly in the presence of salt, form viscoelastic transient networks [63]. Data obtained from an oscillation viscometer as a function of the angular frequency ω, are the complex viscosity η^*, the storage modulus G' and the loss modulus G'', a measure of either elasticity or viscosity, depending on whether the energy supplied to the system is stored or lost in the form of heat. If $G' > G''$, the behavior is said to be predominantly elastic, while in the reverse case the system behaves viscously. As Figure 7.15 shows, the behavior of the network is viscous at low ω. As the frequency increases, G' rises more sharply than G'', the slopes in a double-logarithm plot being 2 and 1, respectively. Hence the surfactant network conforms to the Maxwell model of viscoelasticity, namely a series

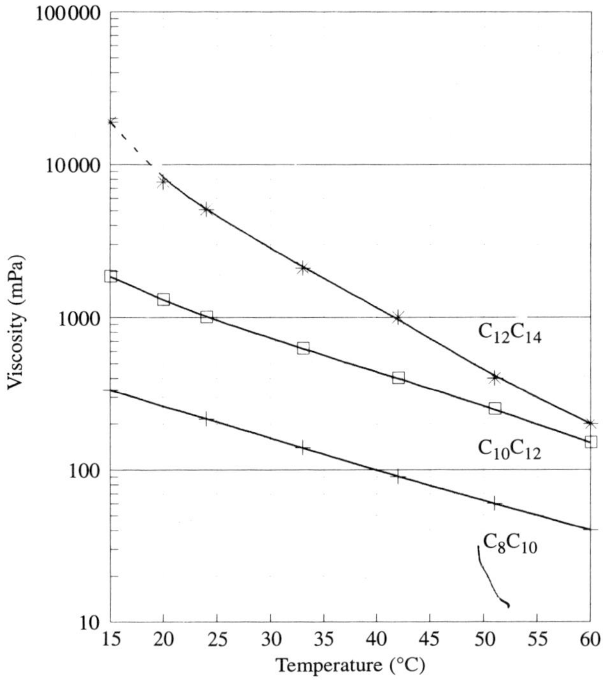

Figure 7.13 Viscosity of alkyl polyglycosides (degree of glucosidation 1.3, active matter 50%) versus temperature, effect of alkyl chain length.

of a spring (for elasticity) and a damping system (for viscosity) [64]. At higher frequencies, the storage modulus outweighs the loss modulus and the liquid is predominantly elastic. G' levels off and passes into a constant G°, which is the elasticity constant in Hook's law. (A more detailed discussion of the properties of the transient networks formed by alkyl polyglucosides and anionic surfactants would be beyond the scope of this section and is presented elsewhere [65].)

(c) Surfactant properties

Micellization and surface activity. The surface tensions of three different alkyl polyglucosides based on C_8C_{10}-, $C_{10}C_{12}$- and $C_{12}C_{14}$-fatty alcohols and with degrees of glucosidation of about 1.2 are shown in Figure 7.16. The curves clearly have two points of inflexion, which should mean that two different micelles are in equilibrium. This situation is quite uncommon for aqueous surfactant solutions, because usually even mixtures of very differently structured surfactants form one type of mixed micelle. For defined isomers like α- and β-alkyl D-glucosides [56, 57, 66] and β-alkyl D-maltosides

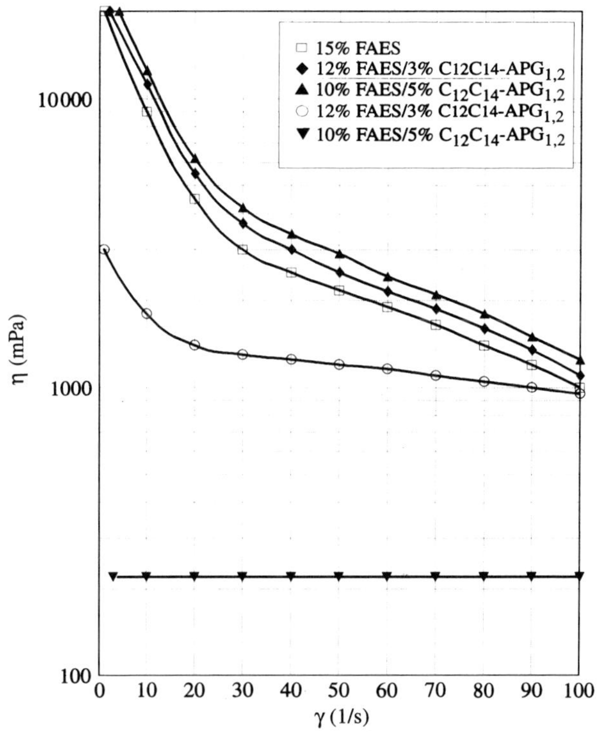

Figure 7.14 Viscosity versus shear rate for FAES/4% NaCl at 25°C, effect of different alkyl polyglucosides.

[67] only normally shaped CMC characteristics are described, or at least any deviation from normality is not mentioned.

This shape (Figure 7.16) or surface tension curve has been observed for all alkyl polyglucosides which have been studied over many years [59]. There is no doubt that CMC_{II} is the concentration at which the rod-shaped micelles already mentioned are formed. But the meaning of CMC_{I} is questionable, and scattering data that would give a closer insight are not obtainable because of the very low concentration range.

The influence of the hydrocarbon chain length on CMC_{I} and CMC_{II} is approximately the same (Figure 7.17). This is also true for the effect of the degree of glucosidation, both CMCs increase similarly in line with increasing c_n. Whereas the required area per molecule is practically not influenced by increasing the hydrocarbon chain length, there is the expected strong effect produced by the voluminous glucosidic head groups. No change in either of the CMCs is brought about by increasing the temperature from 20 to 70°C. Hence CMC_{I} and CMC_{II} behave similarly.

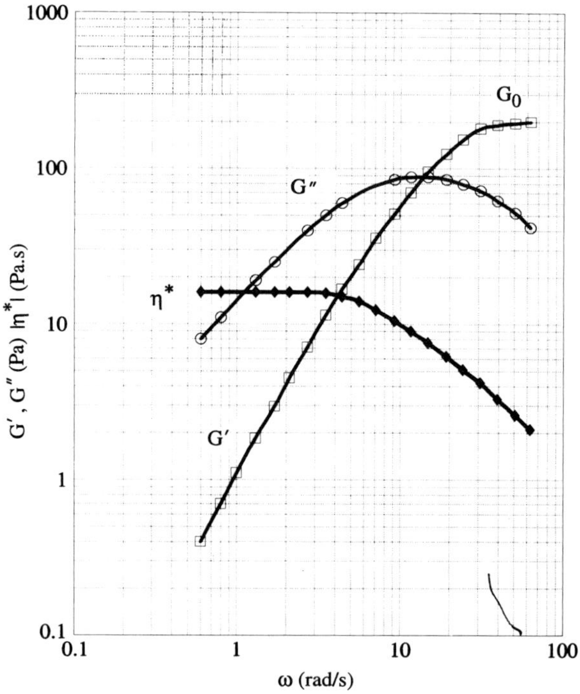

Figure 7.15 Complex viscosity η^*, storage modulus G′ and loss modulus G″ versus angular frequency ω for FAES 7.5%, $C_{12}C_{14}$-$APG_{1.3}$ 7.5%, NaCl 2%, 25°C, ampl. 4°.

A result of adding salt at $\geqslant 1$ g/1 is that only one inflexion point (Table 7.2) exists, i.e. only one type of surfactant micelle remains. The data of this CMC lies between that of CMC_I and CMC_{II} in Figure 7.16. The new CMC should correspond to CMC_{II} because salt favors anisometric micelles. It is surprising that the CMC of non-ionic surfactants decreases on addition of small amounts of salts [66]. However, we have to consider that alkyl polyglucosides are partially anionic surfactants, and because of the electrostatic repulsion of the already voluminous head groups, CMC_I should be the concentration at which globular micelles are formed.

Addition of anionic surfactants such as $C_{13}C_{17}$-alkane sulfonate (PS) or $C_{12}C_{14}$-fatty alcohol ether sulfate with 2 EO (FAES) gives a surface tension curve of the alkyl polyglucosides with only one inflexion point (Table 7.3). The new CMC should correspond to CMC_I, because the birefringence data prove that anisometric micelles have disappeared [68]. As is usual for blends of surfactants with very different CMCs, the CMC of the mixture lies in between [69]. The area required per molecule is greatly enlarged and decreases like the CMC itself in the presence of water hardness, an effect that is well known to be due to the shielding from electrostatic repulsion by alkaline earth ions.

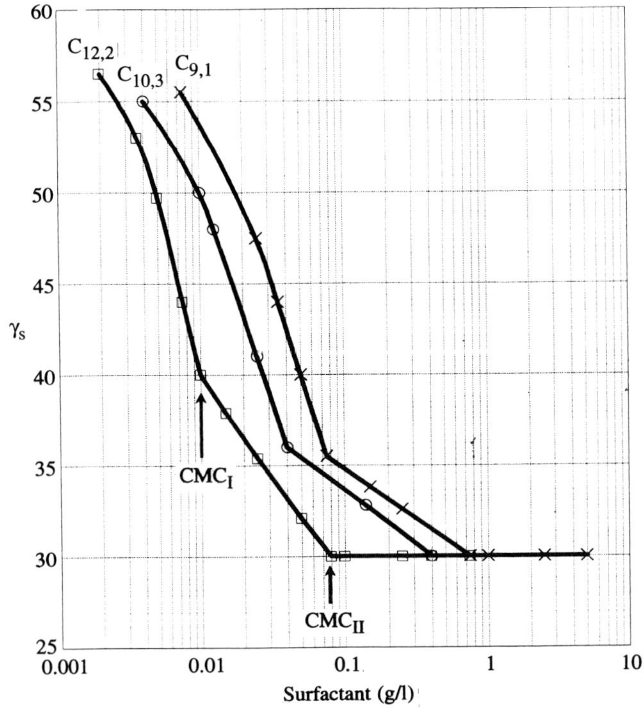

Figure 7.16 Surface tension (mN/m) versus surfactant concentration for C_8C_{10}-, $C_{10}C_{12}$- and $C_{12}C_{14}$-APGl$_{1.2}$ (numbers indicate mean alkyl chain length), 20°C.

The aggregation numbers of the micelles of C_8C_{10}- and $C_{12}C_{14}$-APG$_{1.2}$ are approx. 220 and 420, respectively [60]. It is possible to compare alkyl polyglucosides with other known non-ionic [70], anionic [71] and cationic surfactants [72], as regards surface activity, by plotting the critical micelle concentrations on a semilogarithmic scale as a function of the length of the hydrophobic residue, similar to Traube's rule. This is shown in Figure 7.17. The CMC values of the alkyl polyglucosides correspond to those of the fatty alcohol ethoxylates and are much lower than those of ionic surfactants. What is surprising, however, is the slope, which is lower than for the non-ionic surfactants, but close to that of the ionic surfactants.

Interfacial tension. The interfacial tensions at the water/oil interface—of crucial significance for many processes such as emulsification or detergency—is greatly reduced by alkyl polyglucosides. Figure 7.18 shows this with reference to paraffin oil. As the carbon chain length of the surfactant increases the interfacial tension is sharply decreased. If a cosolvent such as *i*- or *n*-butanol is added, the interfacial tension can be reduced further. At the same time a third phase appears, a middle-phase microemulsion [59],

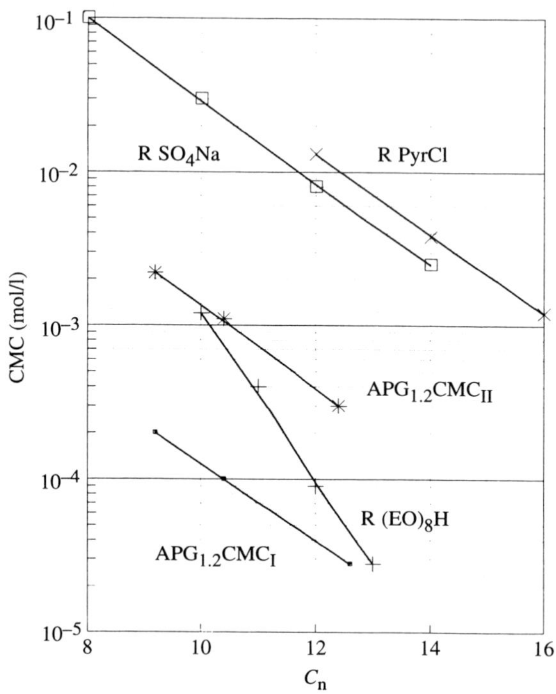

Figure 7.17 CMCs of different surfactant classes versus hydrocarbon chain length. (Figure reproduced with permission from Hanser Verlag, Munich.)

as closer investigations have shown. The appearance of this microemulsion, frequently associated with low interfacial tensions, is observed for many surfactants in the presence of cosolvents, provided the conditions are optimized by salt addition, as in the case of anionic surfactants [73], or via temperature adjustment, as in the case of non-ionic surfactants [74].

Alkyl polyglucosides exhibit different behavior. They are largely invariant to temperature as well as to salinity. The very wide salinity windows for various glucosides and different oil phases are demonstrated in Table 7.4.

Table 7.2 CMC of $C_{10}C_{12}$-$APG_{1.2}$ at 20°C: effect of electrolyte

Electrolyte	CMC_I (g/l)	CMC_I (mmol/l)	$\gamma_s(I)$ (mN/m)	CMC_{II} (g/l)	CMC_{II} (mmol/l)	$\gamma_s(II)$ mN/m
0 g/l NaCl	0.035	0.098	39.0	0.37	1.03	30
1 g/l NaCl	0.033	0.092	36.5	0.29	0.81	29
10 g/l NaCl	—	—	—	0.14	0.39	29
1 g/l $CaCl_2$	–	–	–	0.12	0.33	29

Table 7.3 Critical micelle concentration of $C_{12}C_{14}$-APG$_{1.5}$ blended with anionic surfactants (1:1 by weight): effect of water hardness

Surfactant	Hardness* (°dH)	CMC_I (g/l)	γ_s(I) (g/l)	CMC_{II} (g/l)	γ_s(II) (mN/m)	$-\frac{\partial \gamma s}{\partial \log c}$	A (Å^2)
$C_{12}C_{14}$-APG	0	0.018	36	0.09	30	25	37.4
/PS	0	0.16	31	—	—	17	55.5
/PS	12	0.03	30	—	—	20	46.7
/PS	25	0.03	30	—	—	23	40.6
/FAES	0	0.13	32	—	—	16	58.5
/FAES	12	0.03	31	—	—	23	40.6

*German degree of hardness °dH
1°dH = 1.25°dE (UK) = 1.04°dA (US)
1°dH = 10 mg CaO in 1 litre of water
PS = $C_{13}C_{17}$-paraffinsulfonate-Na
FAES: $C_{12}C_{14}$-fatty alcohol ether (2EO) sulfate-Na.

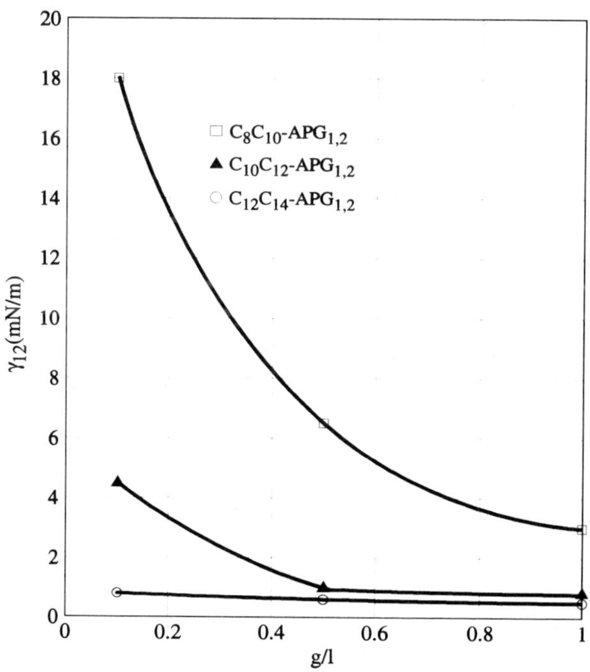

Figure 7.18 Interfacial tension versus surfactant concentration ($C_{10}C_{13}$-paraffin oil/water, 25°C) (for low values $\leqslant 3$ mN/m spinning drop and for high $\geqslant 3$ mN/n ring tensiometer technique were used).

Table 7.4 Salinity requirement for forming a middle-phase micro-emulsion (O/W = 1.4% alkyl polyglucoside, 4% *i*-butanol, 50°C)

APG	Oil (HLB)	Salinity range (%NaCl)
C8C10APG-1.2	*n*-decane (10.5)	6–26*
C12C14-APG1.2	"	0–10
C12C14-APG1.4	"	0–16
C12C14-APG1.2	*iso* propyl myristate (9.5)	1–26*
C9C10-APG1.2	lavender oil (12.5)	0–26*
C8C10-APG1.2	*o*-Xylene (14)	0–26*

*Solubility boundary for NaCl.

More hydrophilic surfactants move the window towards higher salt concentrations, whereas more polar oils shift it to lower salinities. The corresponding solubilization parameters (volumes of oil and water in the middle phase per volume of surfactant) for the alkyl polyglucosides are not very high, as shown in Figure 7.19. With increasing salinity, the oil is more solubilized than the water. For the intersection point of the two quantities V_o/V_s and V_w/V_s, Hue [75] has derived an equation from which the interfacial tension can be estimated*. Using this gives the interfacial tension of $C_{12}C_{14}$-$APG_{1.2}$ in Figure 7.19 of about 0.06 mN/m, which agrees very well with the value of 0.05 mN/m measured by the spinning drop technique. At higher salt concentrations, the three-phase microemulsion is transformed into an upper-phase (i.e. w/o) microemulsion. Similar arguments apply to the corresponding temperature window. Its width (studied between 5 and 95°C) is larger than the usual temperature range of alkyl polyglucoside application [59].

The o/w interfacial tension of the alkyl polyglucosides studied for several oils is independent of the salinity [59] and temperature. There are no deep minima under certain conditions, as is often true of other surfactants. Low interfacial tensions seem to be a general characteristic of alkyl polyglucosides no matter which organic phase is studied. Compared with other emulsifiers such as ethoxylated sorbitan mono fatty acid esters or ethoxylated fatty alcohols the interfacial tensions of alkyl polyglucosides are one to two orders of magnitude lower. This is the basis for their great emulsifying power [10]. Even polysiloxane oils, which are rather difficult to emulsify, can be easily dispersed in water just by employing a small amount of $C_{10}C_{12}$-$APG_{1.2}$ [76].

In a study involving essential oils [10] covering an HLB scale from about 11 to 17 the HLB procedure was successfully applied in order to tailor an efficient emulsifier for each of the oil phases [77]. As shown in Table 7.5, if $C_{12}C_{14}$-APG is employed alone the interfacial tension data are often very low, but perfect emulsification, i.e. bluish or even translucent, finely dispersed emulsions, is only observed under conditions in which the HLB value

*$\gamma_{12} \cdot (V/V_s)2/\cos \pi/4 \approx 0.34$

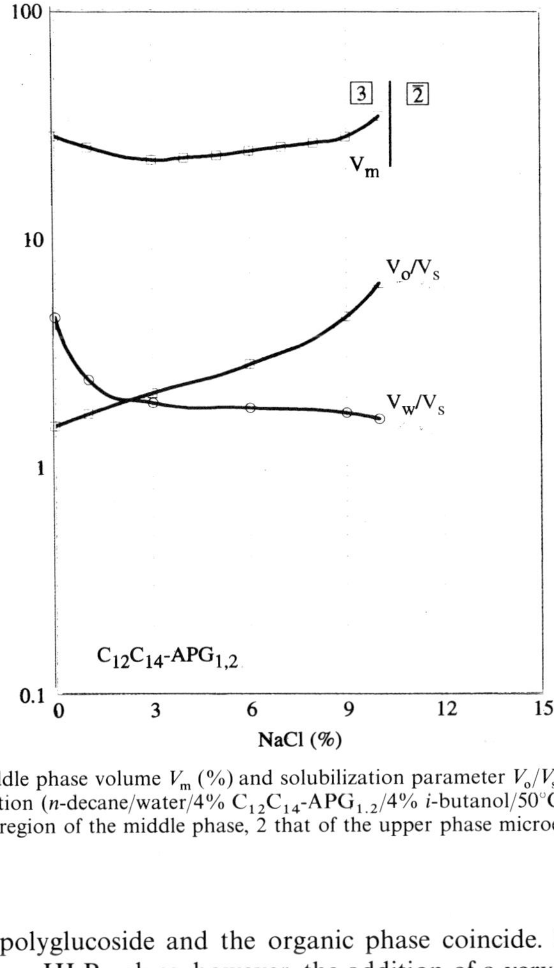

Figure 7.19 Middle phase volume V_m (%) and solubilization parameter V_o/V_s and V_w/V_s versus NaCl concentration (*n*-decane/water/4% $C_{12}C_{14}$-$APG_{1.2}$/4% *i*-butanol/50°C/O/W = 1), 3 indicates region of the middle phase, 2 that of the upper phase microemulsion.

of the alkyl polyglucoside and the organic phase coincide. In the case of higher oil-phase HLB values, however, the addition of a very small amount of any anionic, cationic or betaine surfactant in order to adjust the HLB very often gives perfect emulsification.

Adsorption. As the literature indicates not much work has been done on the adsorption of alkyl polyglucosides on solid surfaces. Studying the dispersing action on titanium dioxide pigments by alkyl polyglucosides in aqueous solution, Smith *et al.* [78] have also studied adsorption. The isotherms for C_8C_{10}-, C_9C_{11}- and $C_{12}C_{16}$-APG with relatively high degrees of glucosidation ($n \geqslant 1.6$) are more or less S-shaped and do not clearly indicate a plateau, which is presumably due to not employing higher equilibrium concentrations. Recent studies by the author, of alkyl polyglucoside adsorption on oxide surfaces such as alumina or Fe_2O_3 from aqueous solutions, have also generated S-shaped isotherms, but with slightly

Table 7.5 Interfacial tension γ_{12} water/oil of $C_{12}C_{14}$-$APG_{1.2}$ (HLB 13.0) and blends with 5% anionic surfactants and emulsification of essential oils in water

Organic phase	HLB	γ_{12}^* (mN/m)	E**	γ_{12}^* +5% co-surf.	E** +5% co-surf.
$C_{10}C_{13}$–paraffin	10.4±0.5	0.3	--		
Anise oil	11.7	1.9	+	>2	+
Lavender oil	12.8	1.8	—	2.5	+
Orange oil	13.0	0.01	++	0.2	+
Eucalyptus oil	13.2	1.0	—	1.5	+
Lime oil	13.6	0.1	—	0.9	++
Lime oil + $C_{12}C_{14}$-$APG_{1.4}$(HLB 13.6)	13.6	0.1	++		
Fir leaf oil	14.1	0.5	—	1.5	++
Fir leaf oil + $C_{12}C_{14}$-$APG_{1.5}$(HLB 14.1)	14.1	0.3	++		
Pine leaf oil	15.6	0.1	—	0.9	++
Cedar oil	16.7	0.5	—		+

* 0.5% surfactant, spinning drop technique
** Surf./oil 1:4, o/w 1:10, spontaneous emulsification, partly by addition of 5% anionic surfactant relating to emulsifier.
++ Translucent emulsion
\+ Bluish-white
o White (stable 24 h)
— White, light creaming
-- Separated after 24 h.

more hydrophobic glucosides saturation values at equilibrium concentrations higher than CMC were clearly achieved [79]. As demonstrated in Figure 7.20 under conditions far from the cloud point, the adsorption level and also the affinity between surfactant and surface depends greatly on the chemical structure of the alkyl polyglucosides. Whereas the adsorption of C_8C_{10}-$APG_{1.2}$ (HLB 15.5) and $C_{12}C_{14}$-$APG_{1.6}$ (HLB 14.3) is rather low, in the case of $C_{12}C_{14}$-$APG_{1.2}$ (HLB 13.0) a plateau is reached at surface areas per molecule of 50 Å^2, slightly higher than those at the air/water interface with 33 Å^2, (Figure 7.16). A space requirement figure of 33 Å^2 corresponds exactly to the saturation adsorption of this surfactant on the iron oxide surface. Further studies are necessary for an interpretation.

The S-shaped isotherms for the oxide surfaces indicate a cooperative interaction in the adsorption layer. In contrast, the adsorption on more hydrophobic surfaces such as carbon black follows the Langmuir isotherm. The space requirement for $C_{12}C_{14}$-$APG_{1.2}$ at the plateau is about 37 Å^2 per molecule and is similar to the data at the air–liquid interface.

7.4.3 Biological properties

Besides economic criteria such as availability and a low cost performance ratio, modern surfactants must meet environmental compatibility and toxicologic safety standards. Because their chemical structure is similar to that of glycolipids [80] and other biological surfactants that are metabolic

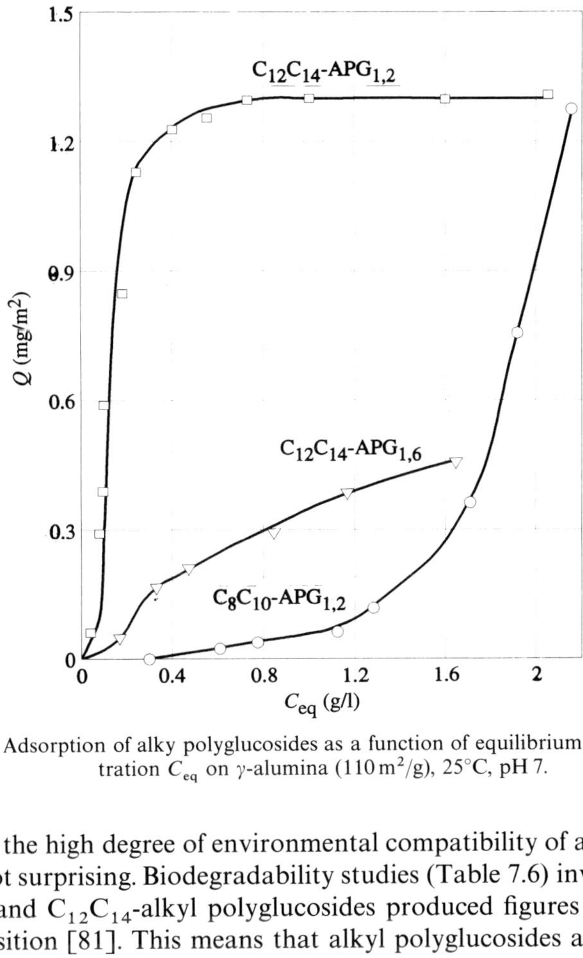

Figure 7.20 Adsorption of alky polyglucosides as a function of equilibrium surfactant concentration C_{eq} on γ-alumina ($110\,m^2/g$), 25°C, pH 7.

products, the high degree of environmental compatibility of alkyl polyglucosides is not surprising. Biodegradability studies (Table 7.6) involving C_8C_{10}-, $C_{10}C_{12}$- and $C_{12}C_{14}$-alkyl polyglucosides produced figures of 95 to 100% decomposition [81]. This means that alkyl polyglucosides are most readily degradable. This also remains true under anaerobic conditions [82].

The data for acute chronic toxicity illustrated by the $C_{12}C_{14}$-alkyl polyglucoside (Table 7.6) demonstrate that they are very readily compatible with the environment. Tests in animal models as well as on humans prove that $C_{12}C_{14}$-APG is far less irritating than conventional surfactants. In blends with more aggressive surfactants it can also greatly mitigate their irritation potential [83].

7.5 Applications

It is not the intention of this section to provide a comprehensive overview of the possible uses of alkyl polyglucosides. In the final analysis, they will be of interest wherever temperatures are not too high and the medium is not

Table 7.6 Biodegradability and toxicity data of $C_{12}C_{14}$-$APG_{1.2}$ [81, 82]

Biodegradability	DOC reduction*
Coupled units test (OECD 303A)	96%
Modified OECD screening test (OECD 301E)	94%
Modified coupled units test (DIN 38412/Teil 26)	101%
Toxicity	
Acute toxicity	
Fish (golden orf, 96 h)	LC_{50} 6 mg/l
Daphnia (EG 84/449, 48 h)	EC_{50} 30 mg/l
Earthworm (EG 88/302, 14 d)	$LC_{50} > 1$ g/kg
Terrestial plants (OECD 208, 16 d)	$EC_{50} > 100$ mg/kg
Chronic toxicity	
Bacteria (DIN 38412 Teil 8, 16 h)	EC_{50} 3400 mg/l
Daphnia (OECD 202, p. II, 21 d)	NOEC/LOEC 1.9/2.6 mg/l
Algae (EG 88/302, 72 h)	E_bC_{50} 7.4 mg/l
Further toxicity data	
Acute oral (rat)	$LD_{50} > 3000$ mg/kg
Skin sensitizing (Guinea-pig, Bühler)	non-sensitizing
Skin irritation (rabbit, Draize)	irritant
Mucous membrane irritation (rabbit, Draize)	slightly irritant
Mutagenicity (Ames, EG 84/449)	non-mutagenic

*DOC: dissolved organic carbon.

too acidic, because they are acetals of a sugar structure. They will be very attractive to use wherever their excellent surfactant performance and favorable ecotoxicological properties are desired, i.e. in cosmetics and in household products; but their very low interfacial tensions and high dispersing power make them attractive for many technical applications. The ability to apply a surfactant depends not only on its own properties but even more on its performance when combined with other surfactants. Being slightly anionic the alkyl polyglucosides are readily compatible with non-ionic, anionic or betaine surfactants. Making allowance for clouding phenomena (section 4.2.2(b)) they are also compatible with cationic surfactants.

In many cases alkyl polyglucosides exhibit favorable synergistic effects in combination with other surfactants, and the practical application of these effects are reflected in the impressive figure of about 400 patent applications for the utilization of alkyl polyglucosides published since 1981. Those cover dishwashing agents, light-duty detergents, heavy-duty detergents, alkaline cleaners, personal care products such as shampoos, shower gels, lotions and emulsions, technical dispersions such as color pastes, formulations for foam inhibitors, demulsifiers, plant protection agents, lubricants, hydraulic liquids, and crude oil production chemicals.

In addition to the basic behavior of the material a very important surfactant property with regard to the application is the foaming power. It

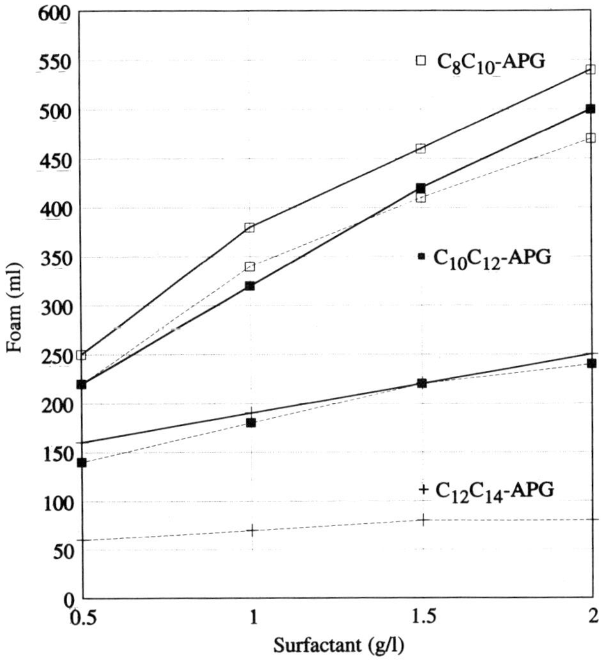

Figure 7.21 Foaming volume versus surfactant concentration, effect of alkyl chain length, degree of glucosidation 1.2, —demin. water, --- 13° German hardness, 25°C.

is desirable in many cases and undesirable in others. Hence foam must be controllable, a criterion that is easy for alkyl polyglucosides to fulfil in most cases.

Alkyl polyglucosides are moderate foamers [79], their foam value decreases in line with increasing hydrocarbon chain length (Figure 7.21). Alkyl polyglucosides with branched short chains such as 2-ethyl hexyl, in contrast to n-octyl with about the same degree of glucosidation, are low foamers although their surface activity is high. Increasing the degree of glucosidation increases the foaming power. In contrast to the behavior of fatty alcohol ethoxylates, the foaming power is often greatly reduced by water hardness. This effect is not limited to hardness ions but is generally due to electrolytes which, with the exception of bases, interact with the partially anionic alkyl polyglucoside in forming coacervates [60] and usually lowering the cloud point. The appearance of the second phase diminishes the foam.

If the negative charge of the alkyl polyglucoside micelle is sharply raised by adding anionic surfactants, clouding disappears, and the foam of the blend often increases even in comparison to the anionics themselves. This is particularly true in the presence of sebum, in which case the volume and stability of the foam of the anionic surfactant are enhanced by adding alkyl polyglucosides [79]. Thus, although moderate foamers by themselves, alkyl polyglucosides are often foam boosters for anionic surfactants.

7.5.1 Manual dishwashing compositions

Manual dishwashing agents are usually blends of anionic surfactants which exhibit synergistic cleansing effects. The use of a non-ionic surfactant as a main component is uncommon. As shown in Figure 7.22 ethoxylates combined with a commonly used anionic do not improve the cleansing. The dependency versus composition is linear. In contrast, when blended with the fatty alcohol ether sulfate, alkyl polyglucosides exhibit a strong synergistic effect in cleansing action. Figure 7.22 indicates a broad maximum which is higher than the effect of the dishwashing liquids available commercially [59]. Similar high and broad synergistic patterns of the dishwashing effect have also been observed for combinations of alkyl polyglucosides with other anionic surfactants such as fatty alcohol sulfate, betaine [84], $C_{13}C_{17}$-alkane

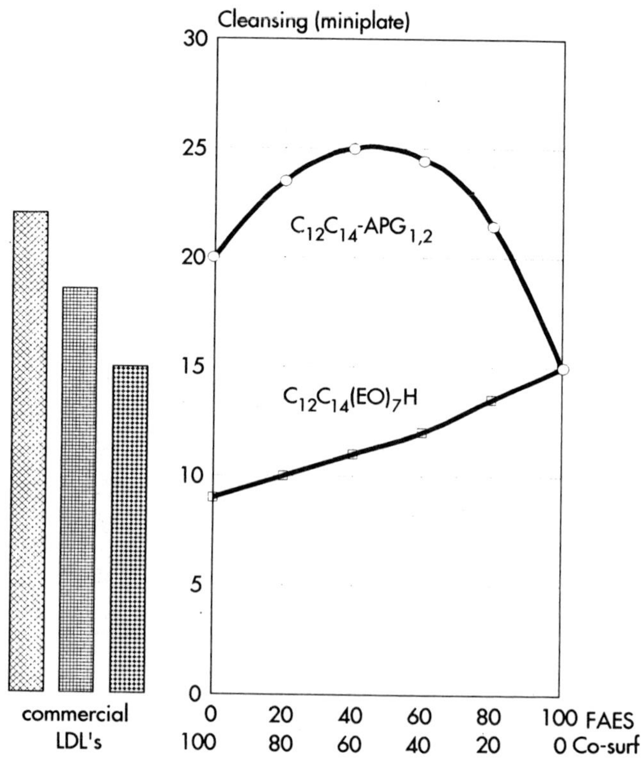

Figure 7.22 Dishwashing effect versus composition, FAES combined with $C_{12}C_{14}$-$(EO)_7H$ or alternatively $C_{12}C_{14}$- $APG_{1.2}$, miniplate test, 0.075 g/l, 13° German hardness (Reproduced with permission from Hanser Verlag, Münich.)

sulfonate, and $C_{11}C_{13}$-alkyl benzene sulfonate [85, 86]. What is also of particular interest is the excellent dishwashing effect of $C_{12}C_{14}$-alkyl polyglucoside alone (Figure 7.22). Adding a low molecular weight alcohol and some salt in order to improve its solubility at low temperatures results in an effective, environmentally mild, low-foaming dishwashing liquid [87]. Adding to this only a tiny amount of a standard anionic surfactant removes the clouding by giving a higher electrical charge to the micelles and results in a high-foaming dishwashing product [88].

7.5.2 *Fabric washing compositions*

The surfactant part of fabric detergents, no matter whether liquid or powder, typically consists of a mixture of anionic and non-ionic surfactants. For sud control soap is usually used. The application of this concept to the use of alkyl polyglucosides creates unexpected results. The combination of alkyl polyglucosides with anionic surfactants such as alkyl benzene sulfonate or fatty alcohol sulfate brings about poor detergency effects. As shown in Figure 7.23, however, combined with a second non-ionic surfactant such as

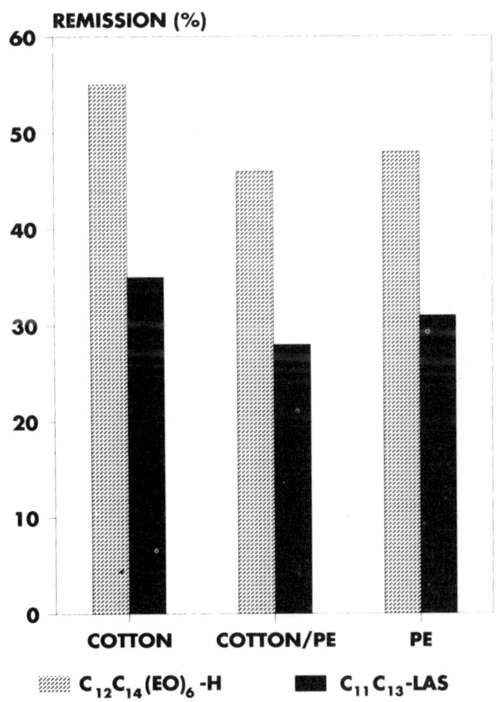

Figure 7.23 Fabric detergency (remission) at 30°C, $C_{12}C_{14}$-$APG_{1.3}$ combined with $C_{11}C_{13}$-alkyl benzene sulfonate or, alternatively, $C_{12}C_{14}$-$(EO)_6H$ + $C_{12}C_{14}$-APG/cosurfactant 1:1, 30% surfactant, silicone defoamer, solvent, pH 8.5, domestic washing machine, detergent 7.5 g/l, demin. water (Reproduced with permission from Hanser Verlag, Münich.)

a fatty alcohol ethoxylate, alkyl polyglucosides perform distinctly better on all fabrics.

Customary washing trials with a surfactant combination of alkyl polyglucosides and fatty alcohol ethoxylate, the ingredients customarily used for heavy duty liquids (HDL) having also been added, confirm the excellent effect in comparison to a commercial product, especially in relation to problem fabrics such as polyester and cotton–polyester (Figure 7.24). An interesting aspect here is also the small effect of the degree of ethoxylation of the fatty alcohol ethoxylate within the range between 4 and 12 [59].

Similar results were also obtained for heavy duty powders (HDP) at 60°C [86]. Addition of an anionic surfactant (besides soap) even in small amounts diminishes the detergency effect. The data also indicate that, with increasing temperatures, the advantage of alkyl polyglucoside containing detergents over conventional formulations decreases. Hence they should preferably be used at lower temperatures, the preferred form being that of liquids. Spray drying of slurries at moderate temperatures is quite feasible without any color deterioration [86]. It is quite interesting that alkyl polyglucosides are beneficial to pumping and spraying of conventional detergent slurries of much higher solid content because they dramatically reduce the slurry viscosity [89].

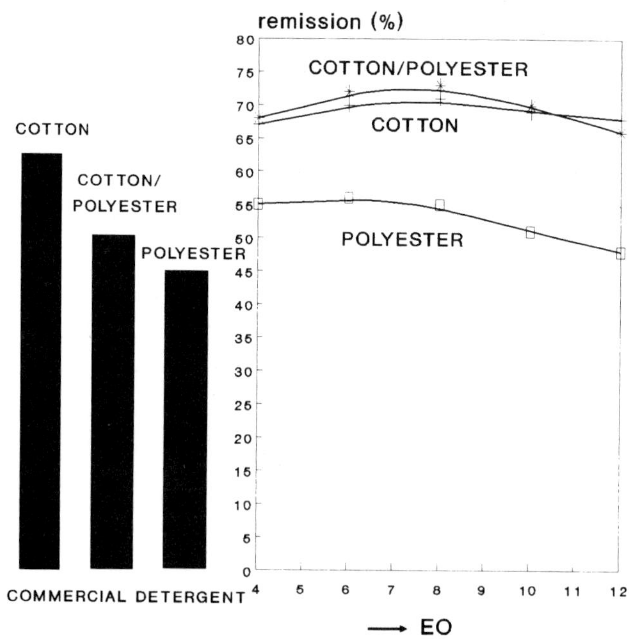

Figure 7.24 Fabric detergency at 30°C versus degree of ethoxylation of a blend $C_{12}C_{14}$-APG/$C_{12}C_{14}$-$(EO)_x$H/soap, 10% each, solvent, domestic washing machine, liquor ratio 1:20, detergent 7.5 g/1, 13° German hardness (Reproduced with permission from Hanser Verlag, Münich.)

7.5.3 *Personal care products*

Alkyl polyglucosides exhibit low interfacial tensions, a basic requirement for emulsification, form long flexible rod-like micelles, which form viscous networks, are foam boosters for surfactants of the kind typically used in toilet formulations, are mild to the skin and of low toxicity. Hence, they are key surfactants for personal care products. As expected, the predominantly non-ionic alkyl polygluconsides do not exhibit any increase in viscosity with added electrolyte of the kind found with some anionic surfactants such as fatty alcohol sulfate or ether sulfate [59]. However, when combined with these, viscosity is strongly increased at much lower salt concentrations than customarily necessary. Similar effects have been observed for combinations with fatty alcohol ether carboxylates or fatty alcohol ether sulfosuccinates, which on their own are viscosified only at extremely high electrolyte concentrations such as 10 to 20% of NaCl, or when combined with surfactants such as alkane sulfonates [10]. As is shown for the very mild carboxymethylated fatty alcohol ethoxylate in Figure 7.25, to produce a gel-like viscosity a blend containing $C_{12}C_{14}$-alkyl polyglucoside needs only 1.5% of NaCl whereas without the alkyl polyglucoside the highly irritating concentration of about 20% NaCl would be necessary [90]. In addition alkyl polyglucosides greatly improve the gentleness to skin and efficiency of personal care formulations [91].

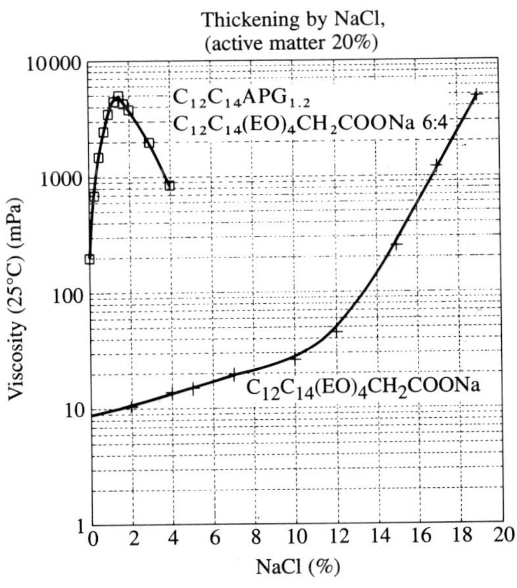

Figure 7.25 Viscosity of $C_{12}C_{14}$-$(EO)_4CH_2COONa$ versus NaCl concentration, effect of $C_{12}C_{14}$-$APG_{1.2}$, surfactant 20%, 25°C.

References

1. Knaut, J., *Seife Öle Fette Wachse*, 1993, **119**, 11.
2. Kosswig, K. In: Kosswig, K. and Stache, H. (eds). *Die Tenside* p. 159. Carl Hanser, München, 1993. Crooks, P.A. *et al.*, *Pharm. Acta Helv.*, 1974, **49**, 274.
3. Biermann, M. *et al.*, In: Falbe, J. (ed). *Surfactants in consumer products*, p. 23. Springer, Berlin, 1987.
4. Kelkenberg, H. *Tenside Surf. Det.*, 1988, **25**, 8.
5. Ruback, W. and Schmidt, S. *3rd International Workshop on Carbohydrates as Organic Raw Materials*, Wageningen, 1994.
6. Scholz, H.J. Symposium *Nachwachsende Rohstoffe* Leverkusen, 1994. Stalmans, M. *et al.* *SÖFW* 1993, **119**, 794.
7. *Chemical carcinogens*, p. 755. ACS-Monograph 182, Washington DC, 1984.
8. Fischer, E. *et al.* *Chem. Ber.*, 1893, **26**, 2400. *Chem. Ber.* 1895, **28**, 1145. *Lieb. Ann.* 1911, **383**, 68.
9. *DRP 593422*, H.Th. Böhme AG, 8.2.1934. *US 2,049,758*, inv. Bertsch, H. and Rauchalles, G. 31.5.1934/4.8.1936.
10. Balzer, D. *Proc. 3rd Cesio International Surf. Congr.* London, 1992. Vol. C, p. 165.
11. Hughes, F.A. and Lew, B.W. *J. Am. Oil. Chem. Soc.*, 1970, **47**, 162.
12. Koenigs, W. and Knorr, E. *Chem. Ber.*, 1901, **34**, 957.
13. De Grip, J.W. and Bovee-Geurts, P.H.M. *Chem. Phys. Lipids*, 1979, **23**, 321.
14. Böcker, Th. Diploma-Thesis, Münster, 1987. Böcker, Th. and Thiem, J. *Tenside Surf. Det.*, 1989, **26**, 5.
15. Thiem, J. and Böcker, Th. *Carbohydrate Res.*, 1992, **230**, 245.
16. Thiem, J. *et al.* *Proc. 2nd World Surf. Congr.*, Paris, 1988. vol. I, p. 197.
17. *US 3,219,956*, inv. Boettner, F.E. 12.8.1963/23.11.1965.
18. *US 3,839,318*, inv. Mansfield, R.C. 27.9.1970/1.10.1974.
19. *US 3,772,269*, inv. Lew, B.W. 24.7.1969/13.11.1973.
20. *US 3,707,535*, inv. Lew, B.W. 24.7.1969/26.12.1972.
21. *US SIR H619*, inv. McDaniel, R.S., Vanderburgh, L.F. and Sommer, S.J. 25.1.1984/4.4.1989.
22. *EP 0252 241*, inv. Lüders, H. 10.7.1986/13.1.1988.
23. *US 4,721,780*, inv. McDaniel, R.S. and Johnson, D.L. 6.2.1985/26.1.1988.
24. *EP 0 482 325*, inv. Oberholz, A., Kahsnitz, J. and Schmidt, S. 26.10.1990/29.4.1992.
25. Balzer, D. and Ripke, N., *Seife Öle Fette Wachse*, 1992, **118**, 894.
26. *DE 19 43 689*, inv. Mansfield, R.C. and Rainey, J.L. 3.9.1968/12.3.1970.
27. *EP 0 092 875*, inv. Mao, M.H.K. 26.4.1982/2.11.1983.
28. *US 4,465,828*, inv. Rau, A.H. and Speckman, D.T. 17.2.1984/14.8.1984.
29. *DE 36 19 796*, inv. Lüders, H. 12.6.1986/17.12.1987.
30. Lüders, H. and Balzer, D., *Proc. 2nd World Surfactant Congress*, Paris, 1988, vol. II, p. 81.
31. *EP 0 253 996*, inv. Lüders, H. 23.7.1986/27.1.1988.
32. *EP 0 306 650*, inv. Lüders, H. 5.9.1987/15.3.1989.
33. *EP 0 521 258*, inv. Ripke, N. 4.7.1991/7.1.1993.
34. *EP 0 531 647*, inv. Müller, B. and Ripke, N. 6.9.1991/17.3.1993.
35. *EP 0 338 151*, inv. McDaniel, R.S. *et al.* 19.4.1988/25.10.1989.
36. *EP 0 418 458*, inv. Fujita, T. *et al.* 19.9.1989/27.3.1991.
37. *EP 0 077 167*, inv. Arnaudis, G. 8.10.1981/20.4.1983.
38. *EP 0 569 682*, inv. Balzer, D. and Ripke, N. 11.5.1992/18.11.1993.
39. *US 4,483,979*, inv. Mao, M.H.K. 8.12.1982/20.1.1984.
40. *EP 0 389 753*, inv. Ripke, N. and Hofmann, P. 30.3.1989/3.10.1990.
41. *EP 0 165 721*, inv. McDaniel, R.S., Glor, P.R. and Kickle, H.L. 25.4.1984/27.12.1985.
42. *US 4,510,306*, inv. Langdon, K.W. 13.9.1983/9.4.1985.
43. Balzer, D. paper presented at VDI/GVC-meeting, Baden-Baden, 19.3.1990.
44. WO 94/02494, inv. McCurry, P.M. *et al.* 15.7.1992/3.2.1994.
45. *DE 40 35 722*, inv. Schulz, P., Eskuchen, R. and Weuthen, M. 9.11.1990/14.5.1992.
46. *EP 0 513 813*, inv. Koike, T. *et al.* 16.5.1991/19.11.1992.
47. *DE 38 33 780*, inv. Hill, K. *et al.* 5.10.1988/12.4.1990.

48. *WO 90/06933,* inv. Rasche, J.F., Pickens, C.E. and McCurry, P.M. 21.12.1988/18.10.1990.
49. *WO 93/24504,* inv. Gibson, M.W., McCurry, P.M. and Pickens, C.E. 29.5.1992/9.12.1993.
50. Schulz, P.T. *Proc. BACS Symp., Chem. Spec. Europe,* Amsterdam, 1991, p. 33.
51. *EP 0 132 043,* inv. Davis, J.E. and Letton, J.C. 15.6.1983/23.1.1985. *EP 0 449 866,* inv. McCurry, P.M., Kozak, W. and Pickens, C.E. 23.12.1988/9.10.1991.
52. Spilker, R. *Tenside Surf. Det.*, 1996, **33**, 21.
53. Noller, C.R. and Rockwell, W.C. *J. Am. Chem. Soc.*, 1938, **60**, 2076.
54. Jeffrey, G.A. *Acc. Chem. Res.* 1986, **19**, 168.
55. Balzer, D. (unpublished).
56. Straathof, A.J.J., van Beckum, H. and Kleboom, A.P.G. Stärke 1988, **40**, 438.
57. Matsumura, S. *et al. J. Am. Oil. Chem. Soc.*, 1990, **67**, 996.
58. Koeltzow, D.E. and Urfer, A.D. *J. Am. Oil. Chem. Soc.*, 1984, **61**, 1651.
59. Balzer, D. *Tenside Surf. Det.*, 1991, **28**, 419.
60. Balzer, D. *Langmuir*, 1993, **9**, 3375.
61. *EP 0 588 024,* inv. Balzer, D. 17.9.1993/23.3.1994.
62. Sadaghiania, A.S. and Khan, A.H. *J. Colloid Interface Sci.*, 1991, **144**, 191.
63. Balzer, D., Schulze, K. and Weihrauch, M. *J. Com. Esp. Deterg.*, 1994, **25**, 81.
64. Hoffmann, H. *et al. Tenside Surf. Deterg.*, 1985 **22**, 290.
65. Balzer, D., Varwig, S. and Weihrauch, M., *Colloids and Surfaces A*, 1995, **99**, 233.
66. Shinoda, K., Yamaguchi, T. and Hori, R., *Bull. Chem. Soc. Japan*, 1961, **34**, 237.
67. Drummond, C.J. *et al. J. Phys. Chem.*, 1985, **89**, 2103.
68. Renner, H. University of Bayreuth, private communication.
69. Lange, H. and Beck, K.H. *Kolloid Z.Z. Polymere*, 1973, **251**, 424.
70. Meguro, K., Ueno, M. and Esumi, K. In: Schick, M.J. (ed). *Surfactant Science Series*, Vol. 23, Nonionic surfactants, p. 109. M. Dekker, New York, 1987.
71. Lange, H. and Schwuger, M.J. *Kolloid Z.Z. Polymere* 1968, **223**, 145.
72. Mukerjee, P. and Mysels, K.J. NSBDS-NBS-36 (1971).
73. Reed, R.L. and Healy, R.N. In: Shah, D.O. and Schechter, R.S. (eds). *Improved oil recovery by surfactant and polymer flooding,* p. 383. Academic Press, New York, 1977.
74. Shinoda, K. and Friberg, S. *Emulsions & solubilization.* John Wiley & Son, New York, 1986.
75. Huh, C. *J. Colloid Interface Sci.*, 1979, **71**, 408.
76. *DE 39 25 846,* inv. Balzer, D. 4.8.1989/12.2.1991.
77. *EP 0 507 047,* inv. Balzer, D. 30.3.1991/7.10.1992.
78. Smith, G.A. *et al. Colloids and Surfaces* 1994, **88**, 67.
79. Balzer, D., *Tenside Surf. Det.,* 1996, **33**, 102.
80. *DE 26 46 506*, inv. Wagner, F. *et al.* 15.10.1976.
81. Schöberl, P., private communication.
82. Scholz, N., private communication.
83. Salka, B. *Cosmetics & Toiletries*, 1993, **108**, 89.
84. *EP 0 513 138*, inv. Guirr, O. and Middelhauve, B. 2.2.1990/8.8.1991.
85. Andree, H. and Middelhauve, B. *Tenside Surf. Det.*, 1991, **28**, 413. Andree, H. and Nitsch, C. *Proc. 3rd Cesio International Surf. Congr., London*, 1992, Vol. B, p. 115.
86. Balzer, D. Tagung Sepawa, Duisburg, 27.11.1992.
87. *EP 0 444 262,* inv. Balzer, D. 26.2.1990/4.9.1991.
88. *EP 0 486 784,* inv. Balzer, D. and Hoffmann, H. 17.11.1990/27.5.1992.
89. *EP 0238 638,* inv. Kickle, H.L. *et al.* 26.9.1985/30.9.1987.
90. *EP 0 572 776,* inv. Balzer, D. 29.5.1992/2.12.1993.
91. Busch, P., Hensen, H. and Tesmann, H. *Tenside Surf. Det.*, 1993, **30**, 116.

8 Low-molecular weight organogelators

P. TERECH

8.1 Introduction

The present chapter deals with a special class of surfactants which have the ability to form viscoelastic fluids or solid-like materials in organic solvents at concentrations lower than about 2%. An increased interest in these systems, considered in the past only as curiosities, has been manifested in recent years. Organogels are now good representatives of the field of the so-called 'soft-matter'. These systems display a variety of physical behaviour ranging from those characteristic of surfactants in solution [1] (micellization, lyotropism, crystallization) to those attributed to polymers in good solvents [2] (swelling, dynamics). Organogels, which are most often binary systems, are usually prepared by dissolution of the solid gelator by heating in appropriate organic liquids and then cooling to room temperature. The resulting materials are named gels or jellies according to their hardening or thickening ability, respectively, and their formal description and classification is given on the basis of the mechanical properties (rheology). The colloidal aggregates in the gel are linked in complex three-dimensional networks immobilizing the liquid component to a variable extent by surface tension to give the typical consistency of the 'gels'. The present review discusses the physical context of organogels, the techniques used to probe these colloidal systems, and the basic thermodynamic concepts involved in the surfactant aggregation phenomenon as they apply to organic solvents.

Organogelator systems are of interest both from the chemical and the physical point of view. One of the 'physical goals' consists of the understanding of the structural correlations of the crystalline solid state and the various types of aggregates which can be found in the gels with some of their macroscopic properties (mainly mechanical and optical). One of the main 'chemical challenges' is the understanding of the molecular requirements which lead to the gelation phenomenon and its specificities. In this context, the various types of molecules known to date to give gels in organic solvents are listed in the present work and some examples are further detailed to draw basic correlations and phenomenological features of the related physical chemistry of organogels.

8.2 Surfactant organogels: physical context

A large number of small molecules are known to spontaneously self-associate in solution to produce various structures. Most of the experimental and theoretical work has been devoted to self-assembly of surfactants in aqueous solutions [3]. In such systems, structural transitions of solute aggregate can be observed by varying the surfactant concentration, the temperature and/or the ionic strength of the dispersing medium. One can thus obtain a spectrum of aggregates ranging from spheres to rods, to discs, to various lyotropic phases [4, 5] (i.e. nematic, hexagonal, lamellar, cubic phases). The related fundamental thermodynamic equations which describe these systems are obtained from the identity of the chemical potential of the surfactant molecules in the different aggregates (equation 8.1).

$$\mu = \mu_N = \mu_N^0 + \frac{kT}{N}\log\left(\frac{X_N}{N}\right) = \text{constant} \tag{8.1}$$

In equation 8.1, aggregates have an aggregation number (N) (which can vary from 1 to infinity) and a mean interaction free energy (μ_N^0) (the standard part of the chemical potential) so that the corresponding mean chemical potential of an amphiphilic molecule in the aggregate is given by (μ_N). (X_N) is the concentration (or activity) of the surfactant in the aggregate. When the condition $\mu_N^0 < \mu_1^0$ holds, stable aggregates are formed.

The most 'efficient' way of obtaining interacting aggregates in gel-like viscoelastic systems (with reference to the volume fraction of aggregated species involved), is for the structure of the assemblies to be rod-like. All examples of compounds discussed below give gel-like materials at concentrations of less than 2%. The structure of the linear aggregates in the gel state is determined by the direction and strength of the binding forces associated with the chemical mechanism of the aggregation of the gelator molecules. In aqueous solutions of amphiphiles, the structure of the aggregates results from a balance between the attractive (interfacial) and repulsive forces (between head groups), optimizing some packing considerations. Simple structural correlations exist between the parent molecule and the resulting aggregate in water [6–8]. Thus, various families of single chain amphiphiles are known to form globules, vesicles, rods, tubes or disks according to the assemblage of their structural elements [9] (flexible tail, rigid segment, hydrophilic head group). The situation in organic liquids is fundamentally different [10–12] as the hydrophobic attraction, which is the driving force for aggregation in water, cannot be invoked. In non-aqueous solvents, the attractive factors are mainly the dipolar interactions and, possibly, intermolecular hydrogen bonds or metal coordination bonds balanced by the repulsive contributions due to the reductions in the translational and rotational freedom of motions.

Aggregation of surfactants in organic solvents [13] has been the subject of some theoretical and quantitative treatments [14–16]. The thermodynamics of micellization in organic solvents is such that normal and reverse micelle formation can be observed for ionic surfactants (see the example of aerosol OT, i.e. sodium bis(2-ethylhexyl) sulfosuccinate [17]). Beyond a critical concentration (C_{cri}), usually much smaller than the equivalent critical micellar concentration, CMC, of the aqueous systems, finite spherical micelles are formed. Such assemblies usually have small (N) values since the packing is driven only by interactions between the polar parts of the molecules while the lipophilic parts remain more or less unchanged. In general, an increase of chain length leads to a decrease of the aggregation number (N). By contrast, when the small amphiphilic molecule is a gelator and/or when the concentration is properly adjusted, the formation of colloidal aggregates can be observed. As long as only the topology of the aggregates is considered and compared to various standard geometries (spherical, lamellar, fibrillar), the rod-like structure appears to exhibit some specificities which determine the thermodynamics of the suspensions. At equilibrium, the optimal configuration is obtained when the standard free energy has reached a minimal value corresponding to a packing which takes into account both the ratio of the cross-sectional areas of the hydrocarbon to the polar moieties of the gelator and the polarity of the solvent. Special mention must be made of the situation at the chain extremities (hemispherical caps) since these ends are energetically unfavourable. Depending on the excess of free energy corresponding to the free ends, either the growth of infinitely long giant micelles or numerous shorter ones (entropically favourable) will be favoured.

The aggregation mechanism involves specific chemical interactions (H-bonding, organometallic coordination bonding, electron transfer between donor and acceptor moieties of the gelator, etc.) with a strength function (ψ) depending on the gelator–gelator bond energy (αkT) in the aggregate. The functional dependence is directly related to the number and type of bonds per gelator molecule and the number of gelators within the cross-sectional area of a given type of morphology of the aggregate. For instance, the simplest situation is that of molecular threads, i.e. strands in which only one gelator molecule per cross-sectional repetition unit along the chain axis is connected by only one type of interaction (or bond). The gelator–gelator bonding energy determines the aggregation threshold ($C_{\text{cri}} \approx e^{-\psi(\alpha)}$) and the size distribution of the aggregates. The size distribution of the chain-like aggregates correlated to the contour length ($\bar{L}$) and flexibility ($\langle l_{\text{p}} \rangle$), is very large and depends strongly on the concentration and temperature. The growing rod-like aggregates are exchanging 'monomeric' species of non-aggregated gelator molecules dispersed in the surrounding bulk and can thus be considered as 'equilibrium aggregates'. Theoretically, the statistical contour length ($\bar{L}$) of such chain-like aggregates can be estimated by equations 8.2 where the exponent (y) is about 0.6 in a scaling approach involving 'blobs' (see below) of size (ξ) with excluded volume effects [2] ($y \approx 1/2$ for a Flory–Huggins or mean field approach [18]).

$$\bar{L} \alpha \phi^{y} \exp(E_{\text{scis}}/2k_{\text{B}}T) \tag{8.2}$$

In equation 8.2, E_{scis} is the scission energy of the chain-like aggregate and ϕ the volume fraction of the gelator. In practice, an explicit form for the micellar size distribution can be a so-called Schulz–Flory distribution [19]. Based upon multiple chemical equilibrium models, an exponential law of the form $n(L) \approx \exp(-L/L_{\text{w}})$ with $L_{\text{w}}/L_{\text{n}} \approx \sigma$, where ($\sigma$) reflects the degree of polydispersity (L_{n} being the number average and L_{w} the weight-average contour length of the linear aggregate) can be conveniently used. The size distribution of the aggregates results from the combination of the entropic decrease on aggregation and the standard free energy change of the surfactant between the non-aggregated and the aggregated states. The profile of this distribution with various physical colligative properties is usually such that only small finite aggregates should be expected through gradual changes of the thermodynamic properties. Obviously, the situation in organogels is rather particular where infinite aggregates (fibre-like) are formed and phase separation (crystallization) is avoided due to the existence both of an interconnected solid-like three-dimensional (3D) organization and a very specific balance of opposite forces with respect to aggregate interactions. The necessity of such a delicate balance of forces contributes to the relative rarity of these systems.

By contrast to some aqueous systems (see for instance cetylpyridinium micelles [20]), a clear sphere to rod phase transition is usually not as well characterized for most of the binary systems of organogelators. In spite of a very polydisperse overall size, experimental results show (see below) that the cross-sectional geometry of the aggregates is rather monodisperse. When the concentration is further increased to the overlap value (C^*), the chains are no longer separated (as in the so-called dilute concentration regime), the rotation becomes hindered and the equivalent spherical volumes of the linear aggregates overlap. A network with a certain average mesh size (ξ) is formed. At a given temperature of the gel-like phase, surfactant organogels can exhibit either permanent rigid networks or transient semi-flexible meshes. The notion of transient mesh size or statistical porosity of a 3D network is similar to a concept used for polymeric solutions where the notion of a 'blob' of size (ξ) refers to a situation where excluded volume correlations prevail so that each individual chain does not interact with any other [2]. Studies summarized in the present article concern mainly the semi-dilute concentration range corresponding to overlapped chain-like aggregates (isotropic gels) but the concentrated range where the chains are more or less ordered (anisotropic gels) is also briefly mentioned.

Most of the systems detailed here are binary and the usual factors affecting the aggregation process are relevant parameters for describing the gelation phenomenon. These parameters include the solvent nature, the polarity and steric nature of the gelator, the temperature and the possible participation of co-surfactants. Some structural and kinetic aspects of the aggregation/gelation phenomenon are also consistent with the aggregation reaction being viewed as a phase separation process involving a nucleation reaction, thus emphasizing the role of the interfacial free energy of the gelators. In this context, the question of the participation of co-surfactants can be of extreme influence on the resulting types of aggregates and phases which can be obtained. In particular, the detergent property of the reverse micellar structure which consists in the solubilization of polar molecules, brings into question the existence of true binary systems. The micellization phenomenon can have two major consequences: it may shift the cooperative aggregation process to concentrations higher than the true C_{cri} and it may be capable of initiating a sphere to rod transition necessary for the subsequent gelation of the mixture at C^*. It is worth noting that the effect of water (or other polar additives) upon reverse micellization and micellar growth reactions is still somewhat controversial [21]. In one case, water has been shown to be a prerequisite for micellization and acts as a 'gluing agent' (as is the case for sodium bis(2-ethyl-hexyl)sulfosuccinate (AOT) in benzene [22]) or in a different example, water has been shown to act as an antimicellization agent (as is the case for sodium bis(2-ethylhexyl) phosphate (NaDEHP) in *n*-heptane [21]). Consequently, great care must always be taken in the composition and preparation method of the gelling mixtures. Systems which are made up of such an important proportion of water that they are better described as microemulsions [11] will not be considered in the present review.

Beyond C^* the individual aggregates of a sol-phase are interconnected in an infinite 3D super-aggregate (network). Gels are typical examples of heterogeneous materials combining two components (solvent and gelator) with very different physical properties resulting in a system exhibiting new behaviours. The macroscopic recognition of a gel state is somewhat tricky and various classifications have been proposed based either on morphology, consistency or chemical and mechanistic considerations [23–25]. So far, rheological properties appear to be most appropriate to characterize the varying connectivity of the aggregates during the sol to gel phase transition [26]. The latter property can be described by statistical physics and, in particular, the percolation models [27, 28] can simulate the mechanical behaviour around the critical gelation threshold. The sol to gel phase transition is generally characterized by a divergence of the connectivity correlation length [29] (as is the case for some second-order phase transitions). As mentioned above, two different types of 3D networks and viscoelastic materials can be obtained in organogel systems (Figure 8.1).

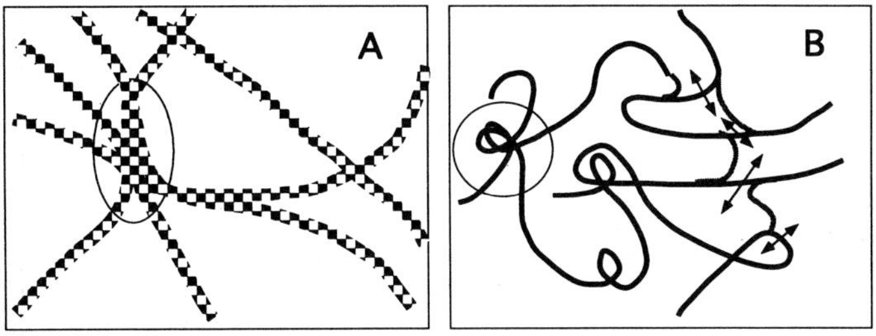

Figure 8.1 The physical properties of organogels are dependent upon the nature of their networks. **A**: 'strong' gels, permanent solid-like networks, the nodes are spatially extended (pseudo)crystalline microdomains. **B**: 'weak' gels, transient networks exhibiting no elasticity over long time periods, the nodes are entanglements or spatially limited organized microdomains. Additional kinetic features such as chain breaking/recombination (arrows) may exist as in the case of the so-called 'living polymers'.

First, materials exhibiting a viscoelastic solid-like mechanical behaviour and which are made of permanent networks obtained through a sharp solution (or sol, i.e. a collection of finite aggregates in a liquid solution) to gel phase transition at a given temperature (T_{SG}) are 'strong' gel-like systems. On the other hand, materials exhibiting a liquid-like viscoelastic behaviour which are made of transient networks (obtained through a much smoother threshold exhibiting some similarities with the glass transition of polymers) are termed 'weak' gel-like systems. Both types of networks are thermally reversible. On heating, the supramolecular architecture is melted and the individual molecules are dispersed back in the bulk solution. In many aspects of the physical properties of surfactant organogels, it can be tempting to adapt and mix some concepts used for semi-dilute polymeric solutions to those used for surfactants in solution. With some special 'weak gel systems', static and dynamic quantities may follow power laws which can be approached by incorporating the concentration and temperature dependencies of the length of the micellar assemblies in the scaling laws for semi-dilute polymers in good solvents (as is the situation of the so-called 'living polymers'). This special class of materials was considered to exist only in salted aqueous solutions [30, 31] of ionic surfactants but is now also represented in organic solvents (see this review).

According to the specific balance between attractive and repulsive forces of the aggregates in the system, various lyotropic organizations can be observed when the concentration is increased far beyond C^*. Bulk phases in the phase diagram or microdomains in the heterogeneous network can develop and produce nematic, hexagonal, rectangular or discotic organizations. Theoretically, the concentration range separating the dilute, semi-dilute and concentrated domains [32] is given by equation 8.3.

$$1/L^3 \ll n \ll 1/(2rL^2) \tag{8.3}$$

In equation 8.3, r is the radius of rigid rods of length (L), and n the number of rods in a unit volume. Equation 8.3 indicates that the solution is thermodynamically but not dynamically ideal since an entanglement of the rods exists. At this stage, it is interesting to keep in mind that the energy input in sample preparations is not minor considering the experimental procedure which always includes either significant shaking and/or heating. The structures formed may have a kinetic stability, since these systems could have been formed in conditions which are far from thermodynamic equilibrium. It has already been proposed that in supercritical conditions of the energy input and entropy export, the structures thus produced could be self-organized and dissipative [33].

8.3 Techniques

It is not the purpose of the present article to recall the basic principles of each of the various techniques applicable to the study of the mechanism of gelation of organic liquids by small amounts of low-molecular weight compounds. Nevertheless, a brief list of the major techniques used in this field is summarized below. Depending upon the chemical specificities of the organogelator molecule, specific spectroscopies can complement the catalogue of the common techniques used in colloid science. Fluorescence spectroscopy, electron paramagnetic resonance (EPR) and circular dichroic absorption spectroscopies constitute such examples.

The first step of any study of gelator systems should be the identification of the gel state. Rheology offers a convenient method [23] to measure the flow properties of the gel-like materials and facilitates the classification among the two above mentioned 'strong' and 'weak' gel types. The mechanical parameters give some strong indication of the nature of the network and the mechanisms of the interactions responsible for its cohesion. Various procedures can probe the mechanical properties of the soft-solids and viscoelastic fluids, such as steady shear, oscillation, stress relaxation and creep experiments. Different measuring geometries enable the shearing of the samples (parallel plates, concentric cylinders, etc.) but in numerous cases a cone-plate geometry is preferred [34]. The elastic and viscous components, mixed in such non-Newtonian materials, are measured in oscillatory or flowing experiments in which a small periodic strain is applied within the appropriate amplitude range defining the linear regime. Oscillatory experiments show that for solid-like gels (soft solids or 'strong' physical gels) the elasticity measured by the storage modulus $G'(\omega)$ is much larger than the dissipated energy (or viscous behaviour as measured by $G''(\omega)$) and extends as a plateau from long times to small ones [26]. Frequently, these systems are characterized by a yield value (σ^*) which is the minimum stress that has to be applied before a significant flow can occur. To identify the gelation time, which can reflect some molecular mechanisms involved in the phenomenon, an accurate method consists in plotting $\tan\delta = G''/G'$ (where δ is the phase angle between the stress and strain mechanical components in dynamic experiments) as a function of the time (t) and to look for the appearance of a frequency independent crossover point [35, 36]. It is interesting to observe that pertinent comparisons and analogies of the structural and mechanical properties of the organogels can be made with the thermoreversible networks of some aqueous biopolymer gels [37–39] as the gel state is frequently involved in natural materials.

After the study of the basic mechanical features, the determination of the phase diagram and the temperature of the sol to gel phase transition (T_{SG}) is usually a compulsory step to specify the system. The methods used for this purpose include calorimetry (DSC), temperature sweep rheological experiments and different spectroscopies (fluorescence, IR, EPR, NMR, etc.) using the amplitude variation of a spectral feature with temperature. The melting process of physical gels at a temperature (T_{GS}) can be analyzed in a simplified framework already utilized with gelatin gels [40]. The related model assumes that the network cohesion comes from crosslinks resulting from binary associations of macromolecular chains. A combination of the mass-action constant, Van't Hoff's law and various simplification hypotheses, leads to equation 8.4 which relates the exothermic heat of association (ΔH^0) to the melting point (T_{GS}).

$$\ln C = -\frac{\Delta H^0}{RT_{GS}} + \text{constant} \tag{8.4}$$

An approach considering the melting of the gels as a dissolution process of crystals would lead to similar trends.

The second level of investigation consists of the structural characterization of the unidirectional aggregates whose interactions are responsible for the viscoelasticity of the materials. Electron microscopy (EM) observations provide an apparently convenient method but will most often require the support of complementary techniques. A good combination is certainly the use of EM and scattering techniques. EM gives a more or less direct way of obtaining the structural information in direct space while scattering studies are carried out in reciprocal distance space necessitating a specific mathematical analysis to extract the information. One of the most convenient experimental procedures for such fragile samples is to prepare the EM specimen by a temperature quenching and subsequent fracture, etching and metal shadowing steps. Direct imaging methods, originally used for biological systems [41, 42], are now used for analogous aqueous solutions of amphiphiles [43] and should be preferred in future with organic media to avoid more perturbative procedures. Still, even with the latter method, all sources of artefactual observations will not completely vanish since confinement of the gel in very thin layers could obscure somewhat the observation of parameters such as the micellar lengths and flexibilities.

The use of scattering techniques for the study of highly dispersed gel systems is facilitated with the use of high flux neutron and synchrotron sources. An additional advantage of the scattering techniques is that the specimens are usually the genuine native samples with their solvents (or deuterated solvents) without any particular perturbative processing. Small angle scattering (SAS) of neutron or X-ray radiation beams provides information on long correlation distances in the distance range typical of colloidal systems ($1000\,\text{Å} < d < 20\,\text{Å}$) and can take advantage of some contrast variation opportunities. Selective deuteration of the gelators, use of deuterated solvents, comparison between X-ray and neutron scattering contrasts are among the most common options of the technique. The angular intensity distribution of long wavelength neutrons, scattered within small angles (small angle neutron scattering SANS), constitutes the scattering curve typical of the sample. It consists of an interference pattern whose reverse Fourier transform provides real-space structural information. The long-range fluctuations depend upon the contrast of the particles in the medium which, for neutron radiation, has a dircct dependence on the isotopic composition of the two-phase system. Consequently, any change in the constitution of the system (solvent or surfactant composition) can lead to a variation of the scattering level in a fashion related to the morphology, internal homogeneity and composition of the particles. Neutron scattering can then be compared to small-angle X-ray scattering (SAXS) whose contrast is a function of the atomic number Z of the constitutive elements in the scatterer. Despite the fact that the structural models deduced from the analyses of the SAS curves cannot be demonstrated to be uniquely correct (due to the lack of phase information and spatial resolution) SAS has proved to be very efficient in characterizing colloidal aggregates [44, 45]. For instance, the calculation of the theoretical scattered intensity for rod-like scatterers which are uncorrelated in position and orientation is simplified in the so-called conditions of particle scattering [46, 47]. The intensity is then a sum of the spherically averaged intensity for each long rod ($L \gg 2r$) and reduces to the simplified equation 8.5.

$$I = \frac{\pi C}{Q}\overline{\Delta b}^2 M_{\mathrm{L}}\left[2\frac{J_1(Qr)}{Qr}\right]^2 \tag{8.5}$$

In equation 8.5, C is the rod concentration (g.cm^{-3}), M_{L} is the mass per unit length of rod (g.Å^{-1}) and $\overline{\Delta b}$ its specific contrast (cm.g^{-1}). The momentum transfer $Q(\text{Å}^{-1})$ is defined as $Q = (4\pi/\lambda)\sin\theta$, where θ is half the scattering angle. Use of appropriate Q-expansions of equation 8.5 can provide information on the unidirectionality of the rod-like aggregates and structural features of their cross-section (shape, dimensions, homogeneity, polydispersity, nature of the interfaces with solvent).

An accurate structural and dynamic study of the aggregates in organogels may involve the determination of the length, flexibility and diffusion coefficients of the chain-like aggregates. Within specific conditions, this can be achieved by the use of the elastic and quasi-elastic light scattering techniques [48]. The so-called Rayleigh ratio measured in a static experiment (i.e. the excess scattering of the solution with respect to the solvent) is a product of an intra-aggregate form-factor and a time-averaged inter-aggregate structure factor. Dynamic light scattering experiments provide estimates of the temporal auto-correlation functions of the field from which self and collective diffusion coefficients can be deduced. Two regimes are distinguished on the basis of the magnitude of the correlation length, which defines the spatial scale of the interactions with respect to the wavelength of the incident light. In the first situation, for small particles in the high dilution limit (sol phase), the Rayleigh ratio is inversely proportional to the osmotic compressibility ($\partial\pi/\partial C$) and the molecular weight. The mean hydrodynamic radius ($\overline{R_h}$), the particle shape and the diffusion coefficient (D) can be extracted with an accuracy mainly dependent upon the polydispersity of the species. The second situation corresponds to that where the particle size is no longer small compared to the incident wavelength. If the scatterers are in the dilute range, the excess Rayleigh ratio is then proportional to the form-factor of the aggregate and its molecular weight and the second virial coefficient, as well as structural and dynamic information, can be extracted. In the semi-dilute concentration range for rigid rods (situation found in stiff gels), the diffusion coefficient is independent of the concentration (equation 8.6) while for flexible chains, D is inversely proportional to a correlation length [2], ξ and is associated with a collective mode of the chains in the transient network of 'weak gels'.

$$\overline{R_h} = k_B T/(6\pi\eta D) \tag{8.6}$$

Besides the usual spectroscopies (NMR, EPR, IR, UV, CD, etc.) and wide-angle X-ray scattering (WAXS), which can provide local structural information, various other techniques can be helpful. For instance, molecular dynamics in thermoreversible networks [49] can be investigated either 'classically' with photon correlation spectroscopy or with dielectric spectroscopy covering a broad frequency band (10^{-1}–10^9 Hz). Such examples of the study of the local and cooperative chain motions are found both for organo [50] and aqueous gels [51]. These types of studies can also be supplemented by the use of inelastic neutron scattering techniques such as neutron spin echo spectroscopy [52].

8.4 Classes of organogelators

The prediction of the gelling or thickening capacity of organic solvents by a given small molecule is not yet attainable and discoveries of new gelators or thickeners remain mainly accidental. In view of defining strategies for the molecular design of new gelators of organic fluids, it is useful to make an inventory of the gelators available up to now. There is a restricted number of chemical classes (about ten) to which gelators (and related derivatives) belong. In the following, the different families are mentioned together with their representative example for which some outstanding features are summarized.

8.4.1 Fatty acid derivatives

*Substituted fatty acids and monovalent metal salts (Scheme 8.1 (**1** and **2**)).* The 12-hydroxyoctadecanoic acid and its related salts (Scheme 8.1) have been known for a number of years for their hardening ability of organic

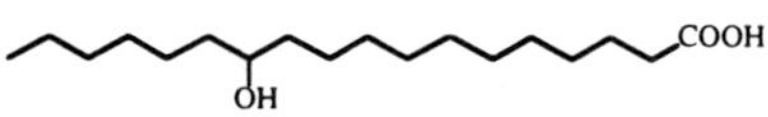

1

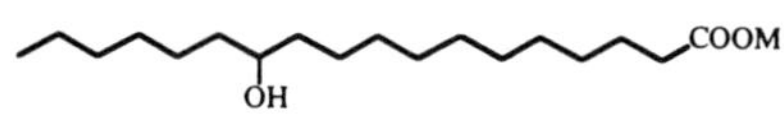

2

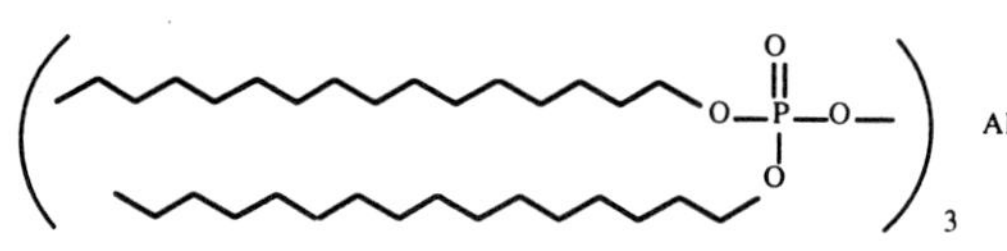

3

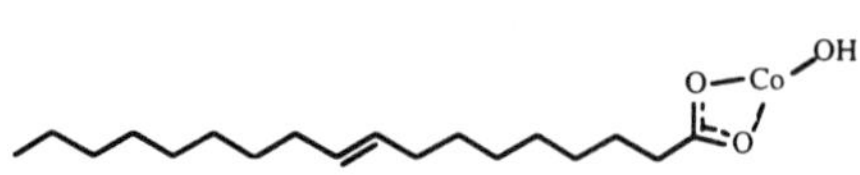

4

Scheme 8.1 Representative structures of the fatty acid organogelators (and their salts). **1**, 12-hydroxyoctadecanoic acid; **2**, monovalent soaps (M = Na, K, Li) of 12-hydroxyoctadecanoic acid; **3**, aluminium salt of dihexadecyl phosphate (C_{16}DP-Al); **4**, cobalt hydroxy monooleate.

fluids [53, 54]. For decades, semi-solid dispersions of the lithium soaps in oils have been used in the lubricating and ointment industries [55]. In many aspects, the history of the studies concerning this gelator (fundamental objectives and use of the techniques) is typical of the evolution of the research in the field. Due to its industrial importance [56], great effort has been devoted to understanding the structural aspects of lubricating greases and how they relate to the mechanical properties. The 3D network of aggregated neutralized fatty acid molecules act as a sponge maintaining the oil close to the friction regions of metallic surfaces [57] in various mechanical systems. One of the first electron micrographs [58], obtained in 1946, has initiated the research on the topic. Electron microscopy was extensively used [59–65] with different procedures (most of them very aggressive with respect to the sample) and were sometimes combined with thermal studies [66]. It has been shown that the dispersion process proceeds through lyotropic phase transitions [67]. Such an observation is comparable to the ability of some optically active acids to form lyotropic liquid crystals in organic solvents [68]. It illustrates the subtle thermodynamic balance which determines the solvation, solubilization or phase separation reactions corresponding respectively to suspensions, solutions or biphasic systems. The first systematic study using pure products (approx. 99%) was performed in the 1980s [69] using complementary techniques, such as circular dichroism, X-ray diffraction, electron microscopy, differential scanning calorimetry (DSC), infra-red spectroscopy (IR) and optical polarizing microscopy to characterize the different levels of organization of the colloidal material. Later on, the use of scattering techniques with powerful radiation beams (neutrons and X-rays) allowed an accurate structural investigation of the native gels [70, 71] within an extended range of distances from the microscopic to the mesoscopic scales. The range of organic solvents which can be gelled by this family of molecules extends from alkanes, carbon tetrachloride (CCl_4) and aromatic solvents to more polar fluids such as nitrobenzene. The kinetics of the sol to gel transition show that the onset of the elastic modulus is sharp. G' and G'' are virtually independent of frequency and the magnitude of G' is large ($G' = 3700$ Pa at $C \approx 0.35\%$). The physical junctions in the network appear as permanent cross-links at a given temperature in the gel domain of the phase diagram. The molecule has an asymmetric carbon, bearing the hydroxyl group (C_{12}), so that the aggregates are chiral and have specific optical properties. The molecular packing in the aggregates of the gels differs from that found in the neat solid by some limited gliding of special reticular planes resulting in a lowered symmetry of the arrangement. Circular dichroism [72] shows a preferential reflection of circularly polarized light which indicates that the fibres exhibit twists related to the optical activity of the amphiphilic molecule (Figure 8.2).

Induced circular dichroism can be generated by adding achiral chromophoric molecules to the gels and confirms the intercalation of the probes in

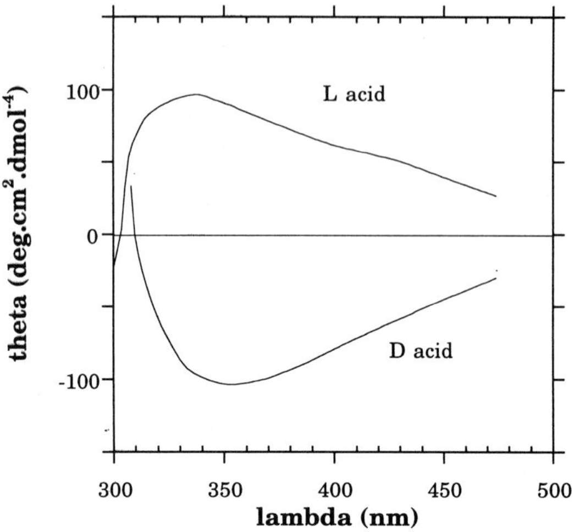

Figure 8.2 Enantiomeric CD bands of 12-hydroxyoctadecanoic acid/CCl_4 gels (10 mm thickness, 25.7 mmol.l^{-1} for the L acid, 35.7 mmol.l^{-1} for the D acid). The location of the CD maximum is dependent on solvents. (Reproduced from Reference 72 with permission from Macmillan Ltd., London.)

the supramolecular helical structures. Enantiomorphic relationships are deduced such that helices are mostly left-handed for the D-acid and right-handed for the L-acid. In addition, super-helicities (the origin of which is not clearly established) can be observed [64] and could be a response of a mechanically stressed chiral fibre during the specimen preparation as has been shown for some steroid-based organogelators [73] or with monolayers at the air–water interface [74]. The racemic (DL) acid produces organogels consisting of fibres interconnected by platelet-like contact zones. Consequently, the enantiomorphic relationships do not fully determine the gelling ability in organic solvents [71]. By contrast, some chiral surfactant systems [75] are observed to form fibres in aqueous solutions, while the racemate derivatives appear to separate not as fibres, but as plate-like crystals without any gel formation. The topology of the cross-sections of the rigid fibres is dependent upon the solvent type and varies from square in benzene (200 Å full side) to very rectangular in nitrobenzene [71] (ribbons of 300 Å thickness). Due to the presence of carboxylic groups on the edges of the fibres, a potential for an anisotropic lateral aggregation or growth of the fibres exists in appropriate solvents which accounts for the development of the 'junction zones' of the gel network. These zones are crystallized microdomains which are responsible for the highly non-Newtonian behaviour of the 'strong' physical gels. The internal structure and morphology of the fibres is obtained [71] from SAXS, SANS and WAXS measurements of gels and xerogels. 'Head to head' contacts between the carboxylic acid groups allow the formation of multiple hydrogen-bond sequences (Figure 8.3).

The organization is monoclinic and is sensitive to the solvent type and thermal/mechanical history of the samples. It is known that this class of compounds exhibits a diversity of structural solid-state variations categorized as polymorphism or polytypism [76]. Xerogels are mesomorphic solids in which helical structures (chiral smectic type) exist in small domains [69] and DSC provides 'thermal signatures' of their local ordering (Figure 8.4).

*(b) Divalent or trivalent metal salts of fatty acids (compounds **1** and **4** in Scheme 8.1).* The thickening properties of various divalent or trivalent metal soaps have been utilized for a long time in some lubricating applications. A comprehensive phenomenological review of the physical properties

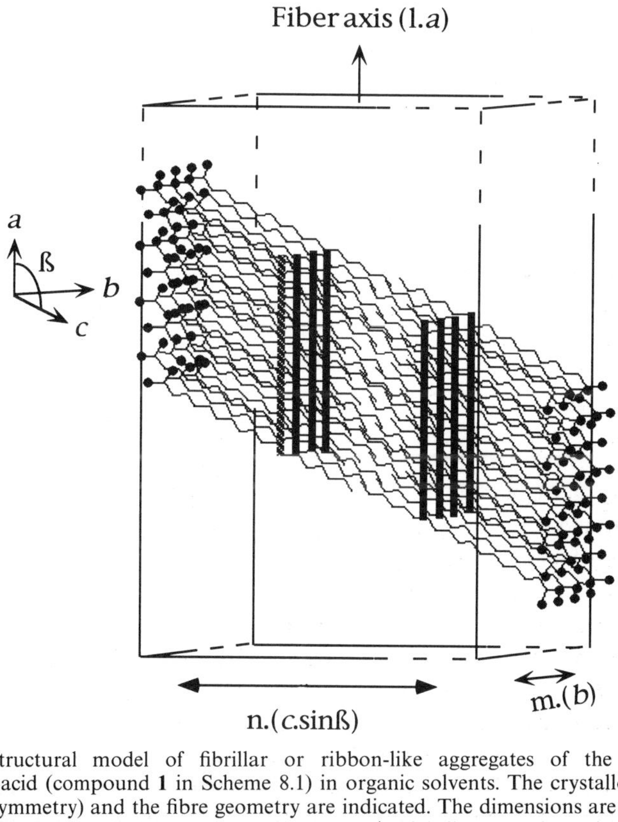

Figure 8.3 Structural model of fibrillar or ribbon-like aggregates of the 12-hydroxyoctadecanoic acid (compound **1** in Scheme 8.1) in organic solvents. The crystallographic axis (monoclinic symmetry) and the fibre geometry are indicated. The dimensions are proportional to the crystallographic cell parameters, as shown. The thick lines are the H-bond sequences connecting the hydroxyl groups at C_{12} and the spheres are carboxylic end groups. (Reproduced from Reference 71 with permission from the American Chemical Society.)

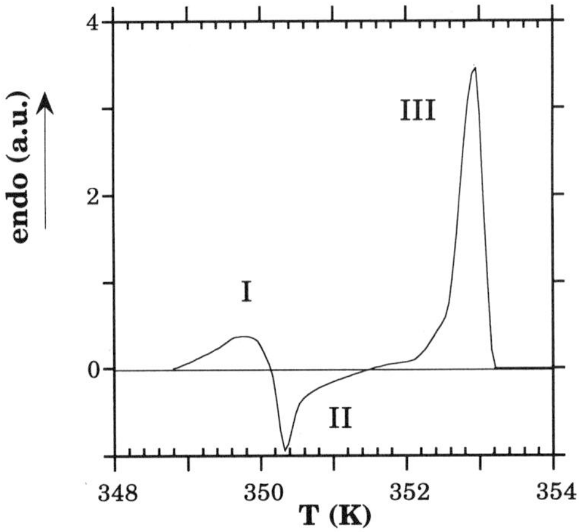

Figure 8.4 DSC traces of the mesomorphic solids made from 12-hydroxyoctadecanoic acid gels in CCl_4 (heating rate: 0.31 $K.min^{-1}$). During heating, a first endothermic process (I) transforms the mesomorphic xerogel to a more disordered solid which is followed by crystal formation (exothermic process II) and finally crystal melting (sharp exothermic process III). Similar traces are observed with steroid xerogels, compound **2** in Scheme 8.2 (see text). (Reproduced from Reference 69 with permission from the Chemical Society of Japan.)

of the related solutions [77] was published more than 30 years ago and reports the various behaviours which were later found not to be restricted to this class of organogelators. The length and degree of unsaturation of the lipophilic part of the molecule, the role of liquid–crystalline phase transitions on the solubility as a function of temperature, the metallic element and the solvent types are some of the parameters that affect the solubility of heavy metal soaps in organic solvents. Viscosity, fluorescence depolarization, osmotic pressure measurements [77] and stress-optical properties [78] have demonstrated that the aluminium soaps form much larger micelles than do the soaps of mono- or divalent metals and have suggested the existence of a specific interaction mechanism. Rheological measurements have shown that traces of water induce marked increases of viscosity through a possible sphere to long thread micellar transition in the dilute concentration range of the gelator. Further increase in water concentration favours the interlinking of the micellar soap through hydrogen bonding up to a proportion of 1 molecule of water per molecule of alkoxide soap. Attempts to model the gel structure as a network of threads interlinked in the 'junction points' of the network and with perpetual breaking and reforming reactions has been initiated in the pioneering work on these systems [77]. More modern theories of the dynamics of reversible networks [79, 80] are now available which aim to correlate the microscopic lifetime of a junction and the macroscopic stress relaxation time. They are usually applicable to polymeric systems or polymer-like solutions having reversible cross-links. Only viscoelastic liquids are concerned by these models and possibly some sol phases of metallic soaps (see below). Very rare examples of systems in organic liquids show rheological properties partially described by these models, one of which being a bicopper tetracarboxylate complex in a cyclohexane solution (section 8.4.6(b) [81, 82]).

More recent structural investigations [83] have shown that the aluminium disoaps form long chains in which the fatty carboxylate groups are almost perpendicular to a linear backbone of aluminium atoms and hydroxyl groups. Fibrous aggregates (studied by scanning electron microscopy, SEM) have also been found for gels made up of the aluminium salt of dihexadecyl phosphate ($C_{16}DP{-}Al$) in hydrocarbons. The fibrous structures, which can be observed for solutions as dilute as 0.01 wt%, result from the stacking of trisubstituted central aluminium atoms and three $C_{16}DP$ groups. The aluminium ions are strongly coordinated by the phosphate bridges (Figure 8.5). The aggregation behaviour is destroyed upon addition

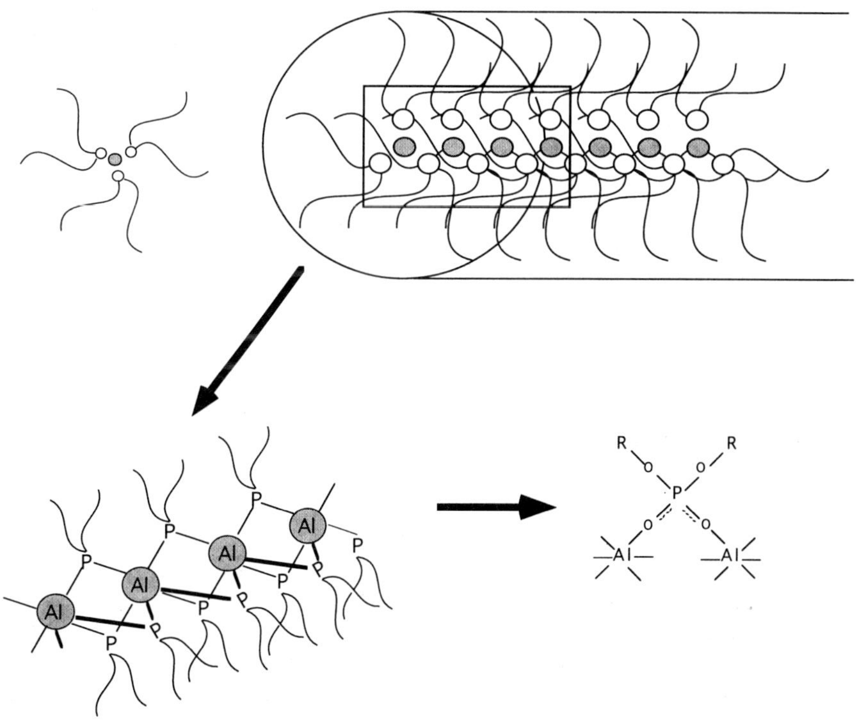

Figure 8.5 Model for the $C_{16}DP$-Al cylindrical aggregates in *n*-hexadecane binary systems. (Reproduced from Reference 83 with permission from Academic Press, San Diego.)

of alkylamines which form links preferentially with the Al atoms. Concentrated solutions of C_{16}DP–Al in *n*-hexadecane (above $C > 18$ wt%) exhibit [83] lyotropic organizations of the inverted hexagonal type H_{II}. The hexagonal array of long cylinders is swelled in *n*-hexadecane such that the intercolumnar spacing varies from 27.5 Å (solid-state) to 54 Å (30 wt% of C_{16}DP–Al).

Dilute solutions of a cobalt hydroxy monooleate derivative (compound 4 in Scheme 8.1) in benzene or heptane ($0.1 < C < 0.4\%$) are studied by capillary viscometry while more concentrated systems ($0.7 < C < 3\%$) are investigated with a cone-plate viscometer [84]. Rheological curves as a function of concentration show non-Newtonian behaviour, exhibiting a shear thinning effect followed by a Newtonian flow behaviour. These solutions, with no yield value, are viscoelastic and show a dramatic decrease of viscosity upon addition of pyridine or acetylacetone. These additives are known to strongly coordinate with cobalt. This observation supports the hypothesis that the cobalt atoms are involved in the coordination bonding process which leads to supramolecular aggregates. Furthermore, the decay curve of the transient electric birefringence recorded for these systems reveals the existence of a broad spectrum of relaxation times of very polydisperse colloidal species. The rotational diffusion coefficient (D_r) of the rod-like chains is estimated to be $32\,s^{-1}$ at infinite dilution. A mechanism of OH bridging around a four-fold coordinated cobalt is proposed, allowing the connection between the 'monomeric' species in linear supermolecular aggregates. Assuming a tetrahedral symmetry of the CoO_4 unit, estimations of the mean length are made and compared to the radius of gyration obtained from static light scattering measurements. Complementary electric birefringence studies of the rotational diffusion lead to an estimation of the mean persistence length of the semi-rigid chain ($l_p \approx 1080$ Å) as well as the weight average number of Kuhn's segments (L/l_p approx. 11).

8.4.2 Steroid derivatives (Scheme 8.2)

Some steroid derivatives [85] (bile acids and salts such as deoxycholic, cholic, apocholic, lithocholic etc.) are known [86] to be thickeners of aqueous salted solutions. Some cholesterol esters and derivatives exhibit thermotropic mesophases [87, 88] which, due to the intrinsic molecular chirality, are frequently cholesteric (i.e. chiral–nematic liquid–crystal). It was thus reasonable to expect a gelling ability from some appropriately functionalized steroid derivatives in organic liquids. The first steroid (androstanyl type) gelator of apolar hydrocarbons was discovered serendipitously during the course of the synthesis of D-homosteroidal nitroxide free radical from isoandrosterone derivatives [89]. Modified steroids (Scheme 8.2) can give remarkably stable gels in apolar hydrocarbons if an hydroxyl group is present at C_3 while the 17a-aza position can be an amine

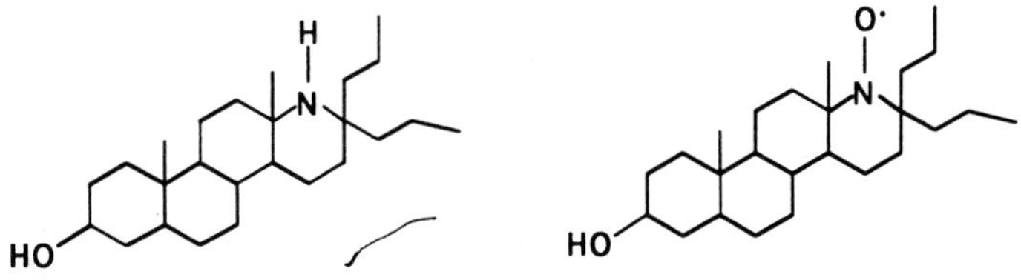

1 2

Scheme 8.2 Representative structures of steroid organogelators; androstanyl derivatives. **1**, D-17,17-dipropyl 17*a* aza-5α-homoandrostan-3β-ol (diamagnetic amine, STNH); **2**, D-17,17-dipropyl-17*a*-aza-5α-homoandrostan-3β-ol 17*a*-oxy (paramagnetic nitroxide, STNO).

or a nitroxide functionality. Depending on position, insertion of unsaturation in the steroid molecular structure can give different gelling abilities. Thus, two allyl groups at C_{17} inhibit the gelation phenomenon while a double bond at $C_{5,6}$ gives efficient organogelators. This class of molecules does not form gels in aromatic or polar solvents and the gel stability in hydrocarbons is very solvent-dependent, with the most stable gels obtained in cyclohexane (stable over decades). Gels formed by the spin-labelled steroid, D-17,17-dipropyl-17a-aza-5α-homoandrostan-3β-ol, 17a-oxy (Scheme 8.1(**2**), STNO) have been studied [90] by electron paramagnetic resonance (EPR). Due to the role played by the nitroxide probe (NO*), acting as a neutral internal reporter, analysis [91] of the EPR spectra allows the nitroxides in a fluid part (three narrow EPR lines) and nitroxides in a solid-like part (broad shifted EPR line) to be distinguished. A partial phase diagram is deduced showing that at a given temperature a small fraction of steroid molecules does not incorporate into the solid-like 3D network. This fraction can be considered as a 'solubility limit' of the system which participates in the multiple equilibrium reactions undergone by the 'equilibrium aggregates'. Details of the kinetics of aggregation have been carefully examined using complementary techniques such as EPR, SANS (Figure 8.6) and IR, UV-visible CD (circular dichroism) spectroscopies [92]. The method used takes advantage of the spectral differences that exist between the isolated 'monomeric' and the aggregated states of the steroid molecules. Once the warmed solution is cooled and equilibrated at a temperature in the gel domain of the phase diagram, there is a time delay before the aggregation and subsequent gelation take place. The induction times as well as the kinetics of growth of the colloidal aggregates follow nucleation and growth laws used for unidirectional aggregates formed from supersaturated solutions [92, 93]. A two-step homogeneous model can be used to simulate the kinetic curves where the first step is autocatalytic and the second one is

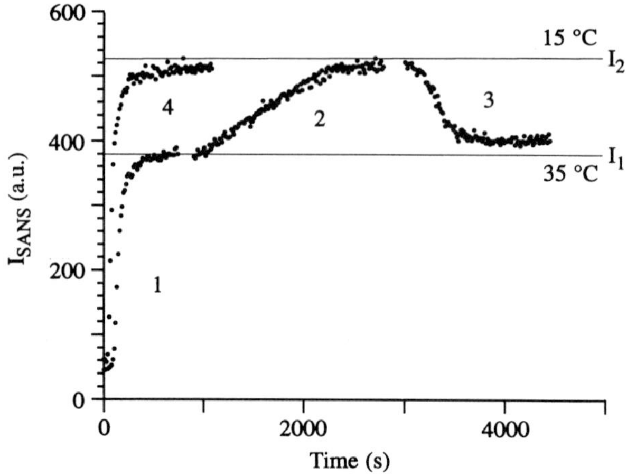

Figure 8.6 SANS curves for the kinetics of aggregation of the steroid STNH (compound **1** in Scheme 8.2, $C_0 = 5.9 \times 10^{-2}$ M) showing the additivity and reversibility of the gelation transition between 35 and 15°C. (Reproduced from Reference 92 with permission from Academic Press, San Diego.) **1**, the sol to gel kinetic transition at 35°C; **2**, transition from 35°C to 15°C in the gel phase resulting from process **1**; **3**, transition from 15°C to 35°C in the gel phase resulting from process **2**; **4**, sol to gel transition at 15°C. I_1, I_2 are the equilibrium integrated scattered intensity within a low-angle window at 35°C and 15°C, respectively. The magnitude of I_{SANS} is correlated to the concentration of large fibrillar aggregates (see equation 8.5).

first order. The cooperativity is based upon the development of multiple hydrogen bond sequences (at the C_3 position) once appropriate specific molecular configuration requirements are satisfied. The kinetics are dependent upon the supersaturation level (σ) of the initial solution so that the resultant gel sample is made up of microdomains, the growth of which is in-phase for high σ values or random for moderate σ values. As shown by neutron scattering experiments, the equilibration temperature of the gel (and not the thermal history) determines the amount of gelator involved in the solid-like network during the aggregation reaction (Figure 8.6).

The kinetic method measures the integrated scattered intensity within a central angular window of the scattering which reflects the amount of growing rod-like aggregates. According to σ, the time delay for a complete aggregation reaction can be varied from a few minutes to several hours. Arrhenius plots of the variation of the kinetic constants with temperature (from SANS, EPR and IR data) provide activation energy values (about 6.5 kcal.mole^{-1}) consistent with a hydrogen bonding mechanism. The SANS technique is also used as a classical method for structure determination and it demonstrates that the steroid aggregates are very long and rigid fibres [94], possibly symmetrical double helices of diameter 99 Å. The cross-section of the fibres, approximately circular, appears to be rather monodisperse as demonstrated by the existence of typical cross-sectional intensity oscillations in the scattering curve at large angles (the so-called Porod Q-range, Figure 8.7).

The 'contact points' or nodes in the network are more realistically 'junction zones' made by a specific overlap through fusion or coiling of two or more individual fibres over long distances compared to their diameter. This picture is fully confirmed by freeze-fracture-etching transmission electron micrographs [95–97] which show 91 Å diameter filaments (Figure 8.8(A)). A helical substructure made up of two or more fibrils with a diameter of 46 Å is also observed. Stereo TEM image pairs show spectacular views of the 3D porous structure of dilute gels. At this point, it is worthwhile mentioning the similarity of images and mechanisms observed with the present steroid gelator system and that obtained with an optically active glycolipid derivative in dilute (1%) aqueous gels [98] known to develop strong intermolecular hydrogen bonds. Similar observations (helical fibres, random networks, kinetic models, hydrogen bonding mechanisms, filament assembly mechanisms) have also been made [99] for some biological and natural systems (actin, tubulin, collagen, DNA).

TEM observations [73] provide new insights on the structural relationship between the native gel state (with solvent) and dried (xerogel) state. Micrographs of entangled quasi-infinite helical filaments (three times the

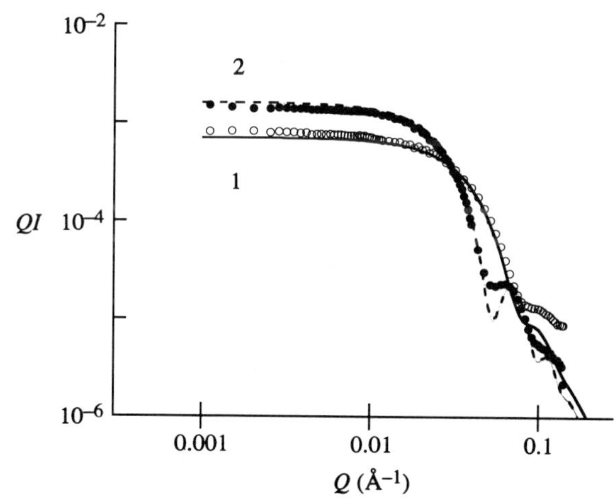

Figure 8.7 SANS of the STNH steroid-based (compound **1** in Scheme 8.2) gels in deuterated cyclohexane (1, ○) ($C = 2.3 \times 10^{-2}$ g.cm^{-3}) and in methyl cyclohexane (2, ●) ($C = 2.1 \times 10^{-2}$ g.cm^{-3}). Cross-sectional intensity QI versus Q is plotted. Points are experimental and full lines are the best adjustments according to the full homogeneous cylindrical rod model described by equation 8.5. The influence of the solvent nature upon the structure of the aggregates is evident. 1, $r_0 = 54$ Å, $\varepsilon = 0.2$; 2, $r_0 = 75$ Å, $\varepsilon = 0.12$. (Reproduced from Reference 104 with permission from the *J. Phys. France*.)

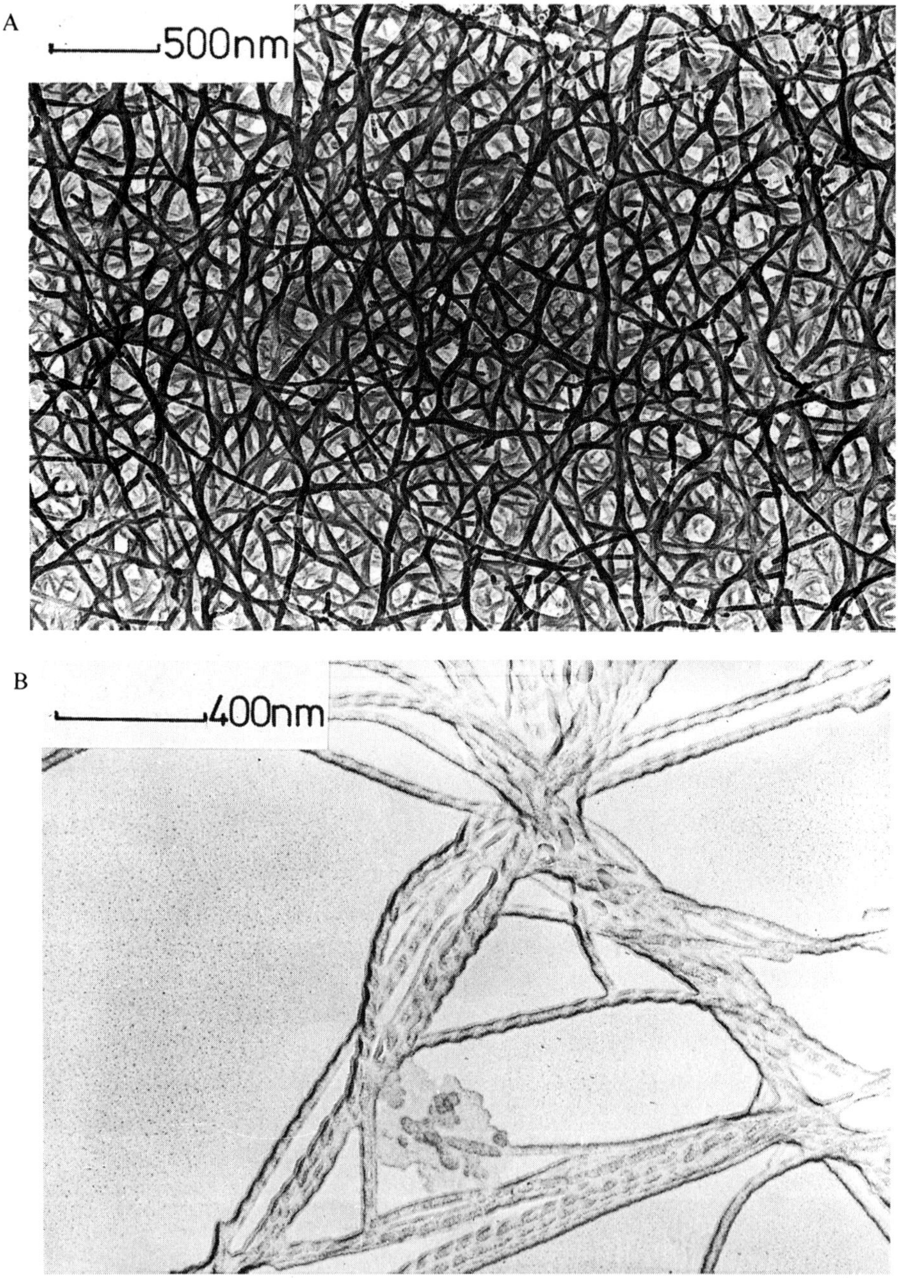

Figure 8.8 Electron microscopy of steroid STNH (compound **1** in Scheme 8.2) gels in cyclohexane. **A**. Porous 3D gel network in cyclohexane. The main filaments of the gel network are about 91 Å in diameter (from References 95 and 97, and Terech, P., Wade, R.G. unpublished data). **B**. Dried gel replica showing long twisted filaments and details of their interactions (from Reference 73, and Terech, P., Wade, R.G. unpublished data).

order of magnitude found in the native gels) with an approx. 580 Å pitch (Figure 8.8(B)) are found along with overtwisted fragments. The shrinking of the gel network with solvent evaporation is regulated by the elasticity and chirality of the fibres. The rearrangement mechanism of the chain trajectory of stress-sensitive chiral filaments involves helical folding of the fibres, collapses in super-aggregates, disruptions of fibres and 'junction zones', and results in increased diameters and rigidities of the chain fragments.

Without external forces, dilute gels are random. Anisotropic gel samples can be obtained either by using strong magnetic fields in appropriate kinetic conditions or at null magnetic field conditions with concentrated solutions confined in thin sample holders [100, 101]. Once a magnetic field (several Teslas) is applied during the first steps of the kinetic process of steroid aggregation, and unstable gel can be obtained which separates out highly birefringent collapsed oriented filaments (Figure 8.9(A)). The examination of the characteristic kinetic times of the experiment (rotational time for the rods and kinetic growth time) shows that only the very first stages of the kinetic sol to gel process are influenced by the application of high magnetic fields so as to produce rod orientation. The ordering process is 'frozen' by the overlap of the growing fibres at C^* but the former oriented rods act as growth precursors for the rest of the growth process. Despite a small anisotropy of diamagnetic susceptibility tensor of the individual steroid molecules, the additivity property obtained in a linear aggregate makes the ordering effect possible. The degree of alignment is evaluated by optical microscopy under crossed polarizers or wide-angle scattering of X-rays.

Clear differences in the molecular organization of the solid, gel and xerogel states have been demonstrated by the different 'signatures' obtained from DSR, IR, optical polarizing microscopy and WAXS techniques [102, 103]. The xerogel is a mesomorphic organization which can be attained only through the solution to gel phase transition. Schlieren textures (Figure 8.9(B)) typical of nematic-like ordering are observed for the diamagnetic compounds while only extinction crosses confined within droplets are observable for the paramagnetic steroid compound (STNO). As observed in most other systems, small changes in the polarity balance of the gelator/solvent couple induce nanoscopic structural variations as well as significant macroscopic changes (stability, mechanical and optical properties). Thus, if the solvent cyclohexane is replaced by methyl-cyclohexane or *cis*-decalin, an increase of the filament diameter (by about 50%) of the resulting steroid aggregates is measured by SAXS [104] indicating that some steric parameter of the lipophilic balance is also involved in the aggregation mechanism. Differentiation between two solvents as similar as *cis* and *trans* decalin is also observed by analysing both the rigidity of the fibres and their ability to grow as oriented helical fibres (*cis*-decalin).

The information on the mechanism of the fibre interactions obtained from EM experiments is complemented by WAXS measurements. Bragg

A

B

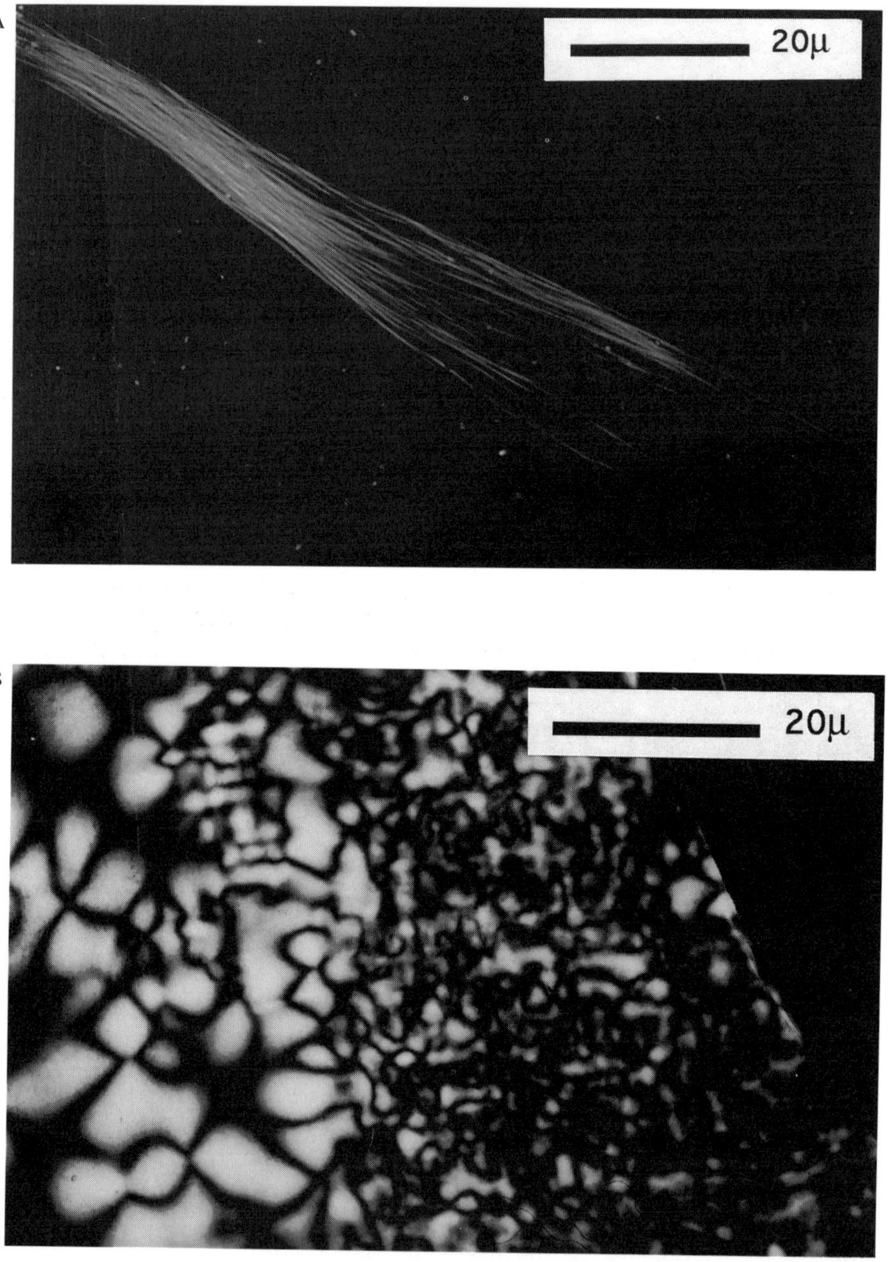

Figure 8.9 Optical microscopy under crossed polarizers of the diamagnetic steroid compound **1** in Scheme 8.2 (STNH) gels in cyclohexane. **A**, Oriented fibres separating from a gel formed within a strong magnetic field (7 T). (Reproduced from Reference 100 with permission from the American Chemical Society.) **B**, Schlieren textures observed in a 1 mm thick gel ($C \approx 4\%$). The black portion is the air–gel interface (Reproduced from Reference 104 with permission from the *J. Phys. France*.)

peaks are observed for the gel samples (for instance, at $Q \approx 0.23\,\text{Å}^{-1}$ and $Q \approx 0.43\,\text{Å}^{-1}$) whose position depends on the gel sample history (age, solvent content and type). The corresponding (pseudo)crystalline (lyotropic) organized regions are assumed to be the 'junction zones' of the heterogeneous network and account for the solid-like mechanical behaviour of the steroid gels. Study of the structural correlations, from the orthorhombic crystallographic organization of the neat solid to the gel and xerogel states, is in progress [105].

8.4.3 Anthryl derivatives (Scheme 8.3)

2,3-bis-*n*-decycloxyanthracene (DDOA, Scheme 8.3(**1**)) forms gels in various alkanes, alcohols, aliphatic amines and nitriles [106]. The solution to gel phase transition is temperature dependent and is approximately in the range of −20 to 40°C depending on the solvent and concentration conditions. DDOA is of special interest due to its photochromic properties which have been characterized by absorption and emission fluorescence spectroscopy measurements [107, 108]. Rheological studies [109] have probed the onset of a 3D elastically connected network of aggregates evidenced by an abrupt increase of the storage modulus (G'), at some characteristic time of the aggregation reaction (Figure 8.10).

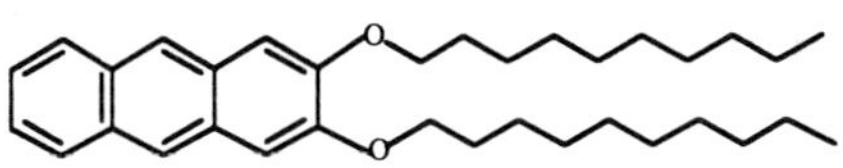

1

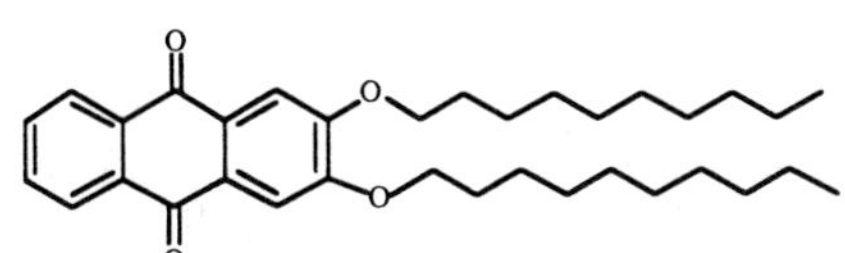

2

Scheme 8.3 Representative structures of anthryl-based organogelators. **1**, 2,3-bis-*n*-decyloxyanthracene (DDOA); **2**, 2,3-bis-*n*-decyloxyanthraquinone.

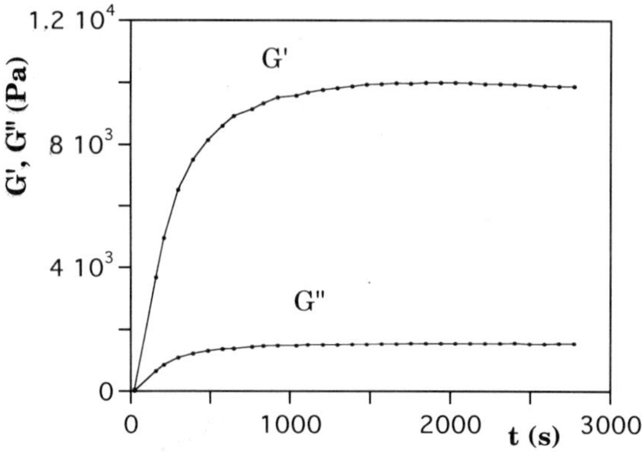

Figure 8.10 Rheological time sweep experiment for a DDOA/octanol gel. The warmed solution ($C = 0.98\%$ wt) is introduced between the cone and plate of the rheometer and cooled to 20°C. Frequency = 1 Hz, angular displacement $\alpha = 6.9 \times 10^{-4}$ rod (strain amplitude $\gamma = 1\%$). (Reproduced from Reference 109 with permission from Academic Press, San Diego.)

The magnitude of the equilibrium value, $G'_{\mathrm{equ.}}$, (approx. 10^4 Pa for a volume fraction $\phi \approx 0.01$) is much larger than the loss modulus ($G''_{\mathrm{equ.}}$) (approx. 1500 Pa). The comparison of the order of magnitude of the elasticity of the gel network with that for other systems suggests that the interaction zones are permanent crystalline microdomains at a given temperature. A temperature sweep experiment performed as a cycle between 20 and 40°C shows that the thermal reversibility of the mechanical properties of the DDOA/1-octanol sample exhibits hysteresis revealing the differences between the melting and formation reactions of the gels. The DDOA/1-octanol system shows, at low frequency, an elasticity which justifies the description of the sample as a solid-like gel with an important yield stress value ($\sigma^* \approx 550$ Pa for $\phi = 0.01$) consistent with the self-supporting ability of the material. This value is much larger than that for many other colloidal materials [110]. The physical junctions of the DDOA network appear as crosslinks with a significant energy associated per unit length of linear DDOA aggregate.

Structural information on the gels (at $\phi \approx 0.02$) is obtained from SANS measurements. DDOA gels in D10-butanol and D-decane exhibit very comparable isotropic scattering patterns showing a monotonous decrease ($I = Q^{-3.7}$) from low-angles to larger angles where diffraction peaks are observed (for instance, at $Q \approx 0.1\ \text{Å}^{-1}$). Since the nuclear contrasts of the DDOA aggregates in the two solvents are very comparable, it is concluded that in the present thermodynamic conditions, the range of distances investigated by the SANS experiments describes part of the morphology of the DDOA aggregate which is insensitive to the solvent type. A Porod analysis shows a broad bump ($0 < Q < 0.075\,Å^{-1}$) attributed to the form-factor of the DDOA aggregates (its width being related to the size polydispersity of the scatterers) while the peaks at larger angles (for instance at $Q \approx 0.1\,Å^{-1}$) are due to the structure factor of ordered heterogeneities in the DDOA gel network. The simulation of the bump using the theoretical form-factor for fibres (equation 8.5), requires the use of several diameters and suggests that the interfacial scattering comes from a distribution of diameters ranging from $D_{min} \approx 100\,Å$ to $D_{max} \approx 500\,Å$.

Freeze–fracture electron micrographs (Figure 8.11) in propanol support the picture of a dense 3D random network of fibrous rigid bundles (600–700 Å). It is worth mentioning that DDOA/acetonitrile gels shrink significantly (syneresis) when submitted to mechanical stress [105]. Such a solvent-dependent behaviour could contribute to the observation of thick fibres in the gels. Comparable 'reactions' of networks to the solvent type are observed at least in two systems: (a) 12-hydroxyoctadecanoic acid which develops more or less extended 'junction zones' along a preferential axis; and (b) an anthryl-appended steroid-based gelator (CAB) whose morphology and helicity of its fibres (section 8.4.4) depend on the solvent 'quality' (by analogy with the terminology used with polymeric systems) or polarity. The network structure results from a balance of forces, among which are its elasticity, the aggregate–aggregate affinity and possible additional terms influencing the osmotic pressure of the gel (such as the ionization degree in highly polar solvents). Modification of this balance of forces is responsible for the swelling or shrinking behaviour as well as for the subsequent phase

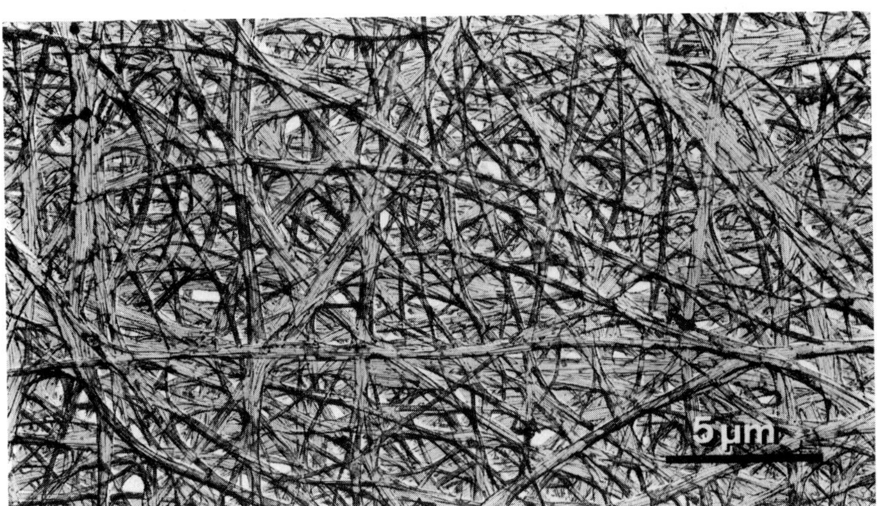

Figure 8.11 Freeze–fracture electron micrograph of gels from DDOA (compound **1** in Scheme 8.3) in propanol (from Reference 108).

transitions observed with some polymeric ionized gels in aqueous systems [111].

A WAXS analysis of the solid powder gives a clearer view of the nature of the 'junction zones'. The diffraction pattern exhibits a sequence of Bragg peaks at 0.095 Å^{-1}, 0.166 Å^{-1}, 0.191 Å^{-1}, 0.254 Å^{-1}, 0.287 Å^{-1}, 0.336 Å^{-1} whose spacings are in the ratios 1, $\sqrt{3}$, $\sqrt{4}$, $\sqrt{7}$, $\sqrt{9}$, $\sqrt{12}$ typical of an hexagonal ordering of the diffracting sample [112]. The DDOA gels are also crystalline as characterized by Bragg peaks in their diffractograms located at the same positions as those for the neat solid. The [11] reflection (associated with a reticular distance $\langle d \rangle \approx 66.1$ Å) corresponds to an intercolumnar distance of 76.3 Å in an hexagonal symmetry which may be that of the crystalline junction zones formed by bundles of aggregates. The 'junction zones' are 'dry' crystalline microdomains in which no swelling by the solvent is involved (by contrast to some steroid gelators (sections 8.4.2 and 8.4.4)). Considering the extended molecular length ($\langle d_{mol} \rangle \approx 24$ Å), various aggregation mechanisms may involve the stacking and partial overlap of the anthryl parts of the DDOA molecules to account for the measured reticular periodicities.

8.4.4 Gelators containing the cholesterol unit and condensed aromatic rings

Despite the fact that the gelation behaviour of the steroids (Scheme 8.2) relies upon the presence of the hydroxyl group at C_3, a new class of organogelators is obtained if the C_3 position is appropriately functionalized. Indeed, an aromatic moiety synthetically linked to a cholesterol part creates a novel family of organogelators. The interest of these systems is both to observe, on a mechanistic level, the result of the combination of two auto-associative parts and also to investigate the possibility of a transfer to the gel of some of the properties intrinsic to the grafted moiety (chirality, luminescence, etc.).

(a) Anthryl and anthraquinone appended steroid-based gelators (Scheme 8.4). Cholesteryl 4-(2-anthryloxy)butanoate (CAB, IV-1) is a member of a family (Scheme 8.4) of molecules (ALS) consisting of a 2-anthryl or 2-anthraquinonyl (CAQ, Scheme 8.4 (**2**)) part (A) connected to a steroid moiety (S) via a linking group (L). The linkage group enables the cholesteryl and anthracenyl groups to adopt a rodlike shape. The polarity of the CAB molecule is low and it does not have any readily available groups capable of developing hydrogen bonds (as is frequently the case for various other gelators). Organic solvents such as dodecane (and various hydrocarbons), alcohols (butanol, octanol etc.), aldehydes (such as heptanal), esters, amines, etc. can be gelled [113] as well as poly(dimethylsiloxane) by CAQ [114]. The gels are luminescent and some of them are metastable, exhibiting a

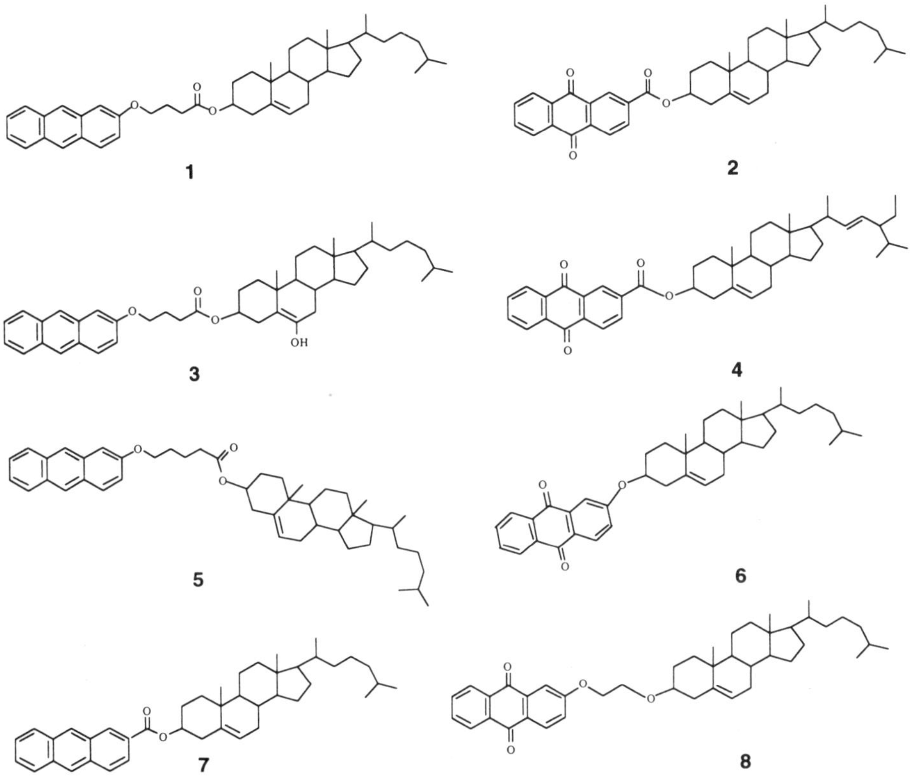

Scheme 8.4 A few examples of anthryl and anthraquinone-steroid based organogelators. **1**, cholesteryl 4-(2-anthryloxy)butanoate (CAB); **2**, cholesteryl anthraquinone-2-carboxylate (CAQ) and their derivatives.

liquid to solid phase separation over variable time periods according to the ALS/solvent pair. A variety of synthetic modifications in the nature of steroid, aromatic and linking groups of the ALS molecules has been made and correlated to the gelation ability [115, 116]. The gelation temperature is detected by a large increase of fluorescence intensity but electronic or dichroic absorption spectroscopies are also suitable methods to follow the gel formation. The gelation temperature (T_{SG}) is not dependent upon the chain length of alkanes while the stability of the gel with respect to solid–liquid phase separation is concentration and solvent dependent (for instance, the stability in cyclohexane is lower than in *n*-alkanes and heptanal).

The fluorescence spectra, arising from the anthryloxy fluorophore, provide information on the local ordering of the CAB molecules in the aggregates of the gel phase. New absorption bands are exhibited in the gel phase (3960 and 3760 Å) which result from the excitation dipole coupling between stacked anthracenyl groups in a face-to-face fashion within the aggregates. In the gel, the CAB molecules overlap partially with their long molecular axes being non-parallel. When the solvent 'quality' or polarity is varied, significant spectral differences of the fluorescent properties of the luminescent CAB gels are observed in hexadecane and 1-octanol [117]. Furthermore, the sol and gel phases of mixtures of CAB/1-octanol/hexadecane have been studied using their spectroscopic (absorption, emission) features (position, shape, intensity of the related bands). It is interesting to note that up to 75/25 1-octanol/hexadecane ratios, T_{SG} of CAB in neat hexadecane samples is observed while with 80/20 mixtures, T_{SG} can be either that of CAB in neat hexadecane or 1-octanol depending upon thermal history (slow or fast cooling rate). It is concluded that two types of morphologies can exist with a configurational transition which is solvent dependent in a stepwise fashion.

Transient optical textures are observed in sample preparations which indicate that a first step towards the dispersed gel state could be a liquid-crystalline-like organization during the dissolution process of the gelator. In addition, CD measurements suggest the presence of helical structures in the gel phase of dodecane while the solvent mobility remains unchanged on the time scale of ^{1}H NMR experiments (300 MHz). In 1-octanol, interactions of the proton donor solvent and specific parts of CAB are suspected to be at the origin of the helicity. Organogels exhibit optical textures, such as faint conic patterns, evolving to fan-like textures while Maltese crosses and other evidence for the existence of lyotropic domains [118] (possibly cholesteric) are also observed. Orientational molecular ordering is also present for neat CAB (and CAQ) which exhibits a thermotropic cholesteric liquid-crystalline phase [119].

Scattering techniques (SANS and SAXS) provide some confirmation [120, 121] of the nanoscopic TEM structural data and CD or other optical measurements. Considering the SANS scattering profiles (Figure 8.12) and the similarity of the neutron contrast values of the CAB aggregates in decane and 1-butanol, it is concluded that the solvent type induces significant structural changes. QI *versus* Q scattering curves of CAB gels in 1-butanol and decane exhibit a low-angle plateau typical of the form-factor of long and rigid fibres (equations 8.5). A Guinier analysis provides the cross-sectional radius of gyration (R_c) values which relate to the geometrical diameters. Assuming homogeneous circular cross-sections, the diameters are found to be slightly larger in alcohols (192 Å) than in alkanes (160 Å). A comparison of SANS and SAXS data reveals a slight variation of the R_c values (59 *versus* 54 Å, respectively) for CAB/decane gels which suggests the existence of heterogeneous cross-sectional contrast profiles. The larger Q-value part of the scattering curve with broad peaks comes from the structure factor of ordered heterogeneities in the network. Consequently, the

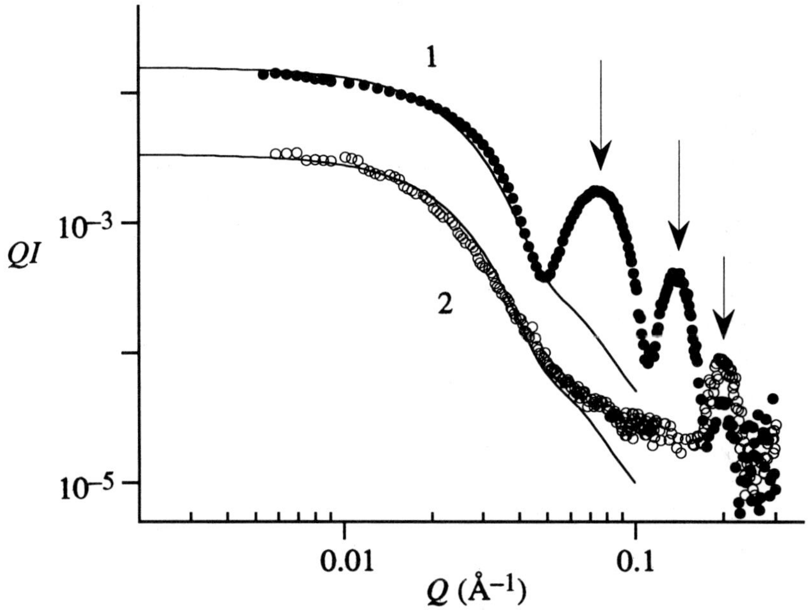

Figure 8.12 SAXS cross-sectional intensity QI versus Q plots for CAB (compound **1** in Scheme 8.4) organogels. 1, (●), decane, $C = 0.0192\,\mathrm{g.cm^{-3}}$; 2, (○), octanol, $C = 0.01058\,\mathrm{g.cm^{-3}}$. Intensity for decane is multiplied by 13 for sake of clarity. Full lines are best fits according to equation 8.5: 1, $r_0 = 82$ Å, $\varepsilon = 0.24$; 2, $r_0 = 85$ Å, $\varepsilon = 0.24$. ε is the radial gaussian dispersity of the cross-sections. The diffraction peaks of the lyotropic microdomains in the decane network are marked by arrows (from Reference 120).

full scattering curve is interpreted as resulting from the mixing of the form-factor of the CAB fibres (low-angle part) and the structure factor (wide-angle part) of ordered domains. The low-angle fit according to equation 8.5, as well as the Guinier and Porod analysis, give a consistent picture of the CAB aggregates in alcohols. Calibrated neutron intensities provide an estimation of the number of CAB molecules per unit length (n_L) of fibre in alcohols [120], $n_L = M_L/M$ about 18 molecules.Å^{-1}. The n_L value is close to that which corresponds to the packing in the solid state since a cylinder of a radius of 81 Å and 1 Å length with $v_{CAB} \approx 1\,\mathrm{cm^3\,g^{-1}}$, has an n_L value $\pi(82)^2.1.1.N_A/M_{CAB} \approx 19.1$ mol.Å^{-1}. The interpretation of the above mentioned SAXS and SANS peaks of the decane gels is made in a crystallographic framework. For the related crystalline powder, the spacings of the Bragg reflections (at 0.096, 0.163, 0.188, 0.248 and 0.281 Å^{-1}) are in the ratio 1, $\sqrt{3}$, $\sqrt{4}$, $\sqrt{7}$, $\sqrt{9}$, indicating its hexagonal ordering. The intercolumnar distance in the neat solids is $D_0 = 2/\sqrt{3}\ 2\pi/Q_{(11)} = 75.57$ Å where $Q_{(11)}$ is the first reflection generated by a two-dimensional hexagonal array of scatterers. The CAB gel network in decane appears to be heterogeneous and made up of fibres 'interconnected' by swollen lyotropic microdomains (intercolumnar distance 102.2 Å). The spacings of the peak positions in the gel are in the ratio of 1/1.92/2.75 characterizing the swollen lyotropic organization of the diffracting units. In 1-butanol, the SAXS diffraction pattern shows that the CAB aggregates are organized differently from both the neat solid and the lyotropic microdomains of CAB/decane. Comparable conclusions are drawn for CAQ organogels [121]. The nodes of the CAQ network are lyotropic microdomains evolving from hexagonal to more compact lamellar-like orderings from decane to 1-octanol. Schlieren optical textures confirm the inhomogeneity of the orientation of the thread-like structures in the birefringent materials while CD spectra confirm the resemblance to CD spectra of the neat CAQ solid cooled from its mesophase.

The comparison of the values of the molecular length of CAB (about 38 Å) and the characteristic lengths calculated on the basis of SAS data (intercolumnar spacings: 75.7 Å in the solid, about 102.2 Å in the decane swollen domains; diameters of the connecting fibres about 160–170 Å) indicate that molecular pairs are involved in the columns of the solid state while in gels, the aggregation mechanism may involve overlapped associated molecules. General trends of the SAS results are supported by TEM experiments [116]. Micrographs of freeze-fractured and etched samples (Figure 8.13) show a 3D network of fibrous bundles emanating from central points. The dimensions of the rectangular cross-sections of the fibres (ribbons) appear as 209 × 104 Å in dodecane (non helical) and 263 × 82 Å (helical) in 1-octanol which correspond approximately to the cross-sectional areas determined by SANS. Other specific behaviour induced by the narrowness of sample holders (typically 50 μm) [117], such as the inhibition of the gel formation, pre-aggregation in sols and crystallization are observed.

Solvent penetrates certain aggregates and the swelling by hydrocarbons may be due to their shape and electrostatic compatibility with the lipophilic external shell of the aggregates (steroid parts). This mechanism is also proposed for some other steroid-based organogelators [122]. When such a swelling is not possible, for instance for gelators lacking a steroid part (such as DDOA), gels with 'dry crystalline nodes' are formed (involving stacks of polarized DDOA anthryl moieties) which lead to almost superimposable diffraction patterns of the solid and gel phases. For networks with nodes involving lyotropic microdomains, the diffraction patterns exhibit broadened and shifted peaks, an extent depending on the solvent type and content [120, 121].

Taking advantage of the large number of ALS gelators available, a 'molecular engineering' approach of the gelation phenomenon has been undertaken [115] and the influence of the different segments of the gelator molecules upon the gelation process identified. For instance, parameters

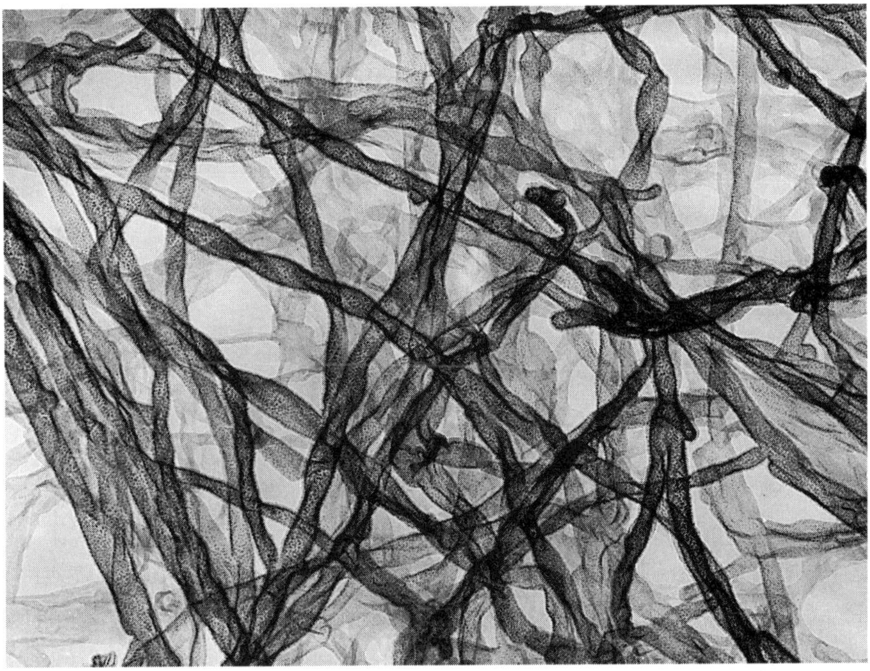

Figure 8.13 Electron micrographs of CAB (compound **1** in Scheme 8.4) in 2-octanol (Lin, Y-c., Kashar, B., Weiss, R.G., unpublished data). The cross-sectional dimensions of the twisted fibres are virtually the same as those for 1-octanol (263 × 82 Å, section 8.4.4(a), (Reproduced from Reference 116 with permission from the American Chemical Society.)

such as the α/β stereochemistry at C-3 of S, the nature of the alkyl chains on C-17, the length and functionality of L have been studied for a series of derivatives. It can be concluded that the gelation ability is favoured by a β stereochemistry at the C-3 of the steroidal part (S) while changes at C-17, albeit minor (unsaturation or ethyl substituent), do not modify it significantly. The contribution of the steroid moiety to the gelation behaviour can be analyzed comparatively in the light of the molecular organization of CAB and DDOA in the solid state. The X-ray diffraction patterns are almost superimposable and suggest that the association mechanism of the aromatic parts (and its more or less flexible arm L) is crucial in determining the structure of the ordered condensed states. In contrast, it is interesting to note that with azobenzene-linked cholesterol organogelators, the primary driving force for aggregation has been postulated to be the one-dimensional stacking of the cholesterol parts [123].

(b) Azobenzene steroid-based gelators (Scheme 8.5). Cholesterol-based gelators having the highly polar azobenzene group at C_3 of the steroid moiety can gel numerous organic liquids such as ethanol, acetic acid,

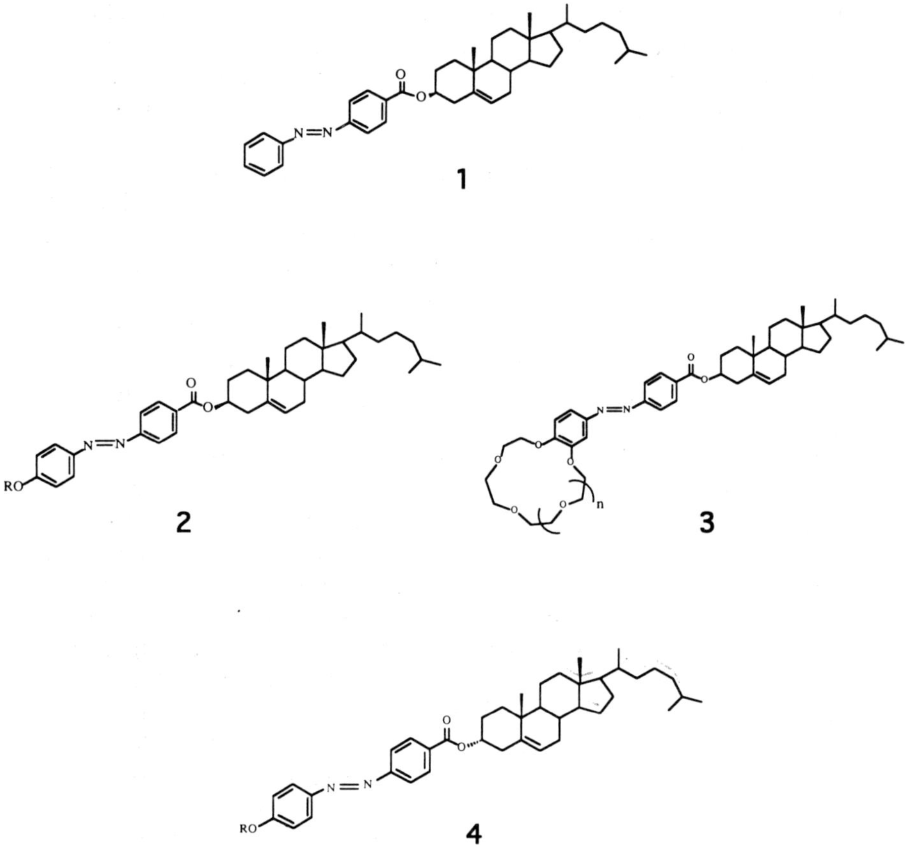

Scheme 8.5 Representative structures of azobenzene steroid-based organogelators. Natural configuration at C-3 ((*S*)-chirality), **1,2,3**. Inverted configuration at C-3 ((*R*)-chirality), **4**.

amines, dichloromethane, ethers, esters in the 'usual' concentration range [123] of about 1%. By using two esterification methods, two families of derivatives have been synthesized with either a natural (*S*)-configuration at C_3 or the inverted (*R*)-configuration. These compounds can gel specifically either polar solvents (*R*-derivatives) or polar solvents (*S*-derivatives). As frequently observed with fibre-like aggregates of optically active surfactants [124], CD can detect the sol to gel phase transition and enables the observation of enantiomorphic relationships. For instance, from the positive exciton coupling band observed for the (*R*)-chirality of compound in Scheme 8.5 (with (*S*) configuration at C_3), it can be deduced that the dipoles of the azobenzene parts in the aggregate are associated in a clockwise fashion (*R*-chirality). The alkane substitution R, located on the azobenzene moieties fully participates in the delicate lipophobic/lipophilic balance governing the gelation phenomenon. For instance, the gelling ability is increased for derivatives with odd numbers of carbons of the alkyl chain. As already mentioned above for other chiral gelators [116], CD signals (free from all possible linear effects due to macroscopic anisotropies within the sample as is frequently observed for orientable rigid-like structures) can be used to track the microscopic aggregation process during which the macroscopic gelation phenomenon occurs. Doing so, exciton coupling bands are identified in the gel phase and ascribed to the asymmetric orientation of the azobenzene chromophore in the aggregate. The cooling speed of the gel samples can affect the sign of the CD spectra showing that metastable structures are formed. Attempts to correlate the strand helicity with the cooling rate and the (*R*)/(*S*)-chirality of the azobenzene-appended, cholesterol-based gelators have been made using SEM observations. Similar correlations have been drawn for the anthryl-appended cholesterol-based organogelators, as discussed above.

Based upon the gelling phenomenology and partial structural information, an aggregation model giving rise to fibre-like helical aggregates is proposed in which the association of the chiral cholesterol moieties constitutes the driving force. The (*S*)-chirality (natural configuration) of the gelator favours additional interactions, strengthening the aggregate through face-to-face orientations of the azobenzene moieties. The absolute configuration at C_3 has also been shown to be determinant for the aggregation modes of monolayers developed at the air–water interfaces [125]. Consequently, the stability of the gels and their aggregative and gelling properties are enhanced while the solubility is reduced, in contrast to the inverted configuration. Part of the phase diagram is determined using three methods—the test-tube tilting method, the ball-drop method and the spectroscopic turbidity method—to determine [123] the gel to sol phase transition temperature (T_{GS}). The related phase transition can be envisaged as a dissolution process of crystals, as far as some thermodynamic parameters are considered, as already shown [92] for the androstanyl and various amino acid derivatives (see below where the hypothesis of the participation of nucleation processes during gelation is analyzed). For instance, the melting energy can be deduced from the measurements of the melting temperature of the gels as a function of concentration (equation 8.4). The insensitivity of the present system to solvent effects is interpreted as an indication that solvent molecules do not participate in the structures of the gelator strands which involve molecular packings close to those found in the neat solid (as measured for some fatty acids derivatives [71] or anthracenyl derivatives [122]). This conclusion cannot be generalized for some other steroid-based derivatives for which lyotropic organizations are present in the gel network (see for instance organogelators depicted in Schemes 8.2(**1**) and 8.4(**1**).

Interestingly, an original reversible control of the sol to gel transition can be performed using photoirradiation of the sample. The azocholesterol derivative undergoes a *trans-cis* photoisomerization of the azobenzene moiety so that the gel transforms reversibly into a solution under appropriate wavelengths of the radiation. The reaction thus provides a photoinduced phase which can be monitored by CD spectroscopy. As already mentioned, variations of the nature of the functional group linked to the cholesterol-based organogelator have been performed in order to investigate the gelation mechanism and to provide new gel systems with original 'grafted' properties. In this context, specific derivatives can be used for chiral recognition as they are capable of forming gels of variable stability (estimated by T_{GS}) when mixed with enantiomers of chiral monosaccharides [126].

8.4.5 Amino acid organogelators (Scheme 8.6)

α-Amino acid oligomers and their close analogues are promising candidates for gelation due to both their amide bonds, which are potential hydrogen-bonding sites, and their relative resistance to crystallization.

(a) 4-[[(1-naphthalenylamino)carbonyl]amino]-benzoic acid, 2-octyldodecylester. 4-[[1-naphthalenylamino)carbonyl]amino]-benzoic acid, 2-octyl dodecyl ester (compound **1** in Scheme 8.6) is known [127] to form nematic-like lyotropic gels at a concentration of about 4%. The gel to liquid transition temperature (T_{GS}), probed by DSC, is slightly higher than that determined by birefringence (Δn) studies which measure the temperature (T_{NI}) at which $\Delta n \to 0$. Schlieren textures are associated with a small birefringence ($\Delta n \approx 10^{-4}$) while shear induced birefringence can be observed with samples that are rapidly quenched. Samples slowly cooled are isotropic but they become uniformly birefringent when annealed at 55°C. It is proposed that the textures with splay deformations arise from finite micelles or breakable/reformable structures. The correlation between the cooling rate of the gelling sample and the nanoscopic structures suggests that metastable states can be formed, as already observed for other systems (such as some steroid-based organogelators (section 8.4.4)).

(b) N-benzyloxycarbonyl-L-alanine 4-hexadecanoyl-2-nitrophenyl ester. N-benzyloxycarbonyl-L-alanine 4-hexadecanoyl-2-nitrophenyl ester (compound **2** in Scheme 8.6) is a gelator [128] of methanol and cyclohexane through intermolecular hydrogen bonding (between N–H and C=O of the urethane bond). Dipole–dipole interactions between NO_2 at the 2-position, C=0 at the 4-position and long hydrocarbon chains are part of the lipophilic/lipophobic balance determining the aggregation process. CD absorption of a dilute gel ($\phi \approx 0.01$), free from linear contributions, demonstrates that the racemic derivative, which is not a gelling agent, does not exhibit any CD absorption.

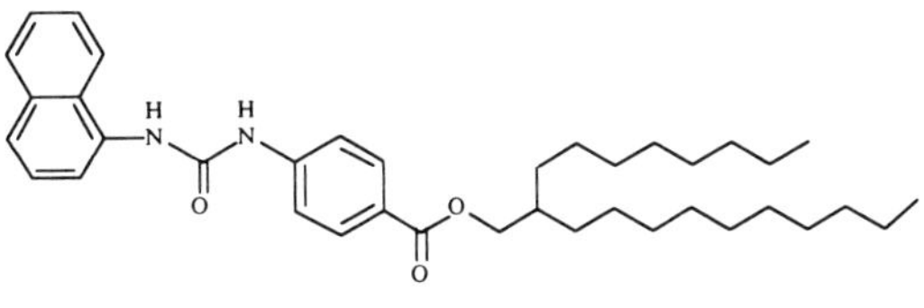

1

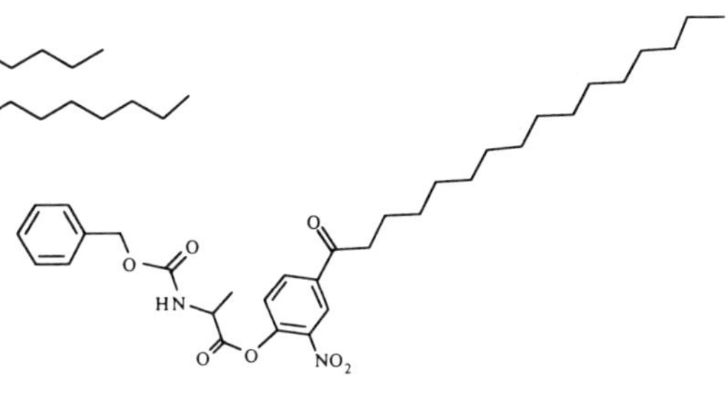

2

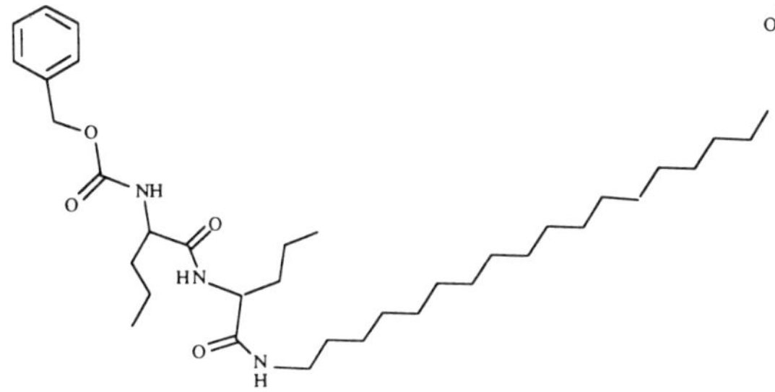

3

$$H\text{-}(NH\text{-}CH(R)\text{-}C(O))_n\ NHC_mH_{2m+1}$$

4

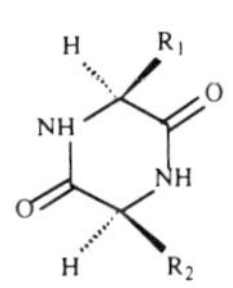

5

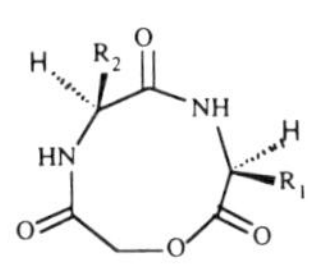

6

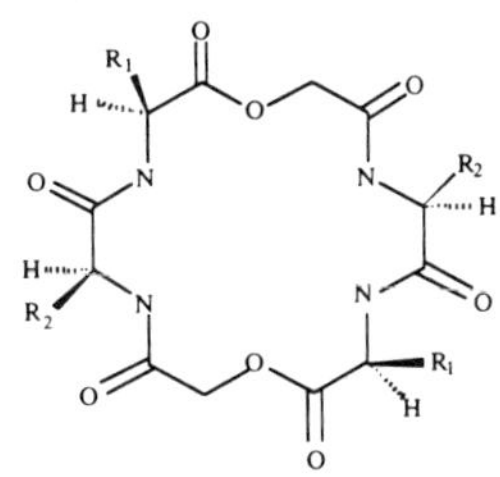

7

Scheme 8.6 Representative structures of amino-acid organogelators. **1**, 4-[[(1-naphthalenylamino)carbonyl]amino]-benzoic acid, 2-octyl dodecyl ester; **2**, *N*-benzyl oxycarbonyl-L-alanine 4-hexadecanoyl-2-nitrophenyl ester; **3**, *N*-benzyloxycarbonyl-L-valyl-L-valine *n*-octadecyl amide; **4**, oligo(α-amino acid) (example R = $CHMe_2$, $n = 5.0$, $m = 18$); **5**, cyclo-(dipeptide), example $R_1 = CH_2Ph$, $R_2 = CH_2CHMe_2$; **6,7**, cyclic depsipeptides; example, cyclo[CH_2-CO-L-leucyl-L-leucine], $R_1 = R_2 = CH_2CHMe_2$).

(c) Long-chain alkylamide derivatives. *N*-benzyl oxycarbonyl-L-valyl-L-valine *n*-octadecyl amide (compound **3** in Scheme 8.6) forms gels in most [129] organic liquids except hexane, tetrahydrofuran, chloroform and pyridine. IR spectroscopy confirms the participation of intermolecular hydrogen bonding (broad band at $\nu \approx 3290\,cm^{-1}$) and the enthalpic contribution deduced from the phase diagram (Figure 8.14) is compatible with the

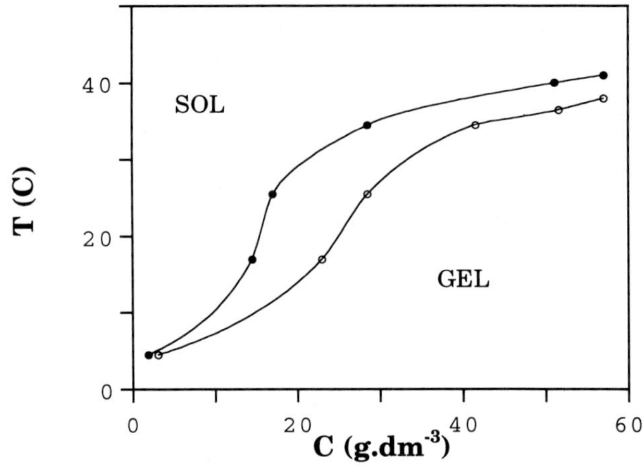

Figure 8.14 Sol to gel phase transition of the *N*-benzyl oxycarbonyl-L-alanine-4-hexadecanoyl-2-nitrophenyl ester (compound **2** in Scheme 8.6) organogelator. (○), methanol; (●), cyclohexane. (Reproduced from Reference 128 with permission from the Royal Society of Chemistry.)

participation of at least two hydrogen bonds, among the three theoretically available between N–H and C=O of an urethane and two amide bonds, in the aggregation reactions. The investigation of the molecular requirements for gelation shows that the structure of the amino acid residue, the optical activity of the molecule, the configuration of the lipophilic tail (dipeptides are more efficient than single amino acids) are some of the important parameters. As already noticed for some aqueous low-molecular weight gelators [75, 130, 131], racemic molecules are found unsuitable for gelation. The latter observation cannot be taken as a general rule since gelation of some racemic materials has been found (such as substituted fatty acids [71]).

(d) α-Amino acid oligomers (*Scheme* 8.6, *compound* **4**). A great variety of organic liquids can be thickened by this class of gelators depending on the degree of oligomerization and functionality of the first carbon position [132]. It appears that the hydrophobicity of the C-end group influences the gelation. From the phase diagram, the standard thermodynamic parameters (ΔH, ΔS) are deduced. Typical values in nitrobenzene are $\Delta H(25°\mathrm{C}) \approx -23.2\,\mathrm{kJ\,mol^{-1}}$ and $\Delta S(25°\mathrm{C}) \approx -21.2\,\mathrm{J.K^{-1}\,mol^{-1}}$ which indicate that two hydrogen bonds are formed. For rather long oligomers (glutamates), CD spectroscopy shows that a β-conformation is adopted through intermolecular hydrogen bonding while the α-helical and random coil conformations are inefficient for gelation.

(e) Cyclo(dipeptide)s. The cyclo(dipeptide)s (Scheme 8.6, compound **5**) have four hydrogen-bonding sites per molecule and those consisting of

different amino acids or having branched alkyl groups are more efficient in their gelling ability of organic liquids than those containing similar amino acids [133]. As already proposed for other gelators, the branching and bulkiness of alkane substitutions are thought to intervene in the stability of the gels as moderators of the aggregate–aggregate interactions which promote crystallization. IR spectroscopy reveals the presence of H-bonds, and TEM studies confirm the existence of fibre-like aggregates. Slow evaporation of the gels produces xerogels in the form of translucent films structured in layers with a spacing of 19.3 Å which may be assimilated to the separation distance in a ladder-like aggregate model [133].

(f) Depsipeptides (Scheme 8.6, compounds **6** *and 7).* Cyclo[CH_2-CO-L-leucyl-L-leucine] is able to gel some polar solvents (MeCN) but does not form gels in apolar ones such as cyclohexane or hexane [134], which is not a very common specificity of organogels. A special preparation procedure is required since the gelator is not soluble in the solvents which are gelled. A presolubilization step in a minimum amount of methanol (or water, dimethyl sulfoxide) is required. The gels exhibit some special characteristics such as surprisingly high T_{GS} (in acetone, the solvent evaporates at 85°C while its boiling point is 57°C), and the gelation is inhibited by some salts ($Zn(NO_3)_2$). As frequently observed, the complexity of the gelation phenomenon prevents a clear identification of the key parameters from the more trivial ones (such as stericity, polarity or solvation ability of the solvents [135]).

8.4.6 *Organometallic compounds (Scheme 8.7)*

(a) Mononuclear copper β-diketonates (Scheme 8.7, compound **1***).* Disc-like molecules have a shape which is favourable to the formation of rod-like aggregates by using an appropriate stacking process. Such an observation has already been reported for solutions of 2,3,6,7,10,11-hexapentoxy-triphenylene in hexadecane [136] or hexakis(*n*-hexyloxy)triphenylene in dodecane [137]. A copper complex substituted by eight paraffinic chains [138, 139] has a particular tendency for aggregation since the neat solid exhibits a thermotropic discotic mesophase and gives organogels containing approx. 99% of cyclohexane [140]. In addition, the cupric atom is responsible for the paramagnetism of the molecule and enables a direct use of EPR (X and Q-bands) without any additional spin labelling procedure. A similar situation was already described for the paramagnetic steroid-based organogelator compound **2** in Scheme 8.2, STNO (section 8.4.2). The polar copper diketonate residue of the molecule interacts with other neighbouring molecules within fibrillar aggregates through a stacking process involving partial recovering of the aromatic moieties. The anisotropy of the hyperfine

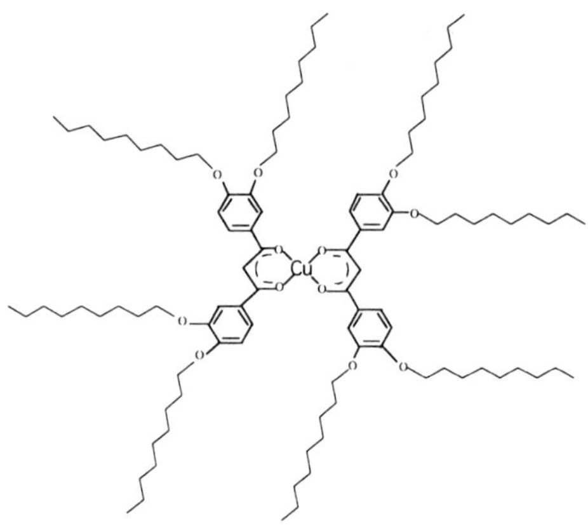

1

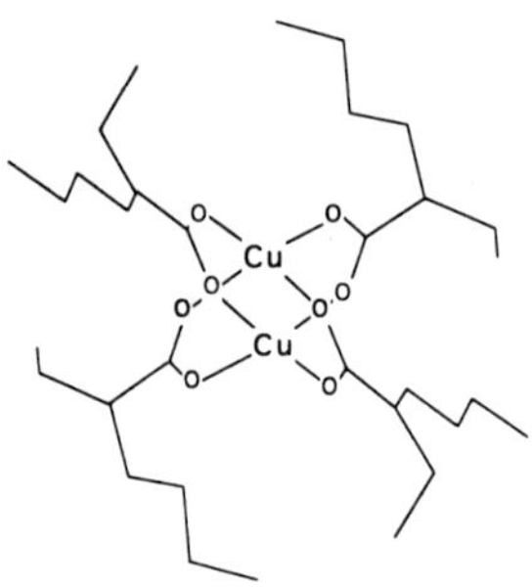

2

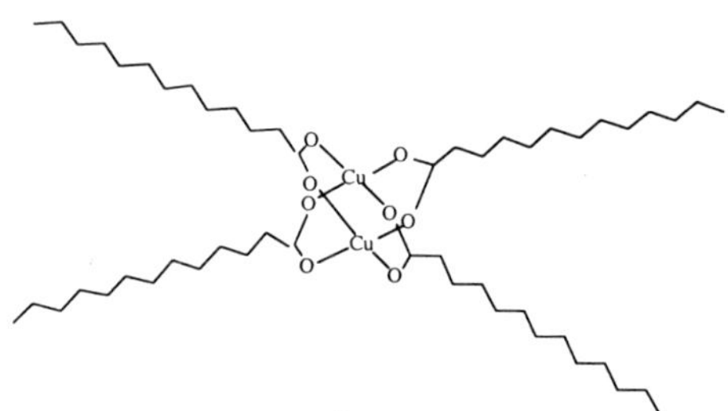

3

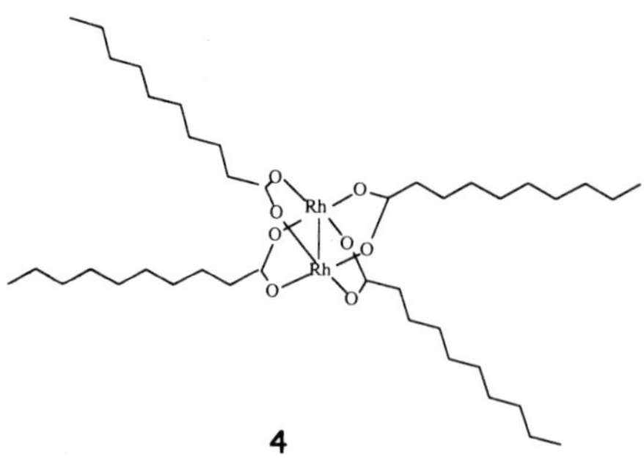

4

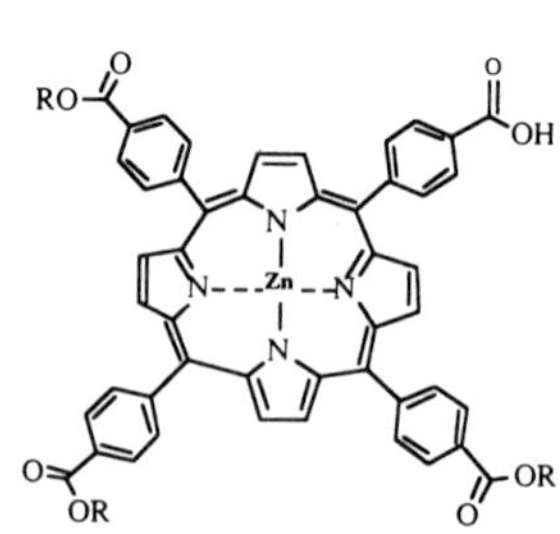

5

Scheme 8.7 Representative structures of organometallic organogelators. **1**, octasubstituted transition metal complexes of the bis(3,4 nonyloxy benzoyl) methanes; **2,3,4,** binuclear copper(II) tetracarboxylate complex of fatty acids; **2**, ethyl-2 hexonate complex; **3**, octanoate complex; **4**, binuclear rhodium(II) tetracarboxylate complex; **5**, Zn(II) complex of a tri-substituted (*para*-carboxyphenyl)porphyrin R = $C_{16}H_{33}$.

structure exhibited by the EPR spectrum of the complex **1** in Scheme 8.7 in the liquid phase is used to estimate the tumbling correlation time ($\tau \approx 4.10^{-10}$ s) of the molecules. In the gel phase, the electron exchange between the stacked molecules averages out the hyperfine structure and the frequency dependency of the spectral anisotropy is analyzed in terms of a reduction of the tumbling rate ($\tau \approx 8.10^{-9}$ s). EPR quantifies the weak electronic exchange interaction between the organometallic species and provides a convenient method to study the kinetics of aggregation. Evidence for a network of fibres is obtained by TEM [97] and SANS experiments [105]. The green gel samples exhibit a yield stress correlated to the existence of (pseudo)crystalline microdomains characterized by Bragg peaks in the WAXS scattering patterns.

(b) Binuclear copper tetracarboxylates (Scheme 8.7, compounds **2–4***).* A class of mesogens formed by disc-like molecules containing metal atoms, such as binuclear copper tetracarboxylates (compounds **2–4** in Scheme 8.7), exhibits thermotropic columnar mesophases [141–143] and gel-like materials [81, 144] in hydrocarbons at concentrations of less than 1%. The jelly is a viscoelastic liquid with no yield stress or elasticity at zero frequency thus justifying its classification as belonging to the 'weak gel' type. The thickening property is very significant since the zero shear viscosity is increased by a factor of 150 000 for a concentration of 2% in cyclohexane. SANS experiments demonstrate that the local structure of the bicopper aggregates is rod-like, while over long distances the chain is semi-rigid [81, 82, 144, 145]. The chains are extremely thin (diameter is approx. 17 Å for the derivative **2** in Section 8.7 where the aliphatic chains are 2-ethylhexanoates) and measurements of neutron absolute intensities indicate that only one molecule lies within the cross-section of the 'molecular thread' (Figure 8.15).

The molecules are stacked on top of each other along the axial direction of the chain and connected through Cu–O coordination bonds so that the symmetry around the copper atoms is a deformed octahedron. The axial packing utilizes sequences of connection sites between neighbouring molecules which are similar to those found in the crystalline solid (layers with a 5.2 Å Cu–Cu interdimer spacing, Figure 8.16, Mode 2) or in discotic liquid crystalline phases (crankshaft-like trajectory with a 4.7 Å Cu–Cu interdimer spacing, Figure 8.16, Mode 1). Any statistical switch between the two possible connection modes generates curvatures in the chain-like assemblies which are responsible for the chain flexibility (Figure 8.16).

This aggregation mechanism is also present in gels of the derivatives with linear aliphatic substitution (octanoates, laurates, etc.) but the viscoelastic jellies are metastable and evolve more or less rapidly towards phase separation (while the compound **2** in Scheme 8.7 in cyclohexane is extremely stable). The bulkiness of the alkane substitution induces a flexion stress in the aggregate during the stacking process which, in turn, generates chain flexibility.

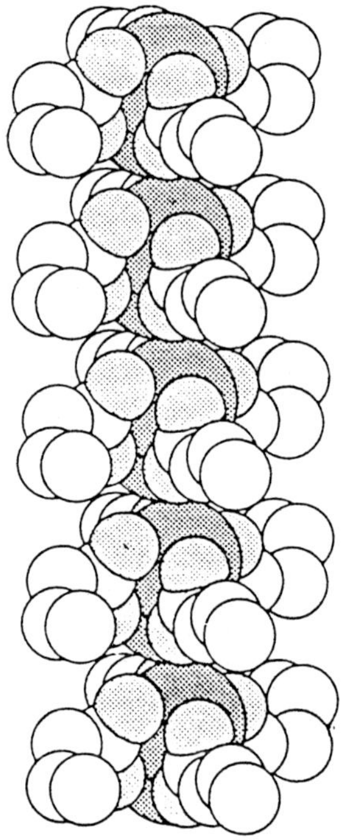

Figure 8.15 Local structure of the molecular thread formed during the aggregation process of a binuclear copper(II) carboxylate (compounds **2**–**4** in Scheme 8.7) in cyclohexane. Atoms: copper (■■), oxygen (▩▩), carbon (□□). The VDW diameter of carbon is 1.57 Å. For the sake of clarity, only the first four C atoms of the alkane substitution are represented. The chain-like assembly is rigid over distances lower than the persistence length.

A rheological investigation of the pseudo-plastic materials within the appropriate concentration range shows that relaxation of the dynamic shear modulus is about monoexponential [81, 82, 144–146]. The dynamic shear and loss modulii approximately follow the frequency dependencies described by equation 8.7. A Cole–Cole representation of the dynamical counterpart, $G^*(\omega)$, of the stress relaxation function $G(t)$ is a convenient way of characterizing the so-called viscoelastic Maxwell fluids (Figure 8.17). The corresponding plot of the imaginary part $G''(\omega)$ as a function of the real part $G'(\omega)$ gives more or less perfect semi-circles characterizing the degree of monoexponentiality of the stress relaxation.

$$G'(\omega) = G_0\omega^2\tau^2/(1 + \omega^2\tau^2)$$
$$G''(\omega) = G_0\omega\tau/(1 + \omega^2\tau^2) \qquad (8.7)$$

This behaviour suggests that the theoretical framework [18, 80] of reversibly breakable polymer-like chains is appropriate to describe the dynamics of stress relaxation in these systems. The so-called 'living polymers' are

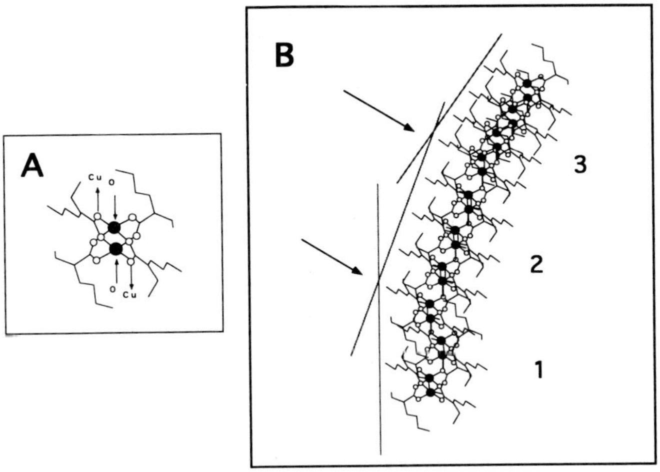

Figure 8.16 A molecular thread (**B**) made up of binuclear copper complex molecules (**A**) connected by copper–oxygen bonds. According to the sequence of connection sites involved, rigid columns or semi-rigid assemblies with different curvatures can be generated. Arrows indicate a change of the curvature of the aggregate.

worm-like equilibrium assemblies which have a molecular weight distribution in thermal equilibrium and exhibit scission/recombination reactions. The dynamics of a chain extremity are modelled by reptation theories [2, 32], at least on the time scale of its recombination with a different chain fragment. The dynamic properties are driven by two characteristic times, the

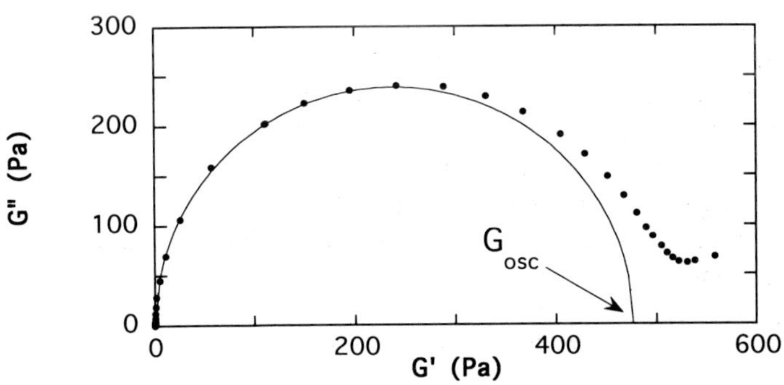

Figure 8.17 Cole–Cole plot for a viscoelastic solution of the compound **2** of Scheme 8.7 ($\phi = 0.71\%$) in *tert*-butylcyclohexane (●) and corresponding osculating semi-circle for a pure Maxwellian behaviour (—). The turn-up observed in the high frequency domain is assigned to a more local relaxation process of the chains (Reproduced from Reference 82 with permission from the *J. Phys. France*.)

lifetime (τ_{break}) of the aggregate of length $\bar{L}$ and the reptation time (τ_{rep}) of the chain. Their relative magnitudes describe the ability of the system to relax through the additional pathways of chain disentanglement provided by the chain scission reactions. Two extreme situations can be envisaged. In the first instance, almost unbreakable and polydisperse chains, corresponding to $\tau_{break} \gg \tau_{rept}$, exhibit a stress relaxation proceeding mainly by reptation and consequently a stretched exponential relaxation spectrum is observed. In the second case, in conditions where fast scissions are dominant ($\tau_{break} \ll \tau_{rept}$), the lifetime of a chain fragment is very short compared to its relaxation through reptation and the stress relaxation is characterized by a single exponential decay (Maxwell liquid).

Scaling behaviour with the volume fraction (ϕ) of quantities such as the plateau modulus (G_0), the zero shear viscosity (η_0) and the terminal stress relaxation time (τ_R) are observed over a small concentration range within the semi-dilute regime. Dynamic and structural information can be obtained from the scaling behaviour as well as from the shape of the stress relaxation spectrum [147–149]. Such procedures have already been successfully applied for 'living polymers' in aqueous electrolyte solutions [150, 151]. Furthermore, the structure of the organometallic aggregates (compound **2** in Scheme 8.7), which is the simplest possible (molecular thread with a polar core of oxygen and metal atoms and an alkanic shell), facilitates the interpretation of some static (or semi-static) and dynamic behaviour [82, 145]. As an example, it can be deduced that the scissions occur through reversible unimolecular reactions instead of contacts between bent portions of chains. The large elastic modulus values at high frequencies (120 Pa at $\phi = 0.01$ and $\nu = 40$ Hz), much larger than those observed for comparable aqueous systems (CTAB: cetyl trimethyl ammonium bromide), are interpreted as being due to both the very thin diameter of the molecular threads (and subsequently the decreased mesh ξ of the network) and the specific nature of the 'crosslinks' (strength of the interactions, flexibility of the connecting chains and their degree of mobility). It is interesting to note that orientational ordering of the aggregates (compound **2** in Scheme 8.7) is observed in the thermotropic discotic mesophase (hexagonal array of rigid columns) and also in the swollen hexagonal and nematic lyotropic phases at concentrations of the order [152] of 50%. In this context, the transient nodes of the heterogeneous 3D network may be transient swollen nematic microdomains whose spatial extension is restricted by the chain flexibility. When the chain concentration is further increased, a topological conflict between the persistence length of the chains and the mesh size of the transient network can take place. In such conditions, the tightness of the net prevents the recombination with different fragments and leads to an apparent increase of the aggregate lifetime (τ_{break}) typical of apparently less breakable polydisperse chains. This cross-over concentration regime where the spatial correlations become important may also correspond to a regime where the recombinations are diffusion controlled [153]. The great versatility of the system is such that both: (a) the chain flexibility can be modified by the type of solvent or alkane substitution used with the tetracarboxylate complex; and (b) the energy of scission of the chain relying on metal–oxygen bonds which can be modified by a change of the metal type (ruthenium [154] or rhodium [152, 155]).

(c) Substituted metalloporphyrins (Scheme 8.7, compound **5***).* A dilute viscoelastic gel-like material has been obtained [156] during the course of the investigation of the thermotropic properties of the metal (Zn) complex of a long chain ester of meso-tetrakis(*para*-carboxyphenyl)porphyrin [157]. This gelling agent is prepared from the tetra-substituted ester eluted on a silica chromatographic column with an appropriate hydrocarbon/alcohol mixture and has one ester group converted into an acidic group. A thixotropic gel-like material is obtained with a weak yield value, not large enough to be self-supporting. The presence of both the metal and acidic group appears to be essential for the gelation phenomenon. Preliminary SAS experiments [156] show that the most probable morphology is the rod-like local structure of moderate length with one molecule per cross-sectional repeating unit.

The aggregation of metalloporphyrins is a well known phenomenon either in the pure thermotropic liquid-crystalline phases [158] or in aqueous [159] solutions where interesting aggregation behaviour has been reported. Tetrasubstituted phenylporphyrins (metallated or not) are known to develop structures with intersecting channels in the solid state [160, 161] so that solvation or intercalation of numerous ‘guest’ molecules is observed. The great variety of clathrates which can be formed qualifies the compounds as ‘sponges’. Furthermore, this class of compounds has received a great deal of attention in the fields of photoelectronic conversion, non-linear optics, O_2 transport and biomimetic studies. So far, very few examples of porphyrin aggregation in organic solvents are known. For instance, preliminary investigations of cationic tetrapyridylporphyrin complexes (Cu^{II}, Mn^{II}, Co^{II}) [162] using NMR and EPR techniques can be mentioned. In the liquid-crystalline state, modifications of the flat core of the porphyrinic rings affect their tendency to aggregate in the discotic (disc-like) or calamitic (rod-like) materials [158].

More extensive studies of aggregation of metallo-porphyrins in aqueous solutions have been reported since they may have a potential ability to act as mimetic models of proteins and redox systems. As an example, a zinc protoporphyrin having two alkylphosphocholine groups aggregates into fibres with a diameter of about 70 Å (molecular length being about 35 Å) [163]. Visible absorption spectra allow the assignment of the stacking mechanism since a blue-shifted Soret band is reported [164, 165] to come from transition dipole interactions of face-to-face stacked porphyrin molecules. The aggregation mechanisms can be even richer depending on the chemical functionality of the porphyrins and, in addition to face-to-face stacking mechanisms, edge-to-edge interactions can also be found in some cases. As an example, a non-metallated amphiphilic protoporphyrin forms ribbons in water by dimerization of the porphyrin macrocycles with a lateral shift and extended edge-to-edge connected chains due to cooperative hydrogen bonding between glycosamide head groups [165]. Exciton models are used to propose the related structural arrangements. Edge-to-edge aggregation of amphiphilic metalloporphyrins is also observed for the so-called 'octopus-porphyrins' (having four alkyl phosphocholine groups on each side of the ring plane) in water [166] and is characterized by a red-shifted Soret band corresponding to an in-line ordering of one of the transition moments of the molecules. Some studies concerning the solid-state [167] may also constitute a basis for interpretation of the properties of the dispersed lyotropic states found in certain gels [168]. In this context, systems providing unidirectional assemblies of organometallic arrays such as coordination polymers [169] or metallomesogens [170, 171] are expected to contribute to the emerging class of organometallic organogelators.

8.4.7 Miscellaneous derivatives (Scheme 8.8)

(a) Macrocycle-based gelators (calix[n]arenes) (Scheme 8.8, compound 1). A representative of the macrocycle-based gelators is a special family of calix[*n*]arenes having long alkane substitutions in the *p*-position (abbreviated as 2_nC_m). These molecules are [1_n]metacyclophanes, obtainable in ring sizes containing 4–8 aryl moieties in the macrocyclic array, which are conformationally flexible to a certain extent [172]. They are frequently utilized in host–guest chemistry [173] and Langmuir monolayer studies [174, 176]. The calix[4]arenes can exist in conformations with extended conical areas while the calix[8]arenes are more flexible. The temperature of gel [176, 177] formation (T_{SG}), determined by DSC or by a rough visual test (inverted test-tube method) of 2_8C_{12}, shows that T_{SG} is dependent upon both the solvent type (alkanes, carbon disulfide, toluene, selected alcohols) and gelator concentration. For the concentration range $0.02 < \phi < 0.15$, T_{SG} is found to be higher in heptane (35–43°C) than in cyclohexane (16–27°C). The thermosensitivity, as defined by the slope (dT_{SG}/dC) of the phase diagram, decreases in the order cyclohexane > hexane > carbon disulfide.

The aggregation mechanism involves the participation of the carbonyl groups in intermolecular hydrogen bonds as demonstrated by the absence of a gelling ability for the analogous derivative 3_8C_{12}. A moderate affinity of the gelator for the solvent is also a prerequisite for gelation. The analysis of gelators of this family (2_4C_{18}, 2_8C_{12}, 2_8C_{18}) and their non-gelator homologues (2_6C_m) adds to the understanding of the appropriate molecular design that a molecule of this class must have to be a gelator. Among

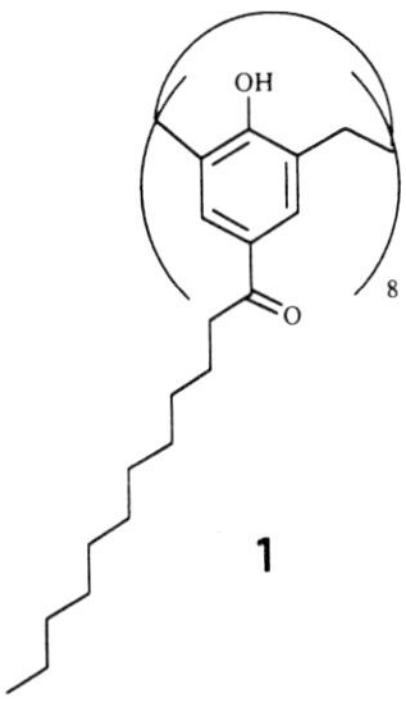

1

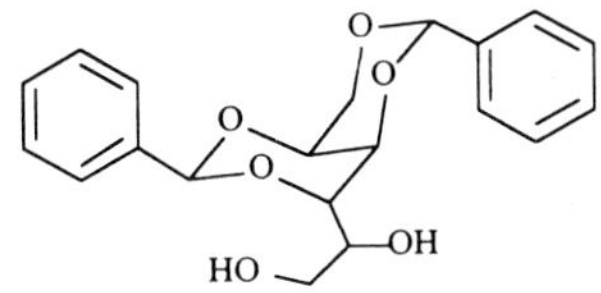

2

$F(CF_2)_n(CH_2)_mH$

3

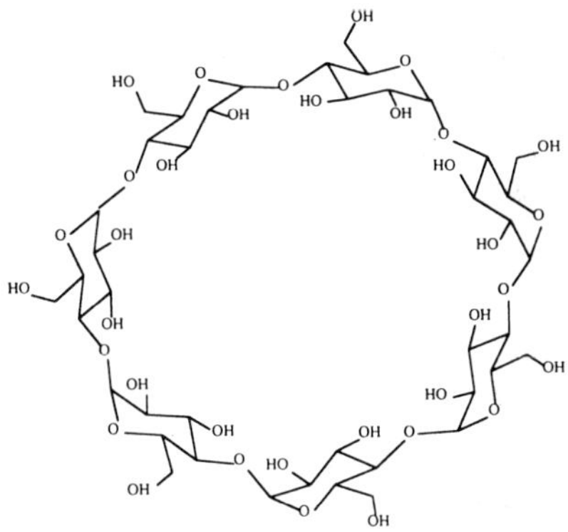

4

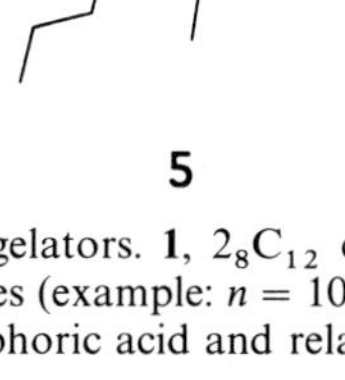

5

Scheme 8.8 Miscellaneous organogelators. **1**, 2_8C_{12} calixarene; **2**, 1,3:2,4-di-*o*-benzylidene-D-sorbitol; **3**, semifluorinated *n*-alkanes (example: $n = 10$, $m = 12$); **4**, *β*-cyclodextrin; **5**, di(2-ethyl hexyl)phosphoric acid and related sodium salt.

corresponding non-cyclic analogues, $2'_2C_{18}$ exhibits a gelling ability restricted to hydrocarbons while $2'_1C_{18}$, with only one carbonyl and one hydroxyl group, is not a gelator. These phenomenological features indicate that a minimum number of connection sites per 'monomeric' species is required to build a 3D network. The gel network contains two pore sizes, one being calibrated molecular holes defined by the calixarene molecules themselves and the other being the statistical nanomolecular holes (interstitial holes of the gel network).

*(b) Sorbitol and polyol derivatives (Scheme 8.8, compound **2**).* 1,3:2,4-di-*o*-benzylidene-D-sorbitol (DBS) is a chiral polyol which can gel a range of solvents from hexadecane, dimethyl phthalate to ethanol and ethylene glycol [178, 179]. The gelation time, as determined by rheological experiments [180] measuring the storage modulus (G') *versus* time during the sol to gel transition, varies strongly with the concentration according to a power law with the exponent 9/2. The kinetics of gelation and the visual appearance of the resulting stiff gels are dependent upon the solvent type. The rigidity of the gels is extremely sensitive to the DBS concentration as demonstrated by the related power-law dependence of the elastic modulus ($G' \approx \phi^4$). Rather dilute clear gels ($\phi < 0.15$) are metastable and crystals separate. The melting of a concentrated gel in ethylene glycol ($\phi = 0.30$) is detected by DSC as an endothermic peak at $T \approx 130°C$. The phase diagram, determined using DSC features, shows three distinct concentration domains (0.05/0.15/0.65). Solid-state NMR studies confirm the bonding of solvent molecules with DBS molecules in the gel state, an interaction which is further demonstrated by the depression of the freezing point at low concentrations ($\phi < 0.1$) due to the fraction of solubilized molecules. Based upon electronic absorption spectroscopy measurements, the relative orientation of the benzene rings in the aggregates of the gel state is estimated. A transition from a linear (corresponding to a hyperchromism absorption feature) to a parallel stacking (corresponding to a hypochromism feature) occurs at $\phi \approx 0.15$ and is accompanied by an increase of crystallinity. CD enhancements are correlated [179] to the gelator concentration increase. Spherulitic textures are observed between crossed-polarizers as extinction crosses (negative optical sign) and rings suggesting the existence of supramolecular helical arrangements. As is the case for various other gelators, the racemic compound (DL-DBS) does not form organogels. IR spectroscopy demonstrates the participation of hydrogen bonds in the aggression reactions while UV spectroscopy suggests that a side by side aggregation of the benzene rings is a probable mechanism [181].

*(c) Semi-fluorinated n-alkanes (Scheme 8.8, compound **3**).* Semifluorinated *n*-alkanes (F_nH_m) can give birefringent gels of hydrogenated alkanes when the gelling derivatives are such that $n = 12$ and $8 < m < 20$. The melting transition proceeds through a gradual disordering as shown by the corresponding broad DSC endothermic peak [182, 183]. The solvent nature (linear and cyclic alkanes) affects the phase diagram in a fashion suggesting

that the interaction parameter between the gelator compound **3** in Scheme 8.8 and the solvent is sensitive to its shape.

(d) β-Cyclodextrin/pyridine gels (*Scheme* 8.8, *compound* ***4***). β-cyclodextrins (β-CD) are cyclic heptasaccharides of α-1,4 linked glucosidic units forming a crown structure with a conical shape. Due to the presence of cages or channels, they can act as hosts. Isotropic gels can be formed with rigorously anhydrous β-CD in dry pyridine [184] or ternary gels in different solvents (toluene, chloroform) can be obtained in the presence of small amounts of water. Syneresis gives rise to crystals which are either free from pyridine or which contain pyridine located in inter and intramolecular channels.

*(e) Sodium bis(2-ethylhexyl) phosphate in n-heptane (Scheme 8.8, compound **5**).* Sodium bis(2-ethylhexyl) phosphate in *n*-heptane [NaDEHP] [21, 185, 186] forms large reverse micelles. Combined dynamic, static light scattering and viscosimetry measurements show that semiflexible rods grow in the absence of water. Observation of large aggregation numbers is surprising for finite micelles in organic solvents with a defined CMC (or C_{cri}) as well as the unusual role of water which contrasts to that observed for instance with AOT in benzene (section 8.4.8(c)). It is known that micellar growth in ternary systems, such as lecithin/water/isooctane [187, 188], usually follows the most common trend in that water promotes the sphere to rod transitions. The presence of large aggregation numbers was already observed for organometallic complexes in organic solutions forming the so-called 'living polymers' [81]. Determination of the mean apparent hydrodynamic radius (R_h) and the mean apparent radius of gyration (R_g), when correlated with viscosity measurements, enables an estimation of the contour length (L_c) of the aggregates and their degree of overlap in the transient network. A molecular packing model is proposed in which three NaDEHP molecules are associated within the cross-section. The main driving force for the growth of the semi-rigid rods in the viscoelastic fluid is thought to be the electrostatic lattice energy with additional contributions coming from dipolar interactions and interfacial energy forces.

8.4.8 Two-component systems (Scheme 8.9)

*(a) Aminopyrimidine + dialkylbarbituric acid mixtures (Scheme 8.9, compound **1**).* Two different components can self-associate through intermolecular hydrogen bonding, constituting a form of molecular recognition. The two complementary subunits associate in a supramolecular structure which may display specific properties. The approach for creating new materials, here developed with a two-component mixture, can be considered as being complementary to that developed for single-component gelators

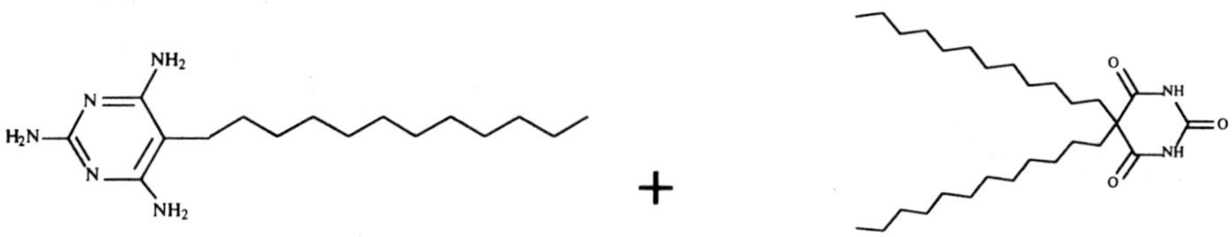

1

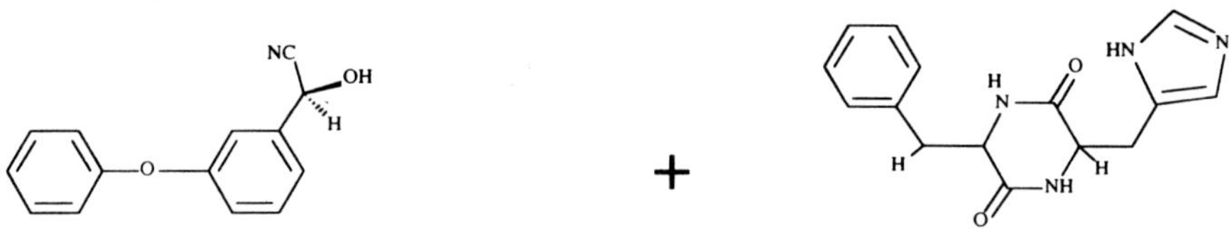

2

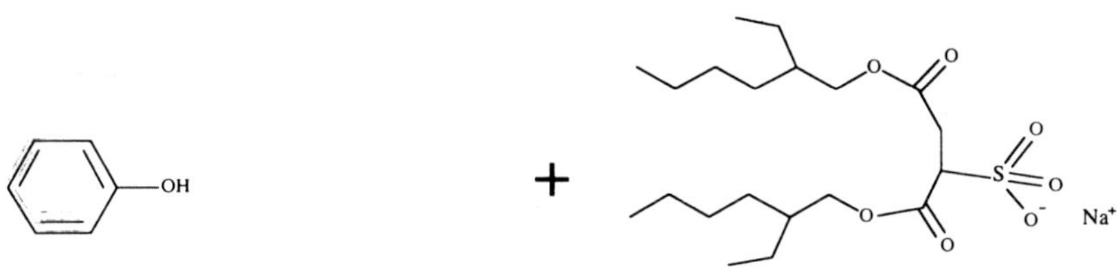

3

Scheme 8.9 Representative structures of two-component organogelator systems. **1**, 5-hexadecyl-2,4,6-triaminopyrimidine + 5,5-didodecylbarbituric acid; **2**, (*S*)-2-hydroxy-2-(3-phenoxyphenyl)acetonitrile + cyclo[(*R*)-phenylalanyl-(*R*)-histidyl]((*R*, *R*)-3); **3**, phenol + sodium bis(2-ethylhexyl)sulfosuccinate (AOT).

having two distinct parts. It offers an original way of conferring specific properties (i.e. optical, electronic, ionic, magnetic properties, etc.) on a macromolecular level. An ordered supramolecular strand can be formed [189] either in solution or in the solid-state (co-crystallization) with all residues located on the same side of the strand (Figure 8.18). This configuration allows a sorting response and a specific orientation of the components in the superstructure. The gelation phenomenon appears as a 'mechanical consequence' due to the overlap of the strands when their concentration is increased.

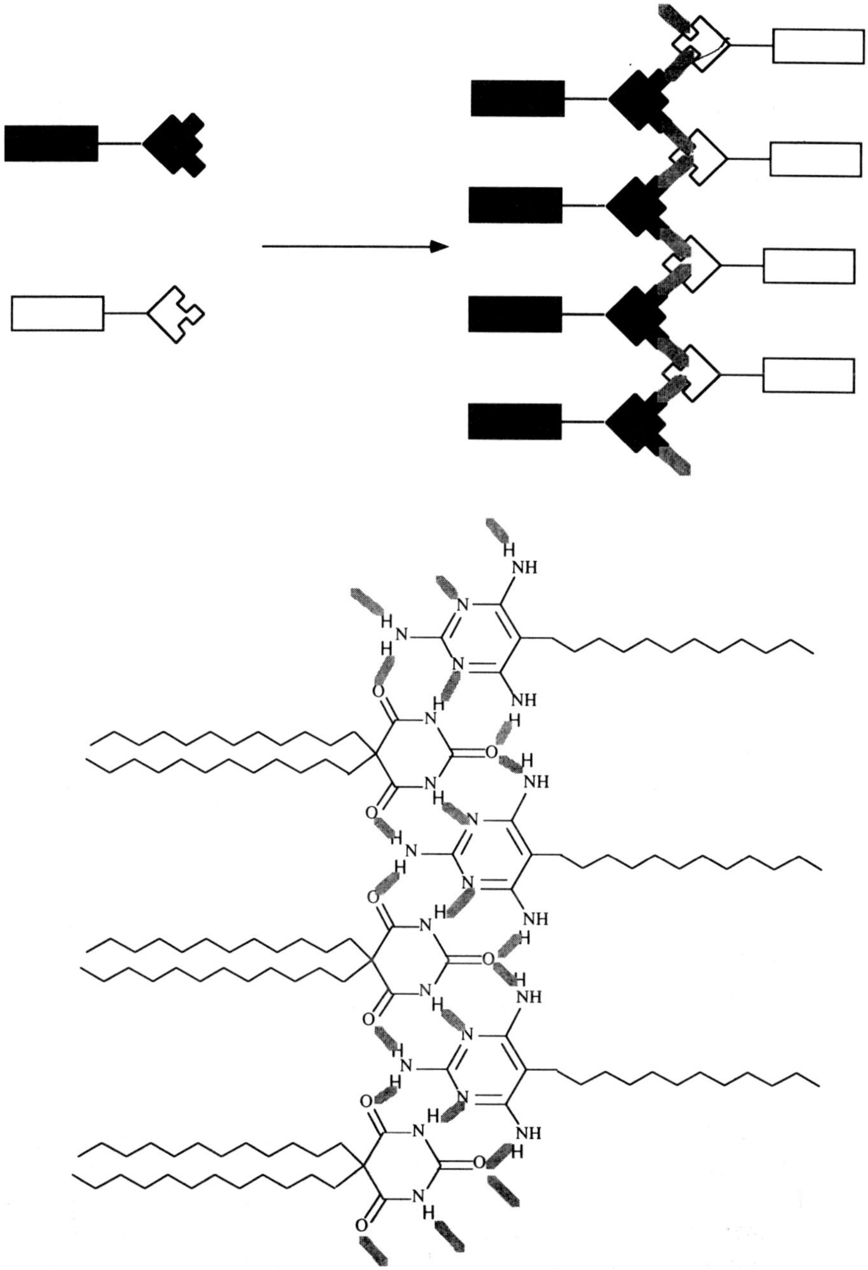

Figure 8.18 Top: schematic representation of the formation of an ordered supramolecular strand by the molecular recognition-induced association of two different molecular units. Each unit contains a group that possesses two identical binding sites complementary to those of the other unit (from Reference 189). Bottom: illustrative example of a two-component gelling system (compound **1** in Scheme 8.9).

As an example [134] mixtures (1:1 molar ratio) of 2,4,6-triaminopyrimidine and 5,5-dialkylbarbituric acid produce stable organogels. The minimum concentration of the mixture necessary for gelation at 25°C is 0.04 $mol.l^{-1}$ in cyclohexane. FTIR spectra confirm that N–H and C=O intermolecular hydrogen bonds develop in the gel phase (as observed in the related co-crystals). EM observations show thick fibres with a diameter (800 Å) suggesting that a large number of molecules participate in the aggregate or that the thick fibres result from the aggregation or collapse of thinner fibres, as frequently observed when the experimental procedures strain or stress the fragile network [73]. In the xerogel, a layered structure is characterized by four narrow reflections corresponding to a 19.2 Å spacing, smaller (due to a bending of the alkyl chains) than that observed in the related co-crystals.

*(b) Phenoxyphenyl × phenylalanylin derivatives (Scheme 8.9, compound **2**).* During the asymmetric hydrocyanation of 3-phenoxybenzaldehyde, a gel was obtained as a consequence of the formation of a complex between two optically active ingredients. The enantioselective complexation is achieved using a racemic mixture of 2-hydroxy-2-(3-phenoxyphenyl) acetonitrile (in which the (*S*) enantiomer is recognized) and the optically pure cyclo[(*R*)-phenylalanylin] toluene in toluene [190].

*(c) Phenol and AOT mixtures (Scheme 8.9, compound **3**).* The twin-tailed anionic surfactant sodium bis(2-ethylhexyl) sulfosuccinate (AOT) is known to self-associate in aqueous and organic solvents [17, 191]. Gels in non-polar solvents are obtained [192] for AOT concentrations less than 0.5% upon addition of specific phenols. Hydrogen bonding between the phenol and the sulfosuccinate head group of AOT is essential for gelation, which is cancelled out by traces of water. The dry reverse micelles of AOT transform into fibres on addition of suitably substituted phenol. Gels cannot be formed with aliphatic alcohols or phenols with long alkyl chains (larger than about three carbon units). A structural model of the aggregate is proposed on the basis of observed broadenings of the phenolic ^{1}H and ^{13}C NMR resonances while those associated with AOT and the solvent remain well resolved. The phenol molecules stack along with strands made up of hydrogen-bonded AOT adsorbed on the external surface [193, 194].

The phenol acid strength controls the binding of the phenol molecules to the reversed micelles [195] as well as the gel strength estimated by its melting point. The solvent dependency of the gelation phenomenon is related to the degree of solvent swelling within the AOT tails. The role of phenol in the transformation of reverse AOT micelles [196] can be analyzed. The phenol molecules decrease the curvature of the interface of the aggregate with the solvent, a mechanism which is different from that involved for the lecithin/isooctane system in the presence of polar additives such as water [197]. According to their nature, the incorporation of various dopants can be performed within the two parts of the aggregates. Acidic molecules, with a gross planar topology, can be intercalated in the stacked aromatic structure of the aggregates while water is confined to the thin hydrophilic shell enveloping the aromatic stacks. With the addition of 2-naphtol, luminescent gels can also be formed which are of interest as they may aid to the understanding of the mechanisms of energy transfer in such unidirectional structures. This class of luminescent gelling systems makes a connection between true binary luminescent gelators (see DDOA anthracene derivatives and anthryl cholesterol-based organogelators, sections 8.4.3 and 8.4.4) and ternary systems in which a third component promotes structural transitions towards linear structure (section 8.4.8(d)). Further structural studies are needed to improve the understanding of the aggregation mechanism.

(d) Ternary systems with small amounts of water. With the two-component systems above mentioned, two amphiphilic small molecules associate specifically to form strands. In addition, mixtures of self-associating surfactants with appropriate small amounts of polar co-solvent or co-surfactant can also form viscoelastic dispersed systems exhibiting all properties of gels. Different qualifications are required to distinguish the ternary systems. In the context of this review, only a few systems with small amounts of water (or polar co-component) are briefly mentioned. In aqueous solutions, it is known that the ratio of the head group area (a) to the hydrocarbon tail of the surfactant molecule at the interface with the solvent and the related interfacial tension determine the curvature of the micellar aggregates (or cross-sectional shape of the unidirectional aggregates). Naturally, by mixing two surfactants with different areas per head group, different curvatures for the interfaces may be obtained, thus generating different phases and micellar topologies.

Complex microemulsion systems can be transformed into hydrocarbon gels in which the major component remains the alkane solvent as is the situation for mixtures of bis(2-ethylhexyl) sodium sulfosuccinate with aqueous solution of gelatin (5%) in isooctane [198]. A proposed picture of such a network involves mechanisms of micellar percolation and gelatin bridging. These microheterogeneous materials compartmentalize the system into bulk solvent and connected water pools. They can be considered as 'frozen' microemulsions where a fluid gelatin-rich water phase is dispersed in the organic solvent [199, 200]. In such a system, the conformation of the polypeptide is random coil or consists of helical segments depending on its environment or transformation rate of the gelation.

A special mention must be made of transparent gel-like emulsions containing up to 99% hydrocarbon in ternary hydrocarbon/water/surfactant systems. These metastable gels (exhibiting a static shear modulus and a yield stress) are polyhedral hydrocarbon cells obtained by compression of droplets separated by a continuous flat aqueous film of surfactant [201]. Usually, two surfactants are required: an oil soluble one and a water soluble one. Representative examples are mixtures of 2-butanol/water/sodium dodecylsulfate in *n*-hexane [202] or mixtures of a fluorinated neutral amine oxide with zwitterionic phosphocholine derivative in perfluorohydrocarbons [203] (cyclic or linear). For higher concentrations of water (up to 50–60%) and surfactant, the so-called 'ringing gels' (made up of spherical microemulsion droplets) can be obtained which display the ringing phenomenon when they are submitted to appropriate mechanical vibrations [204]. Frequently, phase diagrams of various quaternary systems show gel domains covering highly viscous phases (isotropic or anisotropic) of various structures and compositions (see for instance mixtures of sodium dodecylsulphate, butanol, toluene, water [205]).

Organic solutions of lecithin can also be turned into highly viscous gel-like solutions by addition of a small amount of water [187]. A typical formulation of such a ternary system is lecithin ($\phi = 0.14$) in isooctane ($\phi = 0.80$) with a number of about three molecules of water per molecule of lecithin. Numerous complementary techniques have been used to probe the structural, mechanical and dynamical properties of these systems [188, 206–208]. To summarize the data available, the scaling laws followed by static and dynamic quantities (for instance, the zero shear viscosity, the hydrodynamic correlation length, etc.) prove the existence of an entangled network of long semi-rigid cylindrical reverse lecithin micelles. Water appears to induce a sphere to rod phase transition and subsequent growth of giant micelles (see an analogous example consisting of salted aqueous solutions of cetyltrimethylammonium bromide with specific polar additives [209]). Changes of the effective areas of the aliphatic part of the molecule appear to be determinant in promoting the transition from spherical to cylindrical curvatures. Worm-like micelles are well known to exist in complex salted aqueous solutions [20, 210, 211] while lecithin/water/isooctane is an example of a ternary organic system which also exhibits this type of micelles (a binary example also exists with the bicopper tetracarboxylates derivatives, section 8.4.6(b)). Static and dynamic properties (osmotic compressibility, static and hydrodynamic correlation length probed by static and dynamic light scattering experiments) of the lecithin giant micelles exhibit close analogies to semidilute polymer solutions possessing the kinetic features of the so-called 'living polymers' [18, 31, 80, 188, 212]. Information concerning the micellar size distribution and intermicellar interactions as a function of concentration can also be gained from the analysis of the properties in the context of conformational space renormalization group theory [213].

8.5 Applications

Gels have numerous industrial applications due to the great diversity of the structures that they display on microscopic and mesocopic scales [214]. The thermoreversibility, chemical sensitivity to the solvent type, and structural specificity of the constitutive aggregates (interfacial polarity, chirality) are intrinsic characteristics of organogels which make them excellent candidates for future industrial development. A brief overview of the industrial fields where gels are utilized as host media for applications is given below, together with the specific properties of the materials which are used. Fields where some potential can reasonably be expected on the basis of initial applied developments, are also mentioned.

Hydrometallurgy, cosmetics, food and lubrication are just some of the industries where the utilization of gels is important. Solutions to solve the problem of oil recovery from the environment (crude oil spills or disposal of used cooking oil) frequently involve gels or complex microemulsions systems. A promising field of application for organogels is the use of complex systems (such as microemulsion gels) as vehicles for transportation of drugs. Thus, it has been shown that enzymes and bacteria can be entrapped in apolar gels of gelatin in isooctane without loss of their activity [198]. Highly concentrated oil-in-water (O/W) emulsions can be used as aviation fuels or in cosmetic applications while highly concentrated water-in-oil (W/O) emulsions are used in explosives technology. Some complex mixtures of surface active components (surfactants such as cetyl alcohol and polymers) in water and oil are utilized for ointments and creams [215, 216] but the large amount of water they contain (up to 51%) puts this system beyond the scope of the present chapter.

Many aqueous gels are employed for the purposes of macromolecular separation, protein crystallization, etc. Comparable applications specific to organic fluids, using the properties of hindered transport in the pores of the gel networks [217], may be found in the future. Various applications take advantage of the drag effect due to steric and hydrodynamic interactions between permeating macromolecules and the disordered fibrous media. In this context, calix[n]arenes offer interesting potential for membrane filtration. Attempts are currently made to develop two-dimensional oriented networks of molecular pores (between 2.0 and 6.4 Å for calix[4]arene to calix[7]arene-based surfactants) taking advantage of the conical shape [175] of the molecules. The wide interest in membranes for purification could find some developments with aerogel networks providing such two-mode porosity system (small and large diameter holes).

Gels have been used to produce the so-called Turing patterns which result from non-equilibrium self-organization [218]. Such a mechanism, explained as a spatial symmetry breaking instability, is often discussed as being important for some aspects of morphogenetic developments, even though the role of the gel medium is still unclear [219]. The advantages of using gels instead of liquid systems lie in the fact that the hydrodynamic convection is reduced and various chemical species can be immobilized. These features are also of importance for slow crystallization of various compounds [220, 221]. Reverse micellar systems (microemulsions) find numerous applications in biocatalysis, biomembrane mimetics, extraction processes (universal solvents), and preparation of microparticles [222]. When the solvent evaporates from a gel sample, a meniscus develops around the sample which shrinks to give a xerogel. Capillary forces and liquid–vapour interfaces are responsible for the collapse of the tenuous 3D structure. If a supercritical drying process is used at high temperatures and pressures (above T_c and p_c), an extremely porous structure, called an aerogel, is obtained. Some aerogels exhibit remarkable properties concerning sound velocity and heat transport which allow promising thermal insulation and acoustic applications. Up to now, only organometallic oxides (silica) have been utilized to form aerogels [223], but some other organometallic organogelators could be of interest in the future for these purposes.

Anisotropic gels composed of anisotropic networks mixed with low molecular weight liquid crystal molecules can be used as active or passive optical components [224]. Anisotropic gels can be easily obtained from organogelators forming rods of finite short lengths (organometallic conjugated molecules [105, 156] and could give interesting materials in the field of electronic conduction or semi-conduction when appropriate donor-acceptor conjugated systems are used. The electrical properties of organic solutions of metal soaps have been studied due to their implication in the field of the transmission of electric power [77]. The specific conductance (σ) and dielectric loss (ε'') reflect the degree of heterogeneity of the sample (precipitation, gelation). Despite the fact that thermotropic liquid crystals are preferred to their lyotropic analogues due to the energy consumption necessary for the building of display or storage devices, the solvated systems could be of interest in the future. As an example, the linear aggregates of cofacially stacked macrocycles having an extended π-electron system (for instance phthalocyaninato compounds [167]) are of potential interest [225]. The formation of a conduction band allows charge carriers, generated by a partial oxidation, to develop some conducting properties in appropriate structures avoiding the usual inclined stacks which are not favourable to the necessary π-orbital overlap. In addition, the degree of substitution of the metallophthalocyanines determines the degree of their solubilization in the organic solvents. It is known that a tetra-substitution increases the solubilization ability due to a larger dipole moment arising from the unsymmetrical arrangement of the substituents around the macrocycle. The class of low-dimensional linear systems of stacked organometallic compounds such as phthalocyanine stacks, columnar arrays of cation radicals of tetrathiafulvalene and selenium analogues or more complex structures with two-centre bridging elements of the metal [226] can exhibit interesting long-range electronic and magnetic interactions (molecular-based ferromagnets) in the condensed solid-state. Some of these systems are thermotropic liquid-crystalline phases (phthalocyanine derivatives [227, 228]) and some electrical (ferro and antiferroelectricity, semiconductivity) or optical properties (polarizibility and hyperpolarizibility) can be deduced from the features (symmetry, polarity, redox potentials) of the isolated molecular units. Systems such as substituted metacyclophanes [229] and porphyrins with phenyl rings conjugated through a diacetylene spacer [230] are also known to exhibit intermolecular charge transfer properties.

Some of the gelling systems can be used as temperature or humidity sensors while the ordered microstructures found in their networks can be of interest in templating materials synthesis. Luminescent gels could find some applications due to the possibility of linear pathways for energy transfer in their structures. As mentioned above, some chiral organogels can be used for molecular and chiral recognition in the gel phase. Molecular transducers formed by the cholesterol-based organogelators [126] which can translate some chiral recognition interactions into readable outputs are promising practical examples of host–guest signal-responsive chemistry. In such gels, the gel–solution phase transition temperature differentiates the two optical isomers of a monosaccharide (for instance, xylose) added to the gelling system.

In the perspective of producing high modulus polymeric fibres, some xerogels (DBS, dibenzylidene sorbitol) have shown their ability to act as nucleating agents of polyolefins [231]. Some other DBS derivatives (containing ester groups) are used to gel electrolyte solutions for lithium battery applications [232]. Semiconductor nanostructures can already be formed using AOT/phenol organogels. The growing field of sensors, while not restricted to gel-like materials, is based upon the detection of specific functions by appropriate donor–acceptor interactions or detection of morphologies which takes advantage of spatial recognition by cavities, hollows, channels of the sensor host-material [233]. Nanochemistry, which currently utilizes mostly host systems from solid-state chemistry [234] to yield materials exhibiting novel properties (from electronic to mechanical), could take advantage of specificities of the organogel networks.

The present review has attempted to show some facets of the fascinating variety of behaviours displayed by organogelators. Given the rapidly growing nature of the field, it is clear that the next few years will bring discoveries of novel gelators as well as further the understanding of these systems. In the future, organogels may show behaviour as 'intelligent' as their aqueous analogues [235].

References

1. Hoffman, H. *Adv. Colloid Interface Sci.*, 1990, **32**, 123.
2. De Gennes, P.G. *Scaling Concepts in Polymer Physics*, Cornell University Press, Ithaca, NY, 1979.
3. Israelachvili, J.N. *Intermolecular and surface forces*, 3rd edn, pp. 341–435. Academic Press, London, 1992.
4. Tiddy, G.J.T. *Phys. Reports*, 1980, **57**(1), 1.
5. Chevalier, Y. and Zemb, T. *Rep. Prog. Phys.*, 1990, **53**, 279.
6. Israelachvili, J.N., Mitchell, D.J. and Ninham, B.W. *J. Chem. Soc., Faraday Trans II*, 1976, **72**, 1525.
7. Israelachvili, J.N., Marcelja, S. and Horn, R.G. *Quaterly Rev. Biophys.*, 1980, **13**(2), 121.
8. Mitchell, D.J. and Ninham, B.W. *J. Chem. Soc., Faraday Trans II*, 1981, **77**, 601.
9. Kunitake, T., Okahata, Y., *et al. J. Am. Chem. Soc.*, 1981, **103**, 5401.
10. Eicke, H.F. *Topics in current chemistry*, 1980, **87**, 91–145, Springer-Verlag, Berlin.
11. Bourrel, M. and Schechter, R.S. (1988). *Surfactant Science Series*, **30**, 111, Ed. Dekker, M., New York.
12. Ruckenstein, E. and Nagarajan, R. *J. Phys. Chem.*, 1980, **84**, 1349.
13. Ekwall, P., Danielsson, I. and Stenius, P. (1972). Physical Chemistry Series One, Ed. Kerker, M., Butterworths, London, vol. 7, chapter 4.
14. Kitahara, A. and Kon-no, K. (1975). *Colloidal dispersions and micellar behavior*, Ed. Mittal, K.L., ACS Symposium Series 9, Washington DC.
15. Eicke, H.F. and Christen, H. *J. Colloid Interface Sci.*, 1975, **46**(3), 417.
16. Kon-no, K., Jin-no, T. and Kitahara, A. *J. Colloid Interface Sci.*, 1974, **49**, 383.
17. Mukherjee, K. and Moulik, S.P. *Langmuir*, 1993, **9**, 1727.
18. Cates, M.E. *J. Phys. France*, 1988, **49**, 1593.
19. Peebles, J.L.H. *Molecular weight distributions in polymers.* Interscience, New York, 1971.
20. Porte, G. and Appell, J. *J. Phys. Chem.*, 1981, **85**, 2511.
21. Yu, Z-J., Zhou, N-F. and Neuman, R.D. *Langmuir*, 1992, **8**, 1885.
22. Zulauf, M. and Eicke, H.-F. *J. Phys. Chem.*, 1979, **83**(4), 480.
23. Ferry, J.D. *Viscoelastic properties of polymers,* 3rd edition, Wiley, New York, 1980.
24. Hermans, P.H. (1969). *Colloid Science*, H.R. Kruyt, Ed., vol. II, Reversible Systems, Elsevier, Amsterdam.
25. Russo, P.S. (1987). *Reversible polymeric gels and related systems*, Russo, P.S. Ed., ACS Symposium Series 350, Washington DC, chapter 9.
26. Almdal, K., Dyre, J., *et al. Polymer Gels and Networks*, 1993, **1**, 5.
27. Stauffer, D. *Introduction to percolation*, Taylor and Francis, London, 1985.
28. Stanley, H.E., Family, F and Gould, H. *J. Polym. Sci.: Polym. Symp.*, 1985, **73**, 19.
29. Stauffer, D., Coniglio, A. and Adam, M. *Adv. Polym. Sci.*, 1982, **44**, 103.
30. Rehage, H. and Hoffmann, H. *Mol. Phys.*, 1991, **74**(5), 933.
31. Cates, M.E. and Candau, S.J. *J. Phys.: Condens. Matter*, 1990, **2**, 6869.
32. Doi, M. and Edwards, S.F. (1986). *The theory of polymer dynamics.* Clarendon, Oxford, 1986.
33. Wessling, B. *Adv. Mater.*, 1993, **5**(4), 300.
34. Bohlin, L., Carlson, T.L-G. and Bäckström, G. *J. Colloid Interface Sci.*, 1980, **73**(1), 61.
35. Amis, E.J., Hodgson, D.F. and Yu, Q. *Polym. Preprints*, 1991, **32**, 447.
36. Djabourov, M., Leblond, J. and Papon, P. *J. Phys. (Paris)*, 1988, **49**, 333.
37. Burchard, W. and Ross-Murphy, S.B. *Physical networks, polymers and gels.* Elsevier, London, 1990.
38. Burchard, W., Stadler, R. *et al.* (1986). *Biological and synthetic polymer networks*, Ed. Kramer, O., Elsevier, London and New York.
39. Clark, A. and Ross-Murphy, S.B. (1987). *Biopolymers*, **83**, Adv. Polym. Sci., Springer-Verlag, Berlin, 57.
40. Eldridge, J.E. and Ferry, J.D. *J. Phys. Chem.*, 1954, **58**, 992.
41. Dubochet, J., Adrian, M. *et al. Trends Biochem. Sci.*, 1985, **10**, 143.
42. Wade, R.H. and Chretien, D., *J. Structural Biol.*, 1993, **110**, 1.
43. Clausen, T.M., Vinson, P.K. *et al. J. Phys. Chem.*, 1992, **96**, 474.

44. Cabane, C. (1987). *Surfactant Solutions,* Surfactant Science Series, Ed. Zana, R., Marcel Dekker Inc., New York, **22**, 57.
45. Terech, P. *Croatica Chemica Acta*, 1992, **65**(2), 425.
46. Guinier, A. and Fournet, G. *Small angle scattering of X-rays*, Wiley, New York, 1955.
47. Glatter, O. and Kratky, O. *Small angle X-ray scattering*, Academic Press, London, 1982.
48. Candau, S.J. (1987). *Light scattering.* Surfactant Science Series, Ed. Zana, R., Dekker, M., New York, 147.
49. Muller, M., Kremer, F. *et al. Colloid Polym. Sci.*, 1995, **273**, 38.
50. Aliotta, F., Fontanella, M.E. *et al. J. Phys. Chem.*, 1993, **97**, 733.
51. Dissado, L.A., Rowe, R.C. *et al. J. Colloid Interface Sci.*, 1987, **117**(2), 310.
52. Mezei, F., (1980). *Physics*, Neutron Spin Echo, Springer-Verlag, Berlin, **128**, 3.
53. Polishuk, A.T. *J. Am. Soc. Lubn. Eng.*, 1977, **33**, 133.
54. Uzu, Y. *J. Jpn. Oil Chem. Soc.*, 1975, **24**, 261.
55. Fiero, G.W. *J. Am. Pharm. Ass.*, 1940, **29**, 502.
56. Boner, C.J. *Manufacture and application of lubricating greases.* Reinhold Publishing Corp., New York, 1960.
57. Georges J.M. (1982). *Tribology Series*, **7**, 651 Elsevier, Amsterdam.
58. Ellis, S.G. *Canad. J. Res.*, 1946, **A25**(3), 119.
59. Vinogradov, G.V. and Sinitsyn, V.V. *J. Inst. Petrol.*, 1961, **47**(455), 357.
60. Vitali, R. and Borza, M. *NLGI Spokesman*, 1969, **4**, 126.
61. McClellan, A.L. *J. Chem. Phys.*, 1960, **32**, 1271.
62. Tachibana, T. and Kamabara, H. *J. Colloid Interface Sci.*, 1988, **28**(1), 173.
63. Tachibana, T. and Kambara, H. *Bull. Chem. Soc. Japan*, 1989, **42**, 3422.
64. Tachibana, T., Kitazawa, S. and Takeno, H. *Bull. Chem. Soc. Jpn,* 1970, **43**, 2418.
65. Tachibana, T., Kayama, K. and Takeno, H. *Bull. Chem. Soc. Jpn,* 1972, **45**, 415.
66. Evans, D., Hutton, J. F. and Matthews, J.B. *J. Appl. Chem.*, 1952, **2**, 252.
67. Mansot, J.L., Terech, P. and Martin, J.M. *Colloids and Surfaces*, 1989, **39**, 321.
68. Sakamoto, K., Yoshida, R. *et al. J. Am. Chem. Soc.*, 1978, **100**(22), 6898.
69. Tachibana, T., Mori, T. and Hori, K. *Bull. Chem. Soc. Jpn,* 1980, **53**, 1714–19; *ibid.* (1981), **54**, 73.
70. Terech, P. *Colloid Polym. Sci.*, 1991, **269**, 490.
71. Terech, P., Rodriguez, V. *et al. Langmuir*, 1994, **10**, 3406.
71. Tachibana, T., Mori, T. and Hori, K. *Nature*, 1979, **278**, 578.
73. Terech, P. and Wade, R.H. *J. Colloid Interface Sci.*, 1988, **125**(2), 542.
74. Tachibana, T. and Hori, K. *J. Colloid Interface Sci.*, 1977, **61**(2), 398.
75. Fuhrhop, J-H., Schnieder, P. *et al. J. Am. Chem. Soc.*, 1987, **109**(11), 3387.
76. Kaneko, F., Shirai, O. *et al. J. Phys. Chem.*, 1994, **98**, 2185.
77. Pilpel, N. *Chem. Rev.*, 1963, **63**, 221.
78. Gaskins, F.H., Brodnyan, J.G. and Philippoff, W. *Trans. Soc. Rheol.*, 1969, **13**(1), 17.
79. Leibler, L., Rubinstein, M. and Colby, R.H. *Macromolecules*, 1991, **24**, 4701.
80. Cates, M.E. *Macromolecules*, 1987, **20**, 2289.
81. Terech, P., Schaffhauser, V. *et al. Europhys. Lett.*, 1992, **17**, 515.
82. Terech, P., Maldivi, P. and Dammer, C. *J. Phys. II France*, 1994, **4**, 1799.
83. Fukasawa, J-I. and Tsutsumi, H. *J. Colloid Interface Sci.*, 1991, **143**(1), 69.
84. Zhou, Z., Georgalis, Y. *et al. J. Colloid Interface Sci.*, 1987, **116**(2), 473.
85. Klyne, W. *The chemistry of steroids.* Wiley, New York, 1960.
86. Ramanathan, N., Currie, A.L. and Ross Colvin, J. *Nature*, 1961, **4778**, 779.
87. Hoffmann, S. *Mol. Cryst. Liq. Cryst.*, 1984, **110**, 277.
88. Toyne, K.J. *Critical Reports on Applied Chemistry*, 1987, **22**, p. 28, p. 131 and references cited therein, Ed., Gray, G.W., Wiley, Chichester.
89. Martin-Borrett, O., Ramasseul, R. and Rassat, R. *Bull. Soc. Chim. France*, 1979, **7–8**, II-401.
90. Terech, P., Ramasseul, R. and Volino, F. *J. Colloid Interface Sci.*, 1983, **91**(1), 280.
91. Freed, J.H. (1976). *Spin Labelling: Theory and Applications*, chap. 3, Ed., Berliner, L., Academic Press, New York.
92. Terech, P. *J. Colloid Interface Sci.*, 1985, **107**(1), 244.
93. Avrami, M. *J. Chem. Phys.*, 1939, **7**, 1103; *ibid.* 1940, II **8**, 212; *ibid.* 1941, III, **9**, 177.
94. Terech, P., Ramasseul, R. and Volino, F. *J. Phys. France*, 1985, **46**, 895.

95. Wade, R.H., Terech, P. *et al. J. Colloid Interface Sci.*, 1986, **114**(2), 442.
96. Terech, P. *Reversible polymeric gels and related systems,* Russo, P.S. Ed., ACS Symposium Series 350, Washington DC, 1987, chapter 9.
97. Terech, P. *The living cell in four dimensions,* American Institute of Physics Conference Proceedings, Ed. Paillotin, G., 1991, **226**, 518.
98. Pfannemüller, B. and Welte, W. *Chem. Phys. Lipids*, 1985, **37**(2), 227.
99. Paillotin, G. *The living cell in four dimensions*, American Institute of Physics Conference Proceedings, 1991, **226**, 518.
100. Terech, P. and Berthet, C., *J. Phys. Chem.*, 1988, **92**, 4269.
101. Terech, P. *J. Surface Sci. Technol.*, 1989, **5**(2), 163.
102. Terech, P. *Mol. Cryst. Liq. Cryst.*, 1989, **166**, 29.
103. Terech, P. and Rodriguez, V. *Prog. Colloid Polym. Sci.*, 1994, **97**, 151.
104. Terech, P. *J. Phys. France*, 1989, **50**, 1967.
105. Terech, P., unpublished data.
106. Brotin, T., Utermöhlen, R. *et al. J. Chem. Soc., Chem. Commun.*, 1991, 416.
107. Bouas-Laurent, H. and Desvergne, J.P. (1990). *Photochromism, Molecules and Systems*, Ed. Dürr, H. and Bouas-Laurent, H., Elsevier, Amsterdam, p. 561.
108. Brotin, T., Desvergne, J.P. *et al. Photochem. Photobiol.*, 1992, **55**(3), 349.
109. Terech, P., Desvergnes, J.P. and Bouas-Laurent, H. *J. Colloid Interface Sci.*, 1995, **174**, 174, 258.
110. Hoffmann, H. and Rauscher, A. *Colloid Polym. Sci.*, 1993, **271**, 390.
111. Tanaka, T. *Scientific American*, 1981, **Jan.**, 110.
112. Ladd, M.F.C. and Palmer, R.A. *Structure determination by X-ray Crystallography*, Plenum Press, New York, 1985.
113. Lin, Y-c. and Weiss, R.G. *Macromolecules*, 1987, **20**, 414.
114. Itoh, T., Katsoulis, D.E. and Mita, I. *J. Mater. Chem.*, 1993, **3**(12), 1303.
115. Mukkamala, R. and Weiss, R.G. *J. Chem. Soc., Chem. Commun.*, 1995, 375.
116. Lin, Y-c., Kachar, B and Weiss, R.G. *J. Am. Chem. Soc.*, 1989, **111**, 5542.
117. Furman, I. and Weiss, R.G. *Langmuir*, 1993, **9**, 2084.
118. Demus, D. and Richter, L. *Textures of liquid crystals.* Verlag Chemie, Veinheim, 1978.
119. Lin, Y-c. and Weiss, R.G. *Liquid Crystals*, 1989, **4**(4) 367.
120. Terech, P., Furman, I. and Weiss, R.G. *J. Phys. Chem.*, 1995, **99**, 9558.
121. Terech, P., Ostuni, E. and Weiss, R.G. *J. Phys. Chem.*, 1996, **100**, 3759.
122. Terech, P., Furman, I. *et al. J. Chem. Soc., Faraday Discussions*, **101**, (in press).
123. Murata, K., Aoki, M. *et al. J. Am. Chem. Soc.*, 1994, **116**, 6664.
124. Terech, P. *Liquid Cryst.*, 1991, **9**(1), 59.
125. Kawabata, H., Murata, K. *et al. Langmuir*, 1995, **11**, 623.
126. James, T.D., Kawabata, H. *et al. Tetrahedron*, 1995, **51**(2), 555.
127. Campbell, J., Kuzma, M. and Labes, M. *Mol. Cryst. Liq. Cryst.*, 1983, **95**, 45.
128. Hanabusa, K., Obui, K. *et al. J. Chem. Soc., Chem. Commun.*, 1992, 1371.
129. Hanabusa, K., Tange, J. *et al. J. Chem. Soc., Chem. Commun.*, 1993, 390.
130. Fuhrhop, J-H., Schnieder, P. *et al. J. Am. Chem. Soc.*, 1988, **110**, 2861.
131. Fuhrhop, J-H., Demoulin, C. *et al. J. Am. Chem. Soc.*, 1990, **112**, 2827.
132. Hanabusa, K., Naka, Y. *et al. J. Chem. Soc., Chem. Commun.*, 1994, 2683.
133. Hanabusa, K., Matsumoto, Y. *et al. J. Chem. Soc., Chem. Commun.*, 1994, 1401.
134. Hanabusa, K., Miki, T. *et al. J. Chem. Soc., Chem. Commun.*, 1993, 1382.
135. De Vries, E.J. and Kellogg, R.M. *J. Chem. Soc., Chem. Commun.*, 1993, 238.
136. Sheu, E.Y., Liang, K.S. and Chiang, L.Y. *J. Phys. France*, 1989, **50**, 1279.
137. Gallivan, J.P. and Schuster, G.B. *J. Org. Chem.*, 1995, **60**, 2423.
138. Godquin-Giroud, A.M. *J. phys. Lett.*, 1984, **45**, L-387.
139. Godquin-Giroud, A.M., Gauthier, M.M. *et al. Mol. Cryst. Liq. Cryst.*, 1986, **132**, 35.
140. Terech, P., Chachaty, C. *et al. J. Phys. France*, 1987, **48**, 663.
141. Godquin-Giroud, A.M. and Marchon, J.C. *J. Physique Lett.*, 1984, **45**, L-681.
142. Attard, G.S. and Cullum, P.R. *Liq. Cryst.*, 1990, **8**(3), 299.
143. Abied, H., Guillon, D. *et al. Liq. Cryst.*, 1987, **2**, 269.
144. Terech, P., Schaffhauser, V. *et al. Langmuir*, 1992, **8**, 2104.
145. Terech, P. *Il Nuevo Cimento*, 1994, **16D**(7), 757.
146. Dammer, C., Maldivi, P. *et al. Langmuir*, 1995, **11**, 1500.

147. Turner, M.S. and Cates, M.E. *Langmuir*, 1991, **7**, 1590.
148. Turner, M.S. and Cates, M.E. *J. Phys. II France*, 1992, **2**, 503.
149. Granek, R. and Cates, M.E. *J. Chem. Phys.*, 1992, **96**(6), 4758.
150. Kern, F., Zana, R. and Candau, S.J. *Langmuir*, 1991, **7**, 1344.
151. Berret, J.F., Aspell, J. and Porte, G. *Langmuir*, 1993, **9**, 2851.
152. Ibn-Elhaj, M., Guillon, D. *et al. J. Phys. II France*, 1992, **2**, 2197.
153. O'Shaughnessy, B. and Yu, J. *Phys. Rev. Letters*, 1995, **74**(21), 4329.
154. Bonnet, L., Cukiernik, F.D. *et al. Chem. Mater.*, 1994, **6**, 31.
155. Marchon, J.C., Maldivi, P. *et al. Phil. Trans. R. Soc. Lond.*, 1990, **A330**, 109.
156. Terech, P., Gebel, G. and Ramasseul, R., *Langmuir*, 1996 (in press).
157. Ramasseul, R., Maldivi, P. and Marchon, J.C. *Liquid Crystals*, 1993, **13**(5), 729.
158. Bruce, D.W., Wali, M.A. and Qing Min Wang. *J. Chem. Soc., Chem. Commun.*, 1994, 2089.
159. Fuhrhop, J-H., Bindig, U. and Siggel, U. *J. Chem. Soc., Chem. Commun.*, 1994, 1583.
160. Byrn, M.P., Curtis, C.J. *et al. J. Am. Chem. Soc.*, 1990, **112**, 1865.
161. Byrn, M.P., Curtis, C.J. *et al. J. Am. Chem. Soc.*, 1993, **115**, 9480.
162. Yamamura, T., *Chem. Lett.*, 1977, 773.
163. Tsuchida, E., Komatsu, T. *et al. J. Chem. Soc., Chem. Commun.*, 1993, 1731.
164. Dolphin, D. *The porphyrins*, vol. III, Physical Chemistry, Part A, Academic Press, New York, 1978.
165. Fuhrhop., J-H., Demoulin, C. *et al. J. Am. Chem. Soc.*, 1992, **114**, 4159.
166. Komatsu, T., Nakao, K. *et al. J. Chem. Soc., Chem. Commun.*, 1993, 728.
167. Hanack, M. and Lang, M. *Adv. Mater.*, 1994, **6**(11), 819.
168. Simon, J. and Sirlin, C. *Pure Appl. Chem.*, 1989, **61**, 1625.
169. Loeb, S.J. and Shimizu, G.K.H. *J. Chem. Soc., Chem. Commun.*, 1993, 1395.
170. Godquin-Giroud, A.M. and Maitlis, P.M. *Angew. Chem. Int. Ed. Engl.*, 1991, **30**, 375.
171. Chandrasekhar, S. and Ranganath, G.S. *Rep. Prog. Phys.*, 1990, **53**, 57.
172. Gutsche, C.D. and Bauer, L.J. *J. Am. Chem. Soc.*, 1985, **107**, 6052–6059. *ibid*, 6059.
173. *Gutsche, C.D. and Bauer, LJ. J. Am. Chem. Soc.*, 1985, **107**, 6063.
174. Dedek, P., Webber, A.S. *et al. Langmuir*, 1994, **10**, 3943.
175. Markowitz, M.A., Janout, V. *et al. J. Am. Chem. Soc.*, 1989, **111**, 8192.
176. Aoki, M., Murata, K. and Shinkai, S. *Chem. Soc. Jpn., Chem. Lett.*, 1991, 1715.
177. Aoki, M., Nakashima, K. *et al. J. Chem. Soc., Perkin Trans 2*, 1993, 347.
178. Yamasaki, S. and Tsutsumi, H. *Bull. Chem. Soc. Jpn.*, 1994, **67**, 2053.
179. Yamasaki, S. and Tsutsumi, H. *Bull. Chem. Soc. Jpn.*, 1994, **67**, 906.
180. McKenna, G.B., Kern, F. and Candau, S.J. *Polymer Preprints*, 1991, **32**, 455.
181. Yamasaki, S., Ohashi, Y. *et al. Bull. Chem. Soc. Jpn.*, 1995, **68**, 146.
182. Twieg, R.J., Russell, T.P. *et al. Macromolecules*, 1985, **18**, 1361.
183. Rabolt, J.F., Russell, T.P. *et al. Polym. Preprints*, 1991, **32**, 223.
184. De Rango, C., Charpin, P. *et al. J. Am. Chem. Soc.*, 1992, **114**, 5475.
185. Yu, Z-J. and Neuman, R.D. *Langmuir*, 1994, **10**, 2553.
186. Bhattacharyya, S.N. and Ganguly, B. *J. Colloid Interface Sci.*, 1987, **118**(1), 15.
187. Luisi, P.L., Scartazzini, R. *et al. Colloid Polym. Sci.*, 1990, **268**, 356.
188. Schurtenberger, P., Scartazzini, R. and Luisi, P.L. *Rheol. Acta*, 1989, **28**, 372.
189. Lehn, J-M., Mascal, M. *et al. J. Chem. Soc., Chem. Commun.*, 1990, 479.
190. Danda, H., Nishikawa, H. and Otaka, K. *J. Org. Chem.*, 1991, **56**, 6740.
191. Peri, J.B. *J. Colloid Interface Sci.*, 1969, **29**(1), 6.
192. Xu, X., Ayyagari, M. *et al. J. Phys. Chem.*, 1993, **97**, 11350.
193. Tata, M., John, V.T. *et al. J. Am. Chem. Soc.*, 1994, **116**, 9464.
194. Tata, M., John, V.T. *et al. J. Phys. Chem.*, 1994, **98**, 3809.
195. Magid, L.J., Kon-no, K. and Martin, C.A. *J. Phys. Chem.*, 1981, **85**, 1434.
196. Rao, A.M., John, V.T. *et al. Biotechnol. Bioeng.*, 1993, **41**, 531.
197. Capitani, D., Segre, A.L. and Sarapani, R. *Langmuir*, 1991, **7**, 250.
198. Haering, G. and Luisi, P.L. *J. Phys. Chem.*, 1986, **90**, 5892.
199. Quellet, C. and Eicke, H-F. *Chimia*, 1986, **40**(7–8), 233.
200. Quellet, C., Eicke, H-F. and Sager, W. *J. Phys. Chem.*, 1991, **95**, 5642.
201. Bergeron, V. and Sebba, F. *Langmuir*, 1987, **3**, 857–858. Comments by Princen, H.M. 1988, **4**(2), 486.

202. Ebert, G., Platz, G. and Rehage, H. *Ber. Bunsenges. Phys. Chem.*, 1988, **92**, 1158.
203. Krafft, M-P. and Riess, J.G. *Angew. Chem. Int. Engl.*, 1994, **33**(10), 1100.
204. Gradzielski, M., Hoffmann, H. and Oetter, G. *Colloid Polymer Sci.*, 1990, **268**, 167.
205. Hirsch, E., Wittmann, J.C. and Candau, F. *J. Dispersion Sci. Technology*, 1982, **3**(4), 351.
206. Scartazzini, R. and Luisi, P.L. *J. Phys. Chem.*, 1988, **92**, 829.
207. Schurtenberger, P., Scartazzini, R. *et al. J. Phys. Chem.*, 1990, **94**, 3695.
208. Schurtenberger, P., Magid, L.J. *et al. J. Phys. Chem.*, 1991, **95**(11), 4173.
209. Lin, Z., Cai, J.J. *et al. J. Phys. Chem.*, 1994, **98**, 5984.
210. Candau, S.J. *J. Phys. France*, 1990, **51**, 977.
211. Rehage, H. and Hoffmann, H. *J. Phys. Chem.*, 1988, **92**, 4712.
212. Schurtenberger, P. and Cavaco, C. *Langmuir*, 1994, **10**, 100.
213. Schurtenberger, P. and Cavaco, C. *J. Phys. II France*, 1993, **3**, 1279.
214. Prost, J. and Rondelez, F. *Nature*, 1991, **350**, 11.
215. De Vringer, T., Joosten, J.G.H. and Junginger, H.E. *Colloid Polym. Sci.*, 1987, **265**, 167.
216. Patel, H.K., Rowe, R.C. *et al. Int. J. Pharmaceut.*, 1985, **25**, 13.
217. Phillips, R.J., Deen, W.M. and Brady, J.F. *J. Colloid Interface Sci.*, 1990, **139**, 363.
218. Yamaguchi, T., Kuhnert, L. *et al. J. Phys. Chem.*, 1991, **95**, 5831.
219. Agladze, K., Dulos, E. and De Kepper, P. *J. Phys. Chem.*, 1992, **96**, 2400.
220. Andreazza, P., Lefaucheux, F. and Mutaftschiev, B. *J. Crystal Growth*, 1988, **92**, 415.
221. Robert, M.C. and Lefaucheux, F. *J. Crystal Growth*, 1988, **90**, 358.
222. Rees, G.D. and Robinson, B.H. *Adv. Mater.*, 1993, **5**(9), 608.
223. Fricke, J. and Emmerling, A. *Adv. Mater.* 1991, **3**(10), 504.
224. Rifat, A. and Hikmet, A.M. *Adv. Mater.*, 1992, **4**(10), 679.
225. Piechocki, C., Simon, J. *et al. J. Am. Chem. Soc.*, 1982, **104**, 5245.
226. Rosenblum, M. *Adv. Mater.*, 1994, **6**(2), 159.
227. Weber, P., Guillon, D. and Skoulios, A. *J. Phys. Chem.*, 1987, **87**, 2242.
228. Albouy, P.A. *J. Phys. Chem.*, 1994, **98**, 8543.
229. Ricco, M. and Dalcanale, E. *J. Phys. Chem.*, 1994, **98**, 9002.
230. Imahori, H., Higuchi, H. *et al. Bull. Chem. Soc. Jpn.*, 1994, **67**, 2500.
231. Thierry, A., Straupe, C. *et al. Polymer Commun.*, 1990, **31**, 299.
232. Ue, M., Kaitoh, M. *et al. Electrochimica Acta*, 1993, **38**(9), 1301.
233. Dickert, F.L. and Haunschild, A. *Adv. Mater.*, 1993, **5**(12), 887.
234. Ozin, G. *Adv. Mater.*, 1992, **4**(10), 612.
235. Osada, Y. and Ross-Murphy, S.B. *Scientific American*, 1993, **May**, 82.

Index